Lecture Notes in Mathematics 1953

Editors:
J.-M. Morel, Cachan
F. Takens, Groningen
B. Teissier, Paris

T0207323

Khadiga A. Arwini · Christopher T.J. Dodson

Information Geometry

Near Randomness
and Near Independence

 Springer

Authors

Khadiga A. Arwini
Al-Fateh University
Faculty of Sciences
Mathematics Department
Box 13496
Tripoli, Libya
arwini2001@yahoo.com

Christopher T.J. Dodson
University of Manchester
School of Mathematics
Manchester M13 9PL, United Kingdom
ctdodson@manchester.ac.uk

ISBN 978-3-540-69391-8   ISBN 978-3-540-69393-2 (eBook)
DOI 10.1007/978-3-540-69393-2

Lecture Notes in Mathematics ISSN print edition: 0075-8434
                          ISSN electronic edition: 1617-9692

Library of Congress Control Number: 2008930087

Mathematics Subject Classification (2000): 53B50, 60D05, 62B10, 62P35, 74E35, 92D20

Cover design: SPi Publishing Services

Printed on acid-free paper

9 8 7 6 5 4 3 2 1

springer.com

Preface

The main motivation for this book lies in the breadth of applications in which a statistical model is used to represent small departures from, for example, a Poisson process. Our approach uses information geometry to provide a common context but we need only rather elementary material from differential geometry, information theory and mathematical statistics. Introductory sections serve together to help those interested from the applications side in making use of our methods and results. We have available *Mathematica* notebooks to perform many of the computations for those who wish to pursue their own calculations or developments.

Some 44 years ago, the second author first encountered, at about the same time, differential geometry via relativity from Weyl's book [209] during undergraduate studies and information theory from Tribus [200, 201] via spatial statistical processes while working on research projects at Wiggins Teape Research and Development Ltd—cf. the Foreword in [196] and [170, 47, 58]. Having started work there as a student laboratory assistant in 1959, this research environment engendered a recognition of the importance of international collaboration, and a lifelong research interest in randomness and near-Poisson statistical geometric processes, persisting at various rates through a career mainly involved with global differential geometry. From correspondence in the 1960s with Gabriel Kron [4, 124, 125] on his Diakoptics, and with Kazuo Kondo who influenced the post-war Japanese schools of differential geometry and supervised Shun-ichi Amari's doctorate [6], it was clear that both had a much wider remit than traditionally pursued elsewhere. Indeed, on moving to Lancaster University in 1969, receipt of the latest *RAAG Memoirs Volume 4 1968* [121] provided one of Amari's early articles on information geometry [7], which subsequently led to his greatly influential 1985 Lecture Note volume [8] and our 1987 *Geometrization of Statistical Theory Workshop* at Lancaster University [10, 59].

Reported in this monograph is a body of results, and computer-algebraic methods that seem to have quite general applicability to statistical models admitting representation through parametric families of probability density functions. Some illustrations are given from a variety of contexts for geometric characterization of statistical states near to the three important standard basic reference states: (Poisson) randomness, uniformity, independence. The individual applications are somewhat heuristic models from various fields and we incline more to terminology and notation from the applications rather than from formal statistics. However, a common thread is a geometrical representation for statistical perturbations of the basic standard states, and hence results gain qualitative stability. Moreover, the geometry is controlled by a metric structure that owes its heritage through maximum likelihood to information theory so the quantitative features—lengths of curves, geodesics, scalar curvatures etc.—have some respectable authority. We see in the applications simple models for galactic void distributions and galaxy clustering, amino acid clustering along protein chains, cryptographic protection, stochastic fibre networks, coupled geometric features in hydrology and quantum chaotic behaviour. An ambition since the publication by Richard Dawkins of *The Selfish Gene* [51] has been to provide a suitable differential geometric framework for dynamics of natural evolutionary processes, but it remains elusive. On the other hand, in application to the statistics of amino acid spacing sequences along protein chains, we describe in Chapter 7 a stable statistical qualitative property that may have evolutionary significance. Namely, to widely varying extents, all twenty amino acids exhibit greater clustering than expected from Poisson processes. Chapter 11 considers eigenvalue spacings of infinite random matrices and near-Poisson quantum chaotic processes.

The second author has benefited from collaboration (cf. [34]) with the group headed by Andrew Doig of the Manchester Interdisciplinary Biocentre, the University of Manchester, and has had long-standing collaborations with groups headed by Bill Sampson of the School of Materials, the University of Manchester (cf.eg. [73]) and Jacob Scharcanski of the Instituto de Informatica, Universidade Federal do Rio Grande do Sul, Porto Alegre, Brasil (cf.eg. [76]) on stochastic modelling. We are pleased therefore to have co-authored with these colleagues three chapters: titled respectively, Amino Acid Clustering, Stochastic Fibre Networks, Stochastic Porous Media and Hydrology.

The original draft of the present monograph was prepared as notes for short Workshops given by the second author at Centro de Investigaciones de Matematica (CIMAT), Guanajuato, Mexico in May 2004 and also in the Departamento de Xeometra e Topoloxa, Facultade de Matemáticas, Universidade de Santiago de Compostela, Spain in February 2005.

The authors have benefited at different times from discussions with many people but we mention in particular Shun-ichi Amari, Peter Jupp, Patrick Laycock, Hiroshi Matsuzoe, T. Subba Rao and anonymous referees. However, any overstatements in this monograph will indicate that good advice may have been missed or ignored, but actual errors are due to the authors alone.

Khadiga Arwini, Department of Mathematics, Al-Fateh University, Libya
Kit Dodson, School of Mathematics, the University of Manchester, England

Contents

1

Mathematical Statistics
and Information Theory

There are many easily found good books on probability theory and mathematical statistics (eg [84, 85, 87, 117, 120, 122, 196]), stochastic processes (eg [31, 161]) and information theory (eg [175, 176]); here we just outline some topics to help make the sequel more self contained. For those who have access to the computer algebra package *Mathematica* [215], the approach to mathematical statistics and accompanying software in Rose and Smith [177] will be particularly helpful.

The word stochastic comes from the Greek *stochastikos*, meaning skillful in aiming and *stochazesthai* to aim at or guess at, and *stochos* means target or aim. In our context, stochastic colloquially means involving chance variations around some event—rather like the variation in positions of strikes aimed at a target. In its turn, the later word statistics comes through eighteenth century German from the Latin root *status* meaning state; originally it meant the study of political facts and figures. The noun random was used in the sixteenth century to mean a haphazard course, from the Germanic randir to run, and as an adjective to mean without a definite aim, rule or method, the opposite of purposive. From the middle of the last century, the concept of a random variable has been used to describe a variable that is a function of the result of a well-defined statistical experiment in which each possible outcome has a definite probability of occurrence. The organization of probabilities of outcomes is achieved by means of a probability function for discrete random variables and by means of a probability density function for continuous random variables. The result of throwing two fair dice and summing what they show is a discrete random variable.

Mainly, we are concerned with continuous random variables (here measurable functions defined on some $\mathbb{R}^n$) with smoothly differentiable probability density measure functions, but we do need also to mention the Poisson distribution for the discrete case. However, since the Poisson is a limiting approximation to the Binomial distribution which arises from the Bernoulli distribution (which everyone encountered in school!) we mention also those examples.

K. Arwini, C.T.J. Dodson, *Information Geometry.*
Lecture Notes in Mathematics 1953,
© Springer-Verlag Berlin Heidelberg 2008

1.1 Probability Functions for Discrete Variables

For discrete random variables we take the domain set to be $\mathbb{N} \cup \{0\}$. We may view a probability function as a subadditive measure function of unit weight on $\mathbb{N} \cup \{0\}$

$$p \; : \; \mathbb{N} \cup \{0\} \to [0,1) \quad \text{(nonnegativity)} \tag{1.1}$$

$$\sum_{k=0}^{\infty} p(k) = 1 \quad \text{(unit weight)} \tag{1.2}$$

$$p(A \cup B) \le p(A) + p(B), \; \forall A, B \subset \mathbb{N} \cup \{0\}, \quad \text{(subadditivity)} \tag{1.3}$$
$$\text{with equality} \iff A \cap B = \emptyset.$$

Formally, we have a discrete measure space of total measure 1 with σ-algebra the power set and measure function induced by p

$$sub(\mathbb{N} \cup \{0\}) \to [0,1) : A \mapsto \sum_{k \in A} p(k)$$

and as we have anticipated above, we usually abbreviate $\sum_{k \in A} p(k) = p(A)$.

We have the following expected values of the random variable and its square

$$\mathcal{E}(k) = \overline{k} = \sum_{k=0}^{\infty} k \, p(k) \tag{1.4}$$

$$\mathcal{E}(k^2) = \overline{k^2} = \sum_{k=0}^{\infty} k^2 \, p(k). \tag{1.5}$$

Formally, statisticians are careful to distinguish between a property of the whole population—such as these expected values—and the observed values of samples from the population. In practical applications it is quite common to use the bar notation for expectations and we shall be clear when we are handling sample quantities. With slight but common abuse of notation, we call $\overline{k}$ the mean, $\overline{k^2} - (\overline{k})^2$ the variance, $\sigma_k = +\sqrt{\overline{k^2} - (\overline{k})^2}$ the standard deviation and $\sigma_k / \overline{k}$ the coefficient of variation, respectively, of the random variable k. The variance is the square of the standard deviation.

The moment generating function $\Psi(t) = \mathcal{E}(e^{tX})$, $t \in \mathbb{R}$ of a distribution generates the rth moment as the value of the rth derivative of Ψ evaluated at $t = 0$. Hence, in particular, the mean and variance are given by:

$$\mathcal{E}(X) = \Psi'(0) \tag{1.6}$$
$$Var(X) = \Psi''(0) - (\Psi'(0))^2, \tag{1.7}$$

which can provide an easier method for their computation in some cases.

1.1.1 Bernoulli Distribution

It is said that a random variable X has a Bernoulli distribution with parameter p $(0 \le p \le 1)$ if X can take only the values 0 and 1 and the probabilities are

$$P_r(X = 1) = p \tag{1.8}$$

$$P_r(X = 0) = 1 - p \tag{1.9}$$

Then the probability function of X can be written as follows:

$$f(x|p) = \begin{cases} p^x(1-p)^{1-x} & \text{if } x = 0, 1 \\ 0 & \text{otherwise} \end{cases} \tag{1.10}$$

If X has a Bernoulli distribution with parameter p, then we can find its expectation or mean value $\mathcal{E}(X)$ and variance $Var(X)$ as follows.

$$\mathcal{E}(X) = 1 \cdot p + 0 \cdot (1 - p) = p \tag{1.11}$$

$$Var(X) = \mathcal{E}(X^2) - (\mathcal{E}(X))^2 = p - p^2 \tag{1.12}$$

The moment generating function of X is the expectation of e^{tX},

$$\Psi(t) = \mathcal{E}(e^{tX}) = pe^t + q \tag{1.13}$$

which is finite for all real t.

1.1.2 Binomial Distribution

If n random variables $X_1, X_2, \ldots, X_n$ are independently identically distributed, and each has a Bernoulli distribution with parameter p, then it is said that the variables $X_1, X_2, \ldots, X_n$ form n Bernoulli trials with parameter p.

If the random variables $X_1, X_2, \ldots, X_n$ form n Bernoulli trials with parameter p and if $X = X_1 + X_2 + \ldots + X_n$, then X has a binomial distribution with parameters n and p.

The binomial distribution is of fundamental importance in probability and statistics because of the following result for any experiment which can have outcome only either success or failure. The experiment is performed n times independently and the probability of the success of any given performance is p. If X denotes the total number of successes in the n performances, then X has a binomial distribution with parameters n and p. The probability function of X is:

$$P(X = r) = P(\sum_{i=1}^{n} X_i = r) = \binom{n}{r} p^r(1 - p)^{n-r} \tag{1.14}$$

where $r = 0, 1, 2, \ldots, n$.

We write

$$f(r|p) = \begin{cases} \binom{n}{r} p^r (1-p)^{n-r} & \text{if r=0, 1, 2, } \ldots \text{ , n} \\ 0 & \text{otherwise} \end{cases} \qquad (1.15)$$

In this distribution n must be a positive integer and p must lie in the interval $0 \leq p \leq 1$. If X is represented by the sum of n Bernoulli trials, then it is easy to get its expectation, variance and moment generating function by using the properties of sums of independent random variables—cf. §1.3.

$$\mathcal{E}(X) = \sum_{i=1}^{n} \mathcal{E}(X_i) = np \qquad (1.16)$$

$$Var(X) = \sum_{i=1}^{n} Var(X_i) = np(1-p) \qquad (1.17)$$

$$\Psi(t) = \mathcal{E}(e^{tX}) = \prod_{i=1}^{n} \mathcal{E}(e^{tX_i}) = (pe^t + q)^n. \qquad (1.18)$$

1.1.3 Poisson Distribution

The Poisson distribution is widely discussed in the statistical literature; one monograph devoted to it and its applications is Haight [102].
Take $t, \tau \in (0, \infty)$

$$p : \mathbb{N} \cup \{0\} \to [0, 1) : k \mapsto \left(\frac{t}{\tau}\right)^k \frac{1}{k!} e^{-t/\tau} \qquad (1.19)$$

$$\overline{k} = t/\tau \qquad (1.20)$$

$$\sigma_k = t/\tau. \qquad (1.21)$$

This probability function is used to model the number k of events in a region of measure t when the mean number of events per unit region is τ and the probability of an event occurring in a region depends only on the measure of the region, not its shape or location. Colloquially, in applications it is very common to encounter the usage of 'random' to mean the specific case of a Poisson process; formally in statistics the term random has a more general meaning: probabilistic, that is dependent on random variables. Figure 1.1 depicts a simulation of a 'random' array of 2000 line segments in a plane; the centres of the lines follow a Poisson process and the orientations of the lines follow a uniform distribution, cf. §1.2.1. So, in an intuitive sense, this is the result of the least choice, or maximum uncertainty, in the disposition of these line segments: the centre of each line segment is equally likely to fall in every region of given area and its angle of axis orientation is equally likely to fall in every interval of angles of fixed size. This kind of situation is representative

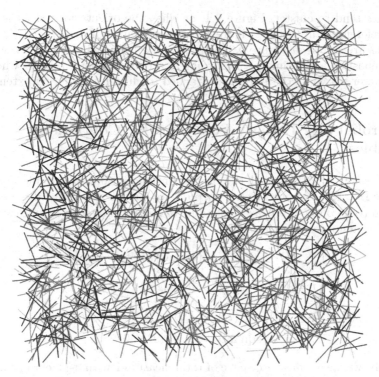

Fig. 1.1. Simulation of a random array of 2000 line segments in a plane; the centres of the lines follow a Poisson process and the orientations of the lines follow a uniform distribution. The grey tones correspond to order of deposition.

of common usage of the term 'random process' to mean subordinate to a Poisson process. A 'non-random' processes departs from Poisson by having constraints on the probabilities of placing of events or objects, typically as a result of external influence or of interactions among events or objects.

Importantly, the Poisson distribution can give a good approximation to the binomial distribution when n is large and p is close to 0. This is easy to see by making the correspondences:

$$e^{-pn} \longrightarrow (1 - (n - r)p) \tag{1.22}$$

$$n!/(n - r)! \longrightarrow n^r. \tag{1.23}$$

Much of this monograph is concerned with the representation and classification of deviations from processes subordinate to a Poisson random variable, for example for a line process via the distribution of inter-event (nearest neighbour, or inter-incident) spacings. Such processes arise in statistics under the term renewal process [150].

We shall see in Chapter 9 that, for physical realisations of stochastic fibre networks, typical deviations from Poisson behaviour arise when the centres of the fibres tend to cluster, Figure 9.1, or when the orientations of their axes have preferential directions, Figure 9.15. Radiographs of real stochastic fibre networks are shown in Figure 9.3 from Oba [156]; the top network consists of fibres deposited approximately according to a Poisson planar process whereas in the lower networks the fibres have tended to cluster to differing extents.

1.2 Probability Density Functions for Continuous Variables

We are usually concerned with the case of continuous random variables defined on some $\Omega \subseteq \mathbb{R}^m$. For our present purposes we may view a probability density function (pdf) on $\Omega \subseteq \mathbb{R}^m$ as a subadditive measure function of unit weight, namely, a nonnegative map on Ω

$$f \; : \; \Omega \to [0, \infty) \quad \text{(nonnegativity)} \tag{1.24}$$

$$\int_\Omega f = f(\Omega) = 1 \quad \text{(unit weight)} \tag{1.25}$$

$$f(A \cup B) \le f(A) + f(B), \; \forall A, B \subset \Omega, \quad \text{(subadditivity)} \tag{1.26}$$

$$\text{with equality} \iff A \cap B = \emptyset.$$

Formally, we have a measure space of total measure 1 with σ-algebra typically the Borel sets or the power set and the measure function induced by f

$$sub(\Omega) \to [0, 1] : A \mapsto \int_A f = \text{integral of } f \text{ over } A$$

and as we have anticipated above, we usually abbreviate $\int_A f = f(A)$. Given an integrable (ie measurable in the σ-algebra) function $u : \Omega \to \mathbb{R}$, the expectation or mean value of u is defined to be

$$\mathcal{E}(u) = \bar{u} = \int_\Omega uf.$$

We say that f is the joint pdf for the random variables $x_1, x_2, \ldots, x_m$, being the coordinates of points in Ω, or that these random variables have the joint probability distribution f. If x is one of these random variables, and in particular for the important case of a single random variable x, we have the following

$$\bar{x} = \int_\Omega xf \tag{1.27}$$

$$\overline{x^2} = \int_\Omega x^2 f. \tag{1.28}$$

Again with slight abuse of notation, we call $\overline{x}$ the mean and the variance is the mean square deviation

$$\sigma_x^2 = \overline{(x - \overline{x})^2} = \overline{x^2} - (\overline{x})^2.$$

Its square root is the standard deviation $\sigma_x = +\sqrt{\overline{x^2} - (\overline{x})^2}$ and the ratio $\sigma_x/\overline{x}$ is the coefficient of variation, of the random variable x. Some inequalities for the probability of a random variable exceeding a given value are worth mentioning.

Markov's Inequality: If x is a nonnegative random variable with probability density function f then for all $a > 0$, the probability that $x > a$ is

$$\int_a^\infty f \le \frac{\overline{x}}{a}. \tag{1.29}$$

Chebyshev's Inequality: If x is a random variable having probability density function f with zero mean and finite variance σ^2, then for all $a > 0$, the probability that $x > a$ is

$$\int_a^\infty f \le \frac{\sigma^2}{\sigma^2 + a^2}. \tag{1.30}$$

Bienaymé-Chebyshev's Inequality: If x is a random variable having probability density function f and u is a nonnegative non-decreasing function on $(0, \infty)$, then for all $a > 0$ the probability that $|x| > a$ is

$$1 - \int_{-a}^a f \le \frac{\overline{u}}{u(a)}. \tag{1.31}$$

The cumulative distribution function (cdf) of a nonnegative random variable x with probability density function f is the function defined by

$$F : [0, \infty) \to [0, 1] : x \mapsto \int_0^x f(t)\, dt. \tag{1.32}$$

It is easily seen that if we wish to change from random variable x with density function f to a new random variable ξ when x is given as an invertible function of ξ, then the probability density function for ξ is represented by

$$g(\xi) = f(x(\xi)) \left| \frac{dx}{d\xi} \right|. \tag{1.33}$$

If independent real random variables x and y have probability density functions f, g respectively, then the probability density function h of their sum $z = x + y$ is given by

$$h(z) = \int_{-\infty}^\infty f(x)\, g(z - x)\, dx \tag{1.34}$$

and the probability density function p of their product $r = xy$ is given by

$$p(r) = \int_{-\infty}^{\infty} f(x) g\left(\frac{r}{x}\right) \frac{1}{|x|} dx. \tag{1.35}$$

Usually, a probability density function depends on a set of parameters, $\theta_1, \theta_2, \ldots, \theta_n$ and we say that we have an n-dimensional family. Then the corresponding change of variables formula involves the $n \times n$ Jacobian determinant for the multiple integrals, so generalizing (1.33).

1.2.1 Uniform Distribution

This is the simplest continuous distribution, with constant probability density function for a bounded random variable:

$$u : [a, b] \to [0, \infty) : x \mapsto \frac{1}{b - a} \tag{1.36}$$

$$\bar{x} = \frac{a + b}{2} \tag{1.37}$$

$$\sigma_x = \frac{b - a}{2\sqrt{3}}. \tag{1.38}$$

The probability of an event occurring in an interval $[\alpha, \beta] \subseteq [a, b]$ is simply proportional to the length of the interval:

$$P(x \in [\alpha, \beta]) = \frac{\beta - \alpha}{b - a}.$$

1.2.2 Exponential Distribution

Take $\lambda \in \mathbb{R}^+$; this is called the parameter of the exponential probability density function

$$f : [0, \infty) \to [0, \infty) : [a, b] \mapsto \int_{[a,b]} \frac{1}{\lambda} e^{-x/\lambda} \tag{1.39}$$

$$\bar{x} = \lambda \tag{1.40}$$

$$\sigma_x = \lambda. \tag{1.41}$$

The parameter space of the exponential distribution is $\mathbb{R}^+$, so exponential distributions form a 1-parameter family. In the sequel we shall see that quite generally we may provide a Riemannian structure to the parameter space of a family of distributions. Sometimes we call a family of pdfs a parametric statistical model.

Observe that, in the Poisson probability function (1.19) for events on the real line, the probability of zero zero events in an interval t is

$$p(0) = e^{-t/\tau}$$

and it is not difficult to show that the probability density function for the Poisson inter-event (or inter-incident) distance t on $[0, \infty)$ is an exponential probability density function (1.39) given by

$$f : [0, \infty) \to [0, \infty) : t \mapsto \frac{1}{\tau} e^{-t/\tau}$$

where τ is the mean number of events per unit interval. Thus, the occurrence of an exponential distribution has associated with it a complementary Poisson distribution, so the exponential distribution provides for continuous variables an identifier for Poisson processes. Correspondingly, departures from an exponential distribution correspond to departures from a Poisson process. We shall see below in §1.4.1 that in rather a strict sense the gamma distribution generalises the exponential distribution.

1.2.3 Gaussian, or Normal Distribution

This has real random variable x with mean μ and variance σ^2 and the familiar bell-shaped probability density function given by

$$f(x) = \frac{1}{\sqrt{2\pi}\sigma} e^{\frac{-(x-\mu)^2}{2\sigma^2}}. \tag{1.42}$$

The Gaussian distribution has the following uniqueness property: For independent random variables $x_1, x_2, \ldots, x_n$ with a common continuous probability density function f, having independence of the sample mean $\bar{x}$ and sample standard deviation S is equivalent to f being a Gaussian distribution [110].

The Central Limit Theorem states that for independent and identically distributed real random variables x_i each having mean μ and variance σ^2, the random variable

$$w = \frac{(x_1 + x_2 + \ldots + x_n) - n\mu}{\sqrt{n}\sigma} \tag{1.43}$$

tends as $n \to \infty$ to a Gaussian random variable with mean zero and unit variance.

1.3 Joint Probability Density Functions

Let f be a probability density function, defined on $\mathbb{R}^2$ (or some subset thereof). This is an important case since here we have two variables, X, Y, say, and we can extract certain features of how they interact. In particular, we define their respective mean values and their covariance, σ_{xy}:

$$\bar{x} = \int_{-\infty}^{\infty} \int_{-\infty}^{\infty} x \, f(x, y) \, dx dy \tag{1.44}$$

$$\bar{y} = \int_{-\infty}^{\infty} \int_{-\infty}^{\infty} y \, f(x, y) \, dx dy \tag{1.45}$$

$$\sigma_{xy} = \int_{-\infty}^{\infty} \int_{-\infty}^{\infty} xy \, f(x, y) \, dx dy \; - \bar{x} \, \bar{y} = \overline{xy} - \bar{x} \, \bar{y}. \tag{1.46}$$

The marginal probability density function of X is f_X, obtained by integrating f over all y,

$$f_X(x) = \int_{v=-\infty}^{\infty} f_{X,Y}(x,v)\, dv \qquad (1.47)$$

and similarly the marginal probability density function of Y is

$$f_Y(y) = \int_{u=-\infty}^{\infty} f_{X,Y}(u,y)\, du \qquad (1.48)$$

The jointly distributed random variables X and Y are called independent if their marginal density functions satisfy

$$f_{X,Y}(x,y) = f_X(x) f_Y(y) \quad for\ all \quad x, y \in R \qquad (1.49)$$

It is easily shown that if the variables are independent then their covariance (1.46) is zero but the converse is not true. Feller [84] gives a simple counterexample: let X take values $-1, +1, -2, +2$, each with probability $\frac{1}{4}$ and let $Y = X^2$; then the covariance is zero but there is evidently a (nonlinear) dependence.

The extent of dependence between two random variables can be measured in a normalised way by means of the correlation coefficient: the ratio of the covariance to the product of marginal standard deviations:

$$\rho_{xy} = \frac{\sigma_{xy}}{\sigma_x \sigma_y}. \qquad (1.50)$$

Note that by the Cauchy-Schwartz inequality, $-1 \le \rho_{xy} \le 1$, whenever it exists, the limiting values corresponding to the case of linear dependence between the variables. Intuitively, $\rho_{xy} < 0$ if y tends to increase as x decreases, and $\rho_{xy} > 0$ if x and y tend to increase together.

A change of random variables from (x, y) with density function f to say (u, v) with density function g and x, y given as invertible functions of u, v involves the Jacobian determinant:

$$g(u,v) = f(x(u,v), y(u,v)) \frac{\partial(x,y)}{\partial(u,v)}. \qquad (1.51)$$

1.3.1 Bivariate Gaussian Distributions

The probability density function of the two-dimensional Gaussian distribution has the form:

$$f(x,y) = \frac{1}{2\pi \sqrt{\sigma_1 \sigma_2 - \sigma_{12}{}^2}}\, e^W. \qquad (1.52)$$

with

$$W = -\frac{1}{2\,(\sigma_1\,\sigma_2 - \sigma_{12}{}^2)}\left(\sigma_2(x-\mu_1)^2 - 2\,\sigma_{12}\,(x-\mu_1)\,(y-\mu_2) + \sigma_1(y-\mu_2)^2\right),$$

where

$$-\infty < x_1 < x_2 < \infty, \quad -\infty < \mu_1 < \mu_2 < \infty, \quad 0 < \sigma_1, \sigma_2 < \infty.$$

This contains the five parameters $(\mu_1, \mu_2, \sigma_1, \sigma_{12}, \sigma_2) = (\xi^1, \xi^2, \xi^3, \xi^4, \xi^5) \in \Theta$. So we have a five-dimensional parameter space Θ.

1.4 Information Theory

Information theory owes its origin in the 1940s to Shannon [186], whose interest was in modelling the transfer of information stored in the form of binary on-off devices, the basic unit of information being one bit: 0 or 1. The theory provided a representation for the corruption by random electronic noise of transferred information streams, and for quantifying the effectiveness of error-correcting algorithms by the incorporation of redundancy in the transfer process. His concept of information theoretic entropy in communication theory owed its origins to thermodynamics but its effectiveness in general information systems has been far reaching. Information theory worked out by the communication theorists, and entropy in particular, were important in providing a conceptual and mathematical framework for the development of chaos theory [93]. There the need was to model the dynamics of adding small extrinsic noise to otherwise deterministic systems. In physical theory, entropy provides the uni-directional 'arrow of time' by measuring the disorder in an irreversible system [164]. Intuitively, we can see how the entropy of a state modelled by a point in a space of probability density functions would be expected to be maximized at a density function that represented as nearly as possible total disorder, colloquially, randomness.

Shannon [186] considered an information source that generates symbols from a finite set $\{x_i | i = 1, 2, \cdots n\}$ and transmits them as a stationary stochastic process. He defined the 'entropy' function for the process in terms of the probabilities $\{p_i | i = 1, 2, \cdots n\}$ for generation of the different symbols:

$$S = -\sum_{i=1}^{i=n} p_i \log(p_i). \tag{1.53}$$

This entropy (1.53) is essentially the same as that of Gibbs and Boltzmann in statistical mechanics but here it is viewed as a measure of the 'uncertainty' in the process; for example S is greater than or equal to the entropy conditioned by the knowledge of a second random variable. If the above symbols are generated mutually independently, then S is a measure of the amount of information available in the source for transmission. If the symbols in a sequence are not mutually independently generated, Shannon introduced the information 'capacity' of the transmission process as $C = \lim_{T\to\infty} \log N(T)/T$, where $N(T)$ is the maximum number of sequences of symbols that can be transmitted in time T. It follows that, for given entropy S and capacity C, the symbols can be encoded in such a way that $\frac{C}{S-\epsilon}$ symbols per second can be transmitted over the channel if $\epsilon > 0$ but not if $\epsilon < 0$. So again, we have a maximum principle from entropy.

Given a set of observed values $< g_\alpha(x) >$ for functions g_α of the random variable x, we seek a 'least prejudiced' set of probability values for x on the assumption that it can take only a finite number of values, x_i with probabilities $p_1, p_2, \cdots, p_n$ such that

$$< g_\alpha(x) > = \sum_{i=1}^{i=n} p_i\, g_\alpha(x_i) \ \text{ for } \alpha = 1, 2, \ldots, N \qquad (1.54)$$

$$1 = \sum_{i=1}^{i=n} p_i. \qquad (1.55)$$

Jaynes [107], a strong proponent of Shannon's approach, showed that this occurs if we choose those p_i that maximize Shannon's entropy function (1.53). In the case of a continuous random variable $x \in \mathbb{R}$ with probability density p parametrized by a finite set of parameters, the entropy becomes an integral and the maximizing principle is applied over the space of parameters, as we shall see below.

It turns out [201] that if we have no data on observed functions of x, (so the set of equations (1.54) is empty) then the maximum entropy choice gives the exponential distribution. If we have estimates of the first two moments of the distribution of x, then we obtain the (truncated) Gaussian. If we have estimates of the mean and mean logarithm of x, then the maximum entropy choice is the gamma distribution.

Jaynes [107] provided the foundation for information theoretic methods in, among other things, Bayes hypothesis testing—cf. Tribus et al. [200, 201]. For more theory, see also Slepian [190] and Roman [175, 176]. It is fair to point out that in the view of some statisticians, the applicability of the maximum entropy approach has been overstated; we mention for example the reservations of Ripley [173] in the case of statistical inference for spatial Gaussian processes.

In the sequel we shall consider the particular case of the gamma distribution for several reasons:

• the exponential distributions form a subclass of gamma distributions and exponential distributions represent Poisson inter-event distances
• the sum of n independent identical exponential random variables follows a gamma distribution

- the sum of n independent identical gamma random variables follows a gamma distribution
- lognormal distributions may be well-approximated by gamma distributions
- products of gamma distributions are well-approximated by gamma distributions
- stochastic porous media have been modelled using gamma distributions [72].

Other parametric statistical models based on different distributions may be treated in a similar way. Our particular interest in the gamma distribution and a bivariate gamma distribution stems from the fact that the exponential distribution is a special case and that corresponds to the standard model for an underlying Poisson process.

Let Θ be the parameter space of a parametric statistical model, that is an n-dimensional smooth family of probability density functions defined on some fixed event space Ω of unit measure,

$$\int_{\Omega} p_{\theta} = 1 \quad \text{for all } \theta \in \Theta.$$

For each sequence $X = \{X_1, X_2, \ldots, X_n\}$, of independent identically distributed observed values, the likelihood function lik_X on Θ which measures the likelihood of the sequence arising from different $p_{\theta} \in S$ is defined by

$$lik_X : \Theta \to [0,1] : \theta \mapsto \prod_{i=1}^{n} p_{\theta}(X_i).$$

Statisticians use the likelihood function, or log-likelihood its logarithm $l = \log lik$, in the evaluation of goodness of fit of statistical models. The so-called 'method of maximum likelihood', or 'maximum entropy' in Shannon's terms, is used to obtain optimal fitting of the parameters in a distribution to observed data.

1.4.1 Gamma Distribution

The family of gamma distributions is very widely used in applications with event space $\Omega = \mathbb{R}^+$. It has probability density functions given by

$$\Theta \equiv \{f(x; \gamma, \kappa) | \gamma, \kappa \in \mathbb{R}^+\}$$

so here $\Theta - \mathbb{R}^+ \times \mathbb{R}^+$ and the random variable is $x \in \Omega = \mathbb{R}^+$ with

$$f(x; \gamma, \kappa) = \left(\frac{\kappa}{\gamma}\right)^{\kappa} \frac{x^{\kappa-1}}{\Gamma(\kappa)} e^{-x\kappa/\gamma} \tag{1.56}$$

Then $\bar{x} = \gamma$ and $Var(x) = \gamma^2/\kappa$ and we see that γ controls the mean of the distribution while κ controls its variance and hence the shape. Indeed, the property that the variance is proportional to the square of the mean, §1.2, actually characterizes gamma distributions as shown recently by Hwang and Hu [106] (cf. their concluding remark).

Theorem 1.1 (Hwang and Hu [106]). *For independent positive random variables with a common probability density function f, having independence of the sample mean and the sample coefficient of variation is equivalent to f being the gamma distribution.*

The special case $\kappa = 1$ in (1.56) corresponds to the situation of the random or Poisson process along a line with mean inter-event interval γ, then the distribution of inter-event intervals is exponential. In fact, the gamma distribution has an essential generalizing property of the exponential distribution since it represents inter-event distances for generalizations of the Poisson process to a 'censored' Poisson process. Precisely, for integer $\kappa = 1, 2, \ldots$, (1.56) models a process that is Poisson but with intermediate events removed to leave only every κ^{th}. Formally, the gamma distribution is the κ-fold convolution of the exponential distribution, called also the Pearson Type III distribution. The Chi-square distribution with $n = 2\kappa$ degrees of freedom models the distribution of a sum of squares of n independent random variables all having the Gaussian distribution with zero mean and standard deviation σ; this is a gamma distribution with mean $\gamma = n\sigma^2$ if $\kappa = 1, 2, \ldots$. Figure 1.2 shows a family of gamma distributions, all of unit mean, with $\kappa = \frac{1}{2}$, 1, 2, 5.

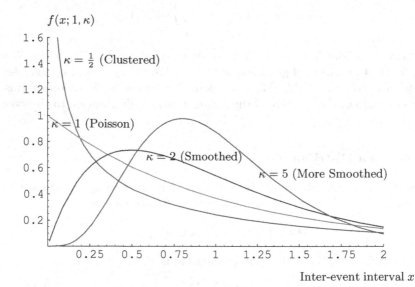

Fig. 1.2. Probability density functions, $f(x; \gamma, \kappa)$, for gamma distributions of inter-event intervals x with unit mean $\gamma = 1$, and $\kappa = \frac{1}{2}$, 1, 2, 5. The case $\kappa = 1$ corresponds to an exponential distribution from an underlying Poisson process. Some organization—clustering ($\kappa < 1$) or smoothing ($\kappa > 1$)—is represented by $\kappa \neq 1$.

Shannon's information theoretic entropy or 'uncertainty' is given, up to a factor, by the negative of the expectation of the logarithm of the probability density function (1.56), that is

$$S_f(\gamma, \kappa) = -\int_0^\infty \log(f(x; \gamma, \kappa)) f(x; \gamma, \kappa) \, dx$$

$$= \kappa + (1 - \kappa) \frac{\Gamma'(\kappa)}{\Gamma(\kappa)} + \log \frac{\gamma \, \Gamma(\kappa)}{\kappa}. \tag{1.57}$$

Part of the entropy function (1.57) is depicted in Figure 1.3 as a contour plot.

At unit mean, the maximum entropy (or maximum uncertainty) occurs at $\kappa = 1$, which is the random case, and then $S_f(\gamma, 1) = 1 + \log \gamma$. So, a Poisson process of points on a line is such that the points are as disorderly as possible and among all homogeneous point processes with a given density, the Poisson process has maximum entropy. Figure 1.4 shows a plot of $S_f(\gamma, \kappa)$, for the case of unit mean $\gamma = 1$. Figure 1.5 shows some integral curves of the entropy gradient field in the space of gamma probability density functions.

We can see the role of the log-likelihood function in the case of a set $X = \{X_1, X_2, \ldots, X_n\}$ of measurements, drawn from independent identically

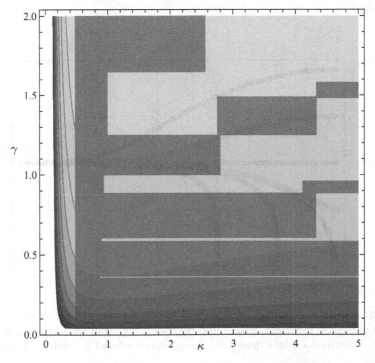

Fig. 1.3. Contour plot of information theoretic entropy $S_f(\gamma, \kappa)$, for gamma distributions from (1.57). The cases with $\kappa = 1$ correspond to exponential distributions related to underlying Poisson processes.

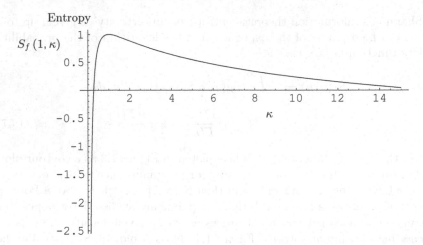

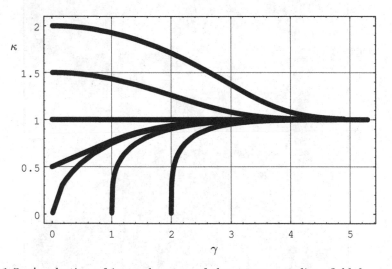

Fig. 1.4. Information theoretic entropy $S_f(\gamma, \kappa)$, for gamma distributions of inter-event intervals with unit mean $\gamma = 1$. The maximum at $\kappa = 1$ corresponds to an exponential distribution from an underlying Poisson process. The regime $\kappa < 1$ corresponds to clustering of events and $\kappa > 1$ corresponds to smoothing out of events, relative to a Poisson process. Note that, at constant mean, the variance of x decays like $1/\kappa$.

Fig. 1.5. A selection of integral curves of the entropy gradient field for gamma probability density functions, with initial points having small values of γ. The cases with $\kappa = 1$ correspond to exponential distributions related to underlying Poisson processes.

distributed random variables, to which we wish to fit the maximum likelihood gamma distribution. The procedure to optimize the choice of γ, κ is as follows. For independent events X_i, with identical distribution $p(x; \gamma, \kappa)$, their joint probability density is the product of the marginal densities so a measure of the 'likelihood' of finding such a set of events is

$$lik_X(\gamma, \kappa) = \prod_{i=1}^{n} f(X_i; \gamma, \kappa).$$

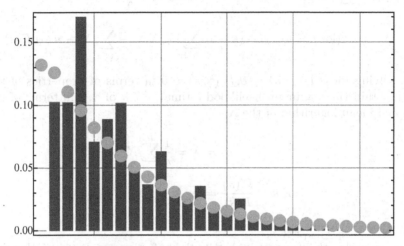

Fig. 1.6. Probability histogram plot with unit mean for the spacings between the first $100,000$ prime numbers and the maximum likelihood gamma fit, $\kappa = 1.09452$, (large points).

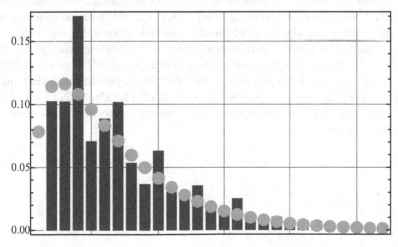

Fig. 1.7. Probability histogram plot with unit mean for the spacings between the first $100,000$ prime numbers and the gamma distribution having the same variance, so $\kappa = 1.50788$, (large points).

We seek a choice of γ, κ to maximize this product and since the log function is monotonic increasing it is simpler to maximize the logarithm

$$l_X(\gamma, \kappa) = \log lik_X(\gamma, \kappa) = \log[\prod_{i=1}^{n} f(X_i; \gamma, \kappa)].$$

Substitution gives us

$$l_X(\gamma, \kappa) = \sum_{i=1}^{n} [\kappa(\log \kappa - \log \gamma) + (\kappa - 1)\log X_i - \frac{\kappa}{\gamma} X_i - \log \Gamma(\kappa)]$$

$$= n\kappa(\log \kappa - \log \gamma) + (\kappa - 1)\sum_{i=1}^{n} \log X_i - \frac{\kappa}{\gamma}\sum_{i=1}^{n} X_i - n \log \Gamma(\kappa).$$

Then, solving for $\partial_\gamma l_X(\gamma, \kappa) = \partial_\kappa l_X(\gamma, \kappa) = 0$ in terms of properties of the X_i, we obtain the maximum likelihood estimates $\hat{\gamma}, \hat{\kappa}$ of γ, κ in terms of the mean and mean logarithm of the X_i

$$\hat{\gamma} = \bar{X} = \frac{1}{n}\sum_{i=1}^{n} X_i$$

$$\log \hat{\kappa} - \frac{\Gamma'(\hat{\kappa})}{\Gamma(\hat{\kappa})} = \overline{\log X} - \log \bar{X}$$

where $\overline{\log X} = \frac{1}{n}\sum_{i=1}^{n} \log X_i$.

For example, the frequency distribution of spacings between the first 100,000 prime numbers has mean approximately 13.0, and variance 112, and 99% of the probability is achieved by spacings up to 4 times the mean. Figure 1.6 shows the maximum likelihood fit gamma distribution with $\kappa = 1.09452$, as points, on the probability histogram of the prime spacings normalized to unit mean; the range of the abscissa is 4 times the mean. Figure 1.7 shows as points the gamma distribution with $\kappa = 1.50788$, which has the same variance as the prime spacings normalized to unit mean. Of course, neither fit is very good and nor is the geometric distribution approximation that might be expected, cf. Schroeder [184] §4.12, in light of The Prime Number Theorem, which says that the average spacing between adjacent primes near n is approximately $\log n$.

2

Introduction to Riemannian Geometry

This chapter is intended to help those with little previous exposure to differential geometry by providing a rather informal summary of background for our purposes in the sequel and pointers for those who wish to pursue more geometrical features of the spaces of probability density functions that are our focus in the sequel. In fact, readers who are comfortable with doing calculations of curves and their arc length on surfaces in $\mathbb{R}^3$ could omit this chapter at a first reading.

A topological space is the least structure that can support arguments concerning continuity and limits; our first experiences of such analytic properties is usually with the spaces $\mathbb{R}$ and $\mathbb{R}^n$. A manifold is the least structure that can support arguments concerning differentiability and tangents–that is, calculus. Our prototype manifold is the set of points we call Euclidean n-space $\mathbb{E}^n$ which is based on the real number n-space $\mathbb{R}^n$ and carries the Pythagorean distance structure. Our common experience is that a 2-dimensional Euclidean space can be embedded in $\mathbb{E}^3$, (or $\mathbb{R}^3$) as can curves and surfaces. Riemannian geometry generalizes the Euclidean geometry of surfaces to higher dimensions by handling the intrinsic properties like distances, angles and curvature independently of any environing simpler space.

We need rather little geometry of Riemannian manifolds in order to provide background for the concepts of information geometry. Dodson and Poston [70] give an introductory treatment with many examples, Spivak [194, 195] provides a six-volume treatise on Riemannian geometry while Gray [99] gave very detailed descriptions and computer algebraic procedures using *Mathematica* [215] for calculating and graphically representing most named curves and surfaces in Euclidean $\mathbb{E}^3$ and code for numerical solution of geodesic equations. Our Riemannian spaces actually will appear as subspaces of $\mathbb{R}^n$ so global properties will not be of particular significance and then the formulae and Gray's procedures easily generalize to more variables.

K. Arwini, C.T.J. Dodson, *Information Geometry.*
Lecture Notes in Mathematics 1953,
© Springer-Verlag Berlin Heidelberg 2008

2.0.2 Manifolds

A smooth n-manifold M is a (Hausdorff) topological space together with a collection of smooth maps (the charts)

$$\{\phi_\alpha : U_\alpha \longrightarrow \mathbb{R}^n \mid \alpha \in A\}$$

from open subsets U_α of M, which satisfy:

i) $\{U_\alpha \mid \alpha \in A\}$ is an open cover of M;
ii) each ϕ_α is a homeomorphism onto its image;
iii) whenever $U_\alpha \cap U_\beta \neq \emptyset$, then the maps between subsets of $\mathbb{R}^n$

$$\phi_\alpha \circ \phi_\beta^{-1} : \phi_\beta(U_\alpha \cap U_\beta) \longrightarrow \phi_\alpha(U_\alpha \cap U_\beta),$$

$$\phi_\beta \circ \phi_\alpha^{-1} : \phi_\alpha(U_\alpha \cap U_\beta) \longrightarrow \phi_\beta(U_\alpha \cap U_\beta),$$

have continuous derivatives of all orders (are C^∞ or smooth).

We call $\{(U_\alpha, \phi_\alpha) \mid \alpha \in A\}$ an atlas of charts for M; the properties of M are not significantly changed by adding more charts. The simplest example is the n-manifold $\mathbb{R}^n$ with atlas consisting of one chart, the identity map.

Intuitively, an n-manifold consists of open subsets of $\mathbb{R}^n$, the $\phi_\alpha(U_\alpha)$, pasted together in a smooth fashion according to the directions given by the $\phi_\alpha \circ \phi_\beta^{-1}$. For example, the unit circle $\mathbb{S}^1$ with its usual structure can be presented as a 1-manifold by pasting together two open intervals, each like $(-\pi, \pi)$. Similarly, the unit 2-sphere $\mathbb{S}^2$ has an atlas consisting of two charts

$$\{(U_N, \phi_N), \; (U_S, \phi_S)\}$$

where U_N consists of $\mathbb{S}^2$ with the north pole removed, U_S consists of $\mathbb{S}^2$ with the south pole removed, and the chart maps are stereographic projections. Thus, if $\mathbb{S}^2$ is the unit sphere in $\mathbb{R}^3$ centered at the origin then:

$$\phi_N : \mathbb{S}^2 \setminus \{n.p.\} \longrightarrow \mathbb{R}^2 : (x, y, z) \longmapsto \frac{1}{1+z}(x, y)$$

$$\phi_S : \mathbb{S}^2 \setminus \{s.p.\} \longrightarrow \mathbb{R}^2 : (x, y, z) \longmapsto \frac{1}{1-z}(x, y).$$

Similar chart maps work also for the higher dimensional spheres.

2.0.3 Tangent Spaces

From elementary analysis we know that the derivative of a function is a linear approximation to that function, at the chosen point. Thus, we need vector spaces to define linearity and these are automatically present in the form of the vector space $\mathbb{R}^n$ at each point of Euclidean point space $\mathbb{E}^n$. At each point x of a manifold M we construct a vector space $T_x M$, called the tangent space to M at x. For this we employ equivalence classes $[T_{\phi_\alpha(x)}\mathbb{R}^n]$ of tangent spaces to the images of x, $\phi_\alpha(x)$, under chart maps defined at x. That is, we borrow the vector space structure from $\mathbb{R}^n$ via each chart (U_α, ϕ_α) with $x \in U_\alpha$, then identify the borrowed copies. The result, for $x \in \mathbb{S}^2$ embedded in $\mathbb{R}^3$, is simply a vector space isomorphic to the tangent plane to $\mathbb{S}^2$ at x. This works here because $\mathbb{S}^2$ embeds isometrically into $\mathbb{R}^3$, but not all 2-manifolds embed in $\mathbb{R}^3$, some need more dimensions; the Klein bottle is an example [70]. Actually, the formal construction is independent of M being embedded in this way; however, the Whitney Embedding Theorem [211] says that an embedding of an n-manifold is always possible in $\mathbb{R}^{2n+1}$.

Once we have the tangent space $T_x M$ for each $x \in M$ we can present it in coordinates, via a choice of chart, as a copy of $\mathbb{R}^n$. The derivatives of the change of chart maps, like

$$\frac{\partial}{\partial x_\beta^i}(\phi_\alpha \circ \phi_\beta^{-1})\,(x_\beta^1, x_\beta^2, \cdots, x_\beta^n)\,,$$

provide linear transformations among the representations of $T_x M$. Next, we say that a map between manifolds

$$f : M \longrightarrow N$$

is differentiable at $x \in M$, if for some charts (U, ϕ) on M and (V, ψ) on N with $x \in U$, $f(x) \in V$, the map

$$\psi \circ f|_U \circ \phi^{-1} : \phi(U) \longrightarrow \psi(V)$$

is differentiable as a map between subsets of $\mathbb{R}^n$ and $\mathbb{R}^m$, if M is an n-manifold and N is an m-manifold. This property turns out to be independent of the choices of charts, so we get a linear map

$$T_x f : T_x M \longrightarrow T_{f(x)} N\,.$$

Moreover, if we make a choice of charts then $T_x f$ appears in matrix form as the set of partial derivatives of $\psi \circ f \circ \phi^{-1}$. The notation $T_x f$ for the derivative of f at x is precise, but in many texts it may be found abbreviated to Df, f_*, f' or Tf, with or without reference to the point of application. When f is a curve in M, that is, a map from some interval

$$f : [0,1] \longrightarrow M : t \mapsto f(t)\,,$$

then $T_t f$ is sometimes denoted by $\dot{f}_t$. This is the tangent map to f at t and the result of its application to the standard unit vector to $\mathbb{R}$ at t, $\dot{f}_t(\hat{1})$, is the tangent vector to f at t. It is quite common for this tangent vector also to be abbreviated to $\dot{f}_t$.

In a natural way we can provide a topology and differential structure for the set of all tangent vectors in all tangent spaces to an n-manifold M:

$$TM = \bigcup_{x \in M} T_x M \, ;$$

details are given in [70]. So, it actually turns out that TM is a $2n$-manifold, called the tangent bundle to M. For example, if $M = \mathbb{R}^n$ then $TM = \mathbb{R}^n \times \mathbb{R}^n$. Similarly, if $M = \mathbb{S}^1$ with the usual structure then TM is topologically (and as a manifold) equivalent to the infinite cylinder $\mathbb{S}^1 \times \mathbb{R}$. The technical term for an n-manifold M that has a trivial product tangent bundle $TM \cong M \times \mathbb{R}^n$ is parallelizable and this property is discussed in the cited texts.

On the other hand, this simple situation is quite rare and it is rather a deep result that for spheres

$$T\mathbb{S}^n \text{ is equivalent to } \mathbb{S}^n \times \mathbb{R}^n \text{ only for } n = 1, 3, 7 \, .$$

For other spheres, their tangent bundles consist of twisted products of copies of $\mathbb{R}^n$ over $\mathbb{S}^n$. In particular, $T\mathbb{S}^2$ is such a twisted product of $\mathbb{S}^2$ with one copy of $\mathbb{R}^2$ at each point. An intuitive picture of a 2-manifold that is a twisted product of $\mathbb{R}^1$ (or an interval from it) over $\mathbb{S}^1$ is a Möbius strip, which we know does not embed into $\mathbb{R}^2$ but does embed into $\mathbb{R}^3$.

A map $f : M \to N$ between manifolds is just called differentiable if it is differentiable at every point of M, and a diffeomorphism if it is differentiable with a differentiable inverse; in the latter case M and N are said to be diffeomorphic manifolds. Diffeomorphism implies homeomorphism, but not conversely. For example, the sphere $\mathbb{S}^2$ is diffeomorphic to an ellipsoid, but only homeomorphic to the surface of a cube because the latter is not a smooth manifold: it has corners and sharp edges so no well-defined tangent space structure. We note one generalisation however, sometimes we want a smooth manifold to have a boundary. For example a circular disc obviously cannot have its edge points homeomorphic to open sets in $\mathbb{R}^2$; so we relax our definition for charts to allow the chart maps to patch together open subsets like $\{(x, y) \in \mathbb{R}^2 | 0 < x \leq 1, 0 < y, < 1\}$ to deal with edge points. This is easily generalized to higher dimensions.

2.0.4 Tensors and Forms

For finite-dimensional real vector spaces it is easily shown that the set of all real-valued linear maps on the space is itself a real vector space, the dual space and similarly multilinear real-valued maps form real vector spaces; multilinear real-valued maps are called tensors. Elementary linear algebra introduces the notion of a real vector space X and its dual space X^* of real-valued linear functions on X; on manifolds we combine these types of spaces in a smooth way using tensor and exterior products to obtain the necessary composite bundle structures that can support the range of multilinear operations needed for geometry. Exterior differentiation, is the fundamental operation in the calculus on manifolds and it recovers all of vector calculus in $\mathbb{R}^3$ and extends it to arbitrary dimensional manifolds.

An m-form is a purely antisymmetric, real-valued, multilinear function on an argument of m tangent vectors, defined smoothly over the manifold. The space of m-forms becomes a vector bundle $\Lambda^m M$ over M with coordinate charts induced from those on M. A 0-form is a real valued function on the manifold. Thus, the space $\Lambda^0 M$ of 0-forms on M consists of sections of the trivial bundle $M \times \mathbb{R}$. The space $\Lambda^1 M$ of 1-forms on M consists of sections of the cotangent bundle $T^* M$, and $\Lambda^k M$ consists of sections of the antisymmetrized tensor product of k copies of $T^* M$. Locally, a 1-form has the local coordinates of an n-vector, a 2-form has the local coordinates of an antisymmetric $n \times n$ matrix. A k-form on an n-manifold has $\binom{n}{k}$ independent local coordinates. It follows that the only k-forms for $k > n$ are the zero k-forms. We summarize some definitions.

There are three fundamental operations on finite-dimensional vector spaces (in addition to taking duals): direct sum $\oplus$, tensor product $\otimes$, and exterior product $\wedge$ on a space with itself. Let F, G be two vector spaces, of dimensions n, m respectively. Take any bases $\{b_1, \cdots, b_n\}$ for $F, \{c_1, \cdots, c_m\}$ for G, then we can obtain bases

$$\{b_1, \cdots, b_n, \ c_1, \cdots, c_m\} \quad \text{for} \quad F \oplus G,$$

$$\{b_i \otimes c_j \mid i = 1, \cdots, n; \ j = 1, \cdots, m\} \quad \text{for} \quad F \otimes G,$$

$$\{b_i \wedge b_j \ = \ b_i \otimes b_j - b_j \otimes b_i \mid i = 1, \cdots, n; \ i < j\} \quad \text{for} \quad F \wedge F.$$

So, $F \oplus G$ is essentially the disjoint union of F and G with their zero vectors identified. In a formal sense (*cf.* Dodson and Poston [70], p. 104), $F \otimes G$ can be viewed as the vector space $L(F^*, G)$ of linear maps from the dual space $F^* = L(F, \mathbb{R})$ to G. Recall also the natural equivalence $(F^*)^* \cong F$. By taking the antisymmetric part of $F \otimes F$ we obtain $F \wedge F$. We deduce immediately:

$$\dim F \oplus G = \dim F + \dim G,$$

$$\dim F \otimes G = \dim F \cdot \dim G,$$

$$\dim F \wedge F = \frac{1}{2} \dim F(\dim F - 1).$$

Observe that only for $\dim F = 3$ can we have $\dim F = \dim(F \wedge F)$. Actually, this is the reason for the existence of the vector cross product $\times$ on $\mathbb{R}^3$ only, giving the uniquely important isomorphism

$$\mathbb{R}^3 \wedge \mathbb{R}^3 \longrightarrow \mathbb{R}^3 : x \wedge y \longmapsto x \times y$$

and its consequences for geometry and vector calculus on $\mathbb{R}^3$.

Each of the operations $\oplus, \otimes$ and $\wedge$ induces corresponding operations on linear maps between spaces. Indeed, the operations are thoroughly universal and categorical, so they should and do behave well in linear algebraic contexts. Briefly, suppose that we have linear maps $f, h \in L(F, J)$ $g \in (G, K)$ then the induced linear maps in $L(F \oplus G, \ J \oplus K)$, $L(F \otimes G, \ J \otimes K)$ and $L(F \wedge F, J \wedge J)$ are

$$f \oplus g : x \oplus y \longmapsto f(x) \oplus g(y) \,,$$
$$f \otimes g : x \otimes y \longmapsto f(x) \otimes g(y) \,,$$
$$f \wedge h : x \wedge y \longmapsto f(x) \wedge h(y) \,.$$

Local coordinates about a point in M induce bases for the tangent vector spaces and their spaces. The construction of the tangent spaces, directly from the choice of the differentiable structure for the manifold, induces a definite role for tangent vectors. An element $v \in T_x M$ turns out to be a derivation on smooth real functions defined near $x \in M$. In a chart about x, v is expressible as a linear combination of the partial derivations with respect to the chart coordinates $x^1, x^2, \ldots, x^n$ as

$$v \ = \ v^1 \partial_1 + v^2 \partial_2 + \cdots + v^n \partial_n$$

with $\partial_i = \frac{\partial}{\partial x^i}$, for some $v^i \in \mathbb{R}$.

This is often abbreviated to $v = v^i \partial_i$, where summation is to be understood over repeated upper and lower indices, the summation convention of Einstein. The dual base to $\{\partial_i\}$ is written $\{dx^i\}$ and defined by

$$dx^j(\partial_i) = \delta_i^j = \begin{cases} 1 & \text{if } i = j \,, \\ 0 & \text{if } i \neq j \,. \end{cases}$$

So a 1-form $\alpha \in T_x^* M$ is locally expressible as

$$\alpha = \alpha_1 dx^1 + \alpha_2 dx^2 + \cdots + \alpha_n dx^n = \alpha_i dx^i$$

for some $\alpha_i \in \mathbb{R}$, but a 2-form γ as

$$\gamma = \sum_{i<j} \gamma_{ij} dx^i \wedge dx^j$$

for some $\gamma_{ij} \in \mathbb{R}$. The common summation convention here is $\gamma = \gamma_{[ij]} dx^i \wedge dx^j$. A symmetric 2-tensor would use (ij).

Since the ∂_i and dx^i are well-defined in some chart (U, ϕ) about x, they serve also as basis vectors [70] at other points in U. Hence, they act as basis fields for the restrictions of sections of $TM \to M$ and $T^*M \to M$ to U, generating thereby local basis fields for sections of all tensor product bundles $T_m^k M \to M$ and exterior product bundles of forms $\Lambda^k M \to M$, restricted to U. The spaces of bases or frames form a structure called the frame bundle over a manifold, details of its geometry may be found in Cordero, Dodson and deLeon [43].

Given two vector fields u, v on M their commutator or Lie bracket is the new vector field $[u, v]$ defined as a derivation on real functions f by

$$[u, v](f) = u(v(f)) - v(u(f)).$$

Locally in coordinates using basis fields, for $u = u^i \partial_i$ and $v = v^j \partial_j$,

$$[u, v] = (u^i \partial_i v^j - v^i \partial_i u^j) \partial_j.$$

The exterior derivative is a linear map on k-forms satisfying

(i) $d : \Lambda^k M \to \Lambda^{k+1} M$ (d has degree $+1$);
(ii) $df = \text{grad } f$ if $f \in \Lambda^0 M$ (locally, $df = \partial_i f \, dx^i$);
(iii) if $\alpha \in \Lambda^a M$ and $\beta \in \Lambda^* M$, then

$$d(\alpha \wedge \beta) = d\alpha \wedge \beta + (-1)^a \alpha \wedge d\beta;$$

(iv) $d^2 = 0$.

This d is unique in satisfying these properties.

2.0.5 Riemannian Metric

We recall the importance of inner products on vector spaces—these allow the definition of lengths or norms of vectors and angles between vectors. The corresponding entity for the tangent vectors to an n-manifold M is a smooth choice of inner products over its family of vector spaces $\{T_x M \mid x \in M\}$. Such a smooth choice is called a Riemannian metric on M. Formally, a Riemannian metric g on n-manifold M is a smooth family of maps

$$g|_x : T_x M \times T_x M \to \mathbb{R}, \ x \in M$$

that is bilinear, symmetric and positive definite on each tangent space. Then we call the pair (M, g) a Riemannian n-manifold. Locally, at each $x \in M$, each $g|_x$ appears in coordinates as a symmetric $n \times n$ matrix $[g_{ij}]$ that is positive definite, so it has positive determinant. For each $v \in T_x M$, the norm of v is defined to be $||v|| = \sqrt{g(v, v)}$.

We can measure the angle θ between any two vectors u, v in the same tangent space by means of

$$\cos \theta = \frac{g(u, v)}{\sqrt{g(u, u) \, g(v, v)}}.$$

For a smooth curve in (M, g)

$$c : [0, 1] \longrightarrow M : t \longmapsto c(t)$$

with tangent vector field

$$\dot{c} : [0, 1] \longrightarrow TM : t \longmapsto \dot{c}(t)$$

the arc length is the integral of the norm of its tangent vector along the curve:

$$L_c(t) = \int_0^1 \sqrt{g_{c(t)}(\dot{c}(t), \dot{c}(t))}\, dt\,.$$

The arc length element ds along a curve can be expressed in terms of coordinates (x^i) by

$$ds^2 = \sum_{i,j} g_{ij}\, dx^i\, dx^j \tag{2.1}$$

which is commonly abbreviated to

$$ds^2 = g_{ij}\, dx^i\, dx^j \tag{2.2}$$

using the convention to sum over repeated indices.

Arc length is often difficult to evaluate analytically because it contains the square root of the sum of squares of derivatives. Accordingly, we sometimes use the 'energy' of the curve instead of length for comparison between nearby curves. Energy is given by integrating the square of the norm of $\dot{c}$

$$E_c(a,b) = \int_a^b ||\dot{c}(t)||^2\, dt. \tag{2.3}$$

A diffeomorphism f between Riemannian manifolds (M, g), (N, h) is called an isometry if its derivative Tf preserves the norms of all tangent vectors: $g(v, v) = h(Tf(v), Tf(v))$. A situation of common interest is when a manifold can be isometrically embedded as a submanifold of some Euclidean $\mathbb{E}^m$ or of $\mathbb{R}^m$ with some specified metric. Note that if we have a Riemannian manifold (M, g) then an open subset X of M inherits a manifold structure using the restriction of chart maps and the metric g induces a subspace metric $g|_X$ so $(X, g|_X)$ becomes a Riemannian submanifold of (M, g). For example, the unit sphere $\mathbb{S}^2$ in $\mathbb{E}^3$ inherits the subspace metric from the Euclidean metric but of course $\mathbb{S}^2$ has spherical not Euclidean geometry. Evidently, the dimension of a submanifold will not exceed the dimension of its host manifold.

2.0.6 Connections

In order to compare tangent vectors at different points along a curve in a manifold M we need to have a procedure that transports tangent space vectors along the curve, so providing a way to 'connect up' unambiguously the tangent spaces passed through. A smooth assignation of tangent vectors along a curve is called a vector field along the curve; one such field is the actual field of tangents to the curve. A suitable connecting entity in the limiting case at a point defines a derivative of a vector field with respect to the tangent to the curve, and gives the result as another tangent vector at the same point. Now, every tangent vector $u \in T_x M$ can be realised as the tangent vector to a curve through x and therefore we finish up with a smooth family of bilinear maps $\nabla = \{\nabla|_x, x \in M\}$ with the property

$$\nabla|_x : T_x \times T_x \to T_x : (u,v) \mapsto \nabla_u v, \quad \text{defined over } x \in M. \qquad (2.4)$$

In coordinates, we have a basis of $T_x M$ given by the derivations (∂_i) and so for some real components (u^i), (v^j), using the summation convention for repeated indices and (∂_i) as basis vector fields $u = u^i \partial_i$, $v = v^j \partial_j$ and then

$$\nabla_u v = (u^i \partial_i v^j + u^k v^m \Gamma^j_{km}) \partial_j \qquad (2.5)$$

for a smooth $n \times n \times n$ array of functions Γ^j_{km} called the Christoffel symbols. It turns out that ∇ provides a derivative for vector valued maps on the manifold, that is of vector field maps $v : M \to TM$, and returns the answer as another vector field; this derivation operator is called the covariant derivative. The smooth family of bilinear maps (2.4) is called a linear connection and there are many ways to formalise its definition [70]. The important theorem here is that for a given Riemannian manifold there is a unique linear connection that preserves the metric and has symmetric Christoffel symbols, this is the Levi-Civita or symmetric metric connection.

Now, we have seen above §2.0.3 that the derivative of a smooth map between manifolds $f : M \to N$ gives a corresponding map $Tf : TM \to TN$. Also, a vector field v on M, is a section $v : M \to TM$ of the tangent bundle projection $\pi : TM \to M$; this means that $\pi \circ v$ is the identity map on M. Therefore the derivative of the vector field will not be another vector field but a map $Tv : TM \to TTM$. This is why we need the connection, it provides a projection of a derivative Tv back onto the the tangent bundle; the covariant derivative of a vector field is precisely the projection of a derivative.

Formally, a linear connection ∇ gives a smooth bundle splitting at each $u \in TTM$ of the space $T_u TM$ into a direct sum

$$T_u TM \cong H_u TM \oplus V_u TM$$

where $V_u TM = \ker(T\pi : T_u TM \to T_{\pi(u)} M)$. We call $H_u TM$ the horizontal subspace (of TTM) at $u \in TM$ and $V_u TM$ the vertical subspace at $u \in TM$. They comprise the horizontal and vertical subbundles, respectively, of TTM.

$$TTM = HTM \oplus VTM.$$

For our purposes, the important role of a connection is that it induces isomorphisms called horizontal lifts from tangent spaces on the base M to horizontal subspaces of the tangent spaces to TM:

$$\uparrow \; : T_{\pi(u)} M \longrightarrow H_u TM \subset T_u TM : v \longmapsto v^\uparrow.$$

Technically, a connection splits the exact sequence of vector bundles

$$0 \longrightarrow VTM \longrightarrow TTM \longrightarrow TM \longrightarrow 0$$

by providing a bundle morphism $TM \to TTM$ with image the bundle of horizontal subspaces.

Along any curve $c : [0, 1) \to M$ in M we can construct through each $u_0 \in \pi^{-1}(c(0)) \subset TM$ a unique curve $c^\uparrow : [0, 1) \longrightarrow TM$ with horizontal tangent vector and $\pi \circ c^\uparrow = c$, $c^\uparrow(0) = u_0$. The map

$$\tau_t : \pi^{-1}(c(0)) \longrightarrow \pi^{-1}(c(t)) : u_0 \longmapsto c^\uparrow(t)$$

defined by the curve is called parallel transport along c. Parallel transport is always a linear isomorphism. An associated parallel transport map satisfies $\tilde{\tau}_t \circ v(c(t)) = v(c(t))$. The covariant derivative of v along c is defined to be the limit, if it exists

$$\lim_{h \to 0} \frac{1}{h} \left(\tilde{\tau}_h^{-1} \circ v(c(t+h)) - v(c(t)) \right)$$

and is usually denoted by $\nabla_{\dot{c}(t)} v$. Using integral curves c, this extends easily to $\nabla_w v$ for any vector field w. Evidently, the operator ∇ is linear and a derivation:

$$\nabla_w(u + v) = \nabla_w u + \nabla_w v \quad \text{and} \quad \nabla_w(fv) = w(f)v + f\nabla_w v \, ;$$

it measures the departure from parallelism. The local appearance of ∇ on basis fields (∂_i) about $x \in M$ is

$$\nabla_{\partial_i} \partial_j = \Gamma_{ij}^k \, \partial_k$$

where the Γ_{ij}^k are the Christoffel symbols defined earlier.

For a linear connection we define two important tensor fields in terms of their action on tangent vector fields: the torsion tensor field T is defined by

$$T(u, v) = \nabla_u v - \nabla_v u - [u, v]$$

and the curvature tensor field is the section of $T_3^1 M$ defined by

$$R(u, v)w = \nabla_u \nabla_v w - \nabla_v \nabla_u w - \nabla_{[u,v]} w \, .$$

The connection is called torsion-free or symmetric when $T = 0$ and flat when $R = 0$.

In local coordinates with respect to base fields (∂_i),

$$T(\partial_j, \partial_k) = (\Gamma_{jk}^i - \Gamma_{kj}^i)\partial_i \, ,$$
$$R(\partial_k, \partial_l)\partial_j = (\partial_k \Gamma_{lj}^i - \partial_l \Gamma_{kj}^i + \Gamma_{lj}^h \Gamma_{kh}^i - \Gamma_{kj}^h \Gamma_{lh}^i)\partial_i \, .$$

The connection form ω is an $\mathbb{R}^{n^2}$-valued linear function on vector fields and is expressible as a matrix valued 1-form with components

$$\omega_j^i = \Gamma_{jk}^i \, dx^k \, . \tag{2.6}$$

Hence

$$d\omega_j^i = d(\Gamma_{jk}^i) \wedge dx^k$$
$$= \partial_r\, \Gamma_{jk}^i\, dx^r \wedge dx^k$$
$$\omega_h^i \wedge \omega_j^h = \Gamma_{hr}^i\, \Gamma_{jk}^h\, dx^r \wedge dx^k$$

The curvature form Ω is an $\mathbb{R}^{n^2}$-valued antisymmetric bilinear function on pairs of vector fields and it has the local expression

$$\Omega_j^i = \frac{1}{2} R_{jrk}^i\, dx^r \wedge dx^k$$
$$= R_{jrk}^i\, dx^r \wedge dx^k\,.$$

2.1 Autoparallel and Geodesic Curves

A curve $c : [0,1) \to M$ that has a parallel tangent vector field $\dot{c} = \dot{c}^j \partial_j$ satisfies:

$$\nabla_{\dot{c}(t)} \dot{c}(t) = 0 \qquad (2.7)$$

which in coordinate components from (2.5) becomes

$$\ddot{c}^i + \Gamma_{jk}^i\, \dot{c}^j \dot{c}^k = 0 \text{ for each } i.$$

It is then called an autoparallel curve . In the case that the connection ∇ is the Levi-Civita connection of a Riemannian manifold (M, g), all the parallel transport maps are actually isometries and then the autoparallel curves c satisfying (2.7) are called geodesic curves (cf. [70] for more discussion of geodesic curves). Geodesic curves have extremal properties—between close enough points they provide uniquely shortest length curves. For example, in Euclidean $\mathbb{E}^3$ the geodesics are straight lines and so provide shortest distances between points; on the standard unit sphere $\mathbb{S}^2 \subset \mathbb{E}^3$ the geodesics are arcs of great circles and so between pairs of points the two arcs provide maximal and minimal geodesic distances.

2.2 Universal Connections and Curvature

A connection, §2.0.6 encodes geometrical choices, and through its curvature, underlying topological information. In some situations, both in geometry and in theoretical physics, it is necessary to consider a family of connections, for example with regard to stability of certain properties [36]. Also, it is common for statisticians to consider a number of linear connections on a given statistical manifold and so it can be important to be able to handle these connections as a geometrical family of some kind.

In general, the space of linear connections on a manifold is infinite dimensional, but Mangiarotti and Modugno [140, 152] introduced the idea of a system (or structure) of connections which gives a representation of the space of linear connections as a finite dimensional bundle. On this system there is a 'universal' connection and corresponding 'universal' curvature; then all linear connections and their curvatures are pullbacks of these universal objects.

A full account of the underlying geometry of jet bundles and their morphisms is beyond our present scope so we refer the interested reader to Mangiarotti and Modugno [140, 152]. Dodson and Modugno [69] provided a universal calculus for this context. An application of universal linear connections to a stability problem in spacetime geometry was given by Canarutto and Dodson [36] and further properties of the system of linear connections were given by Del Riego and Dodson [53]. An explicit set of geometrical examples with interesting topological properties was provided by Cordero, Dodson and Parker [44]. The first application to information geometry was given by Dodson [59] for the system of α-connections.

The technical details would take us too far from our present theme but our recent results on statistical manifolds are given in Arwini, Del Riego and Dodson [16]. There we describe the system of all linear connections on the manifold of exponential families, using the tangent bundle, §2.0.3, to give the system space. We provide formulae for the universal connections and curvatures and give an explicit example for the manifold of gamma distributions, §3.5. It seems likely that there could be significant developments from the results on universal connections for exponential families §3.2, for example in the context of group actions on random variables.

3

Information Geometry

We use the term information geometry to cover those topics concerning the use of the Fisher information matrix to define a Riemannian metric, § 2.0.5, on smooth spaces of parametric statistical models, that is, on smooth spaces of probability density functions. Amari [8, 9], Amari and Nagaoka [11], Barndorff-Nielsen and Cox [20], Kass and Vos [113] and Murray and Rice [153] provide modern accounts of the differential geometry that arises from the Fisher information metric and its relation to asymptotic inference. The Introduction by R.E. Kass in [9] provided a good summary of the background and role of information geometry in mathematical statistics. In the present monograph, we use Riemannian geometric properties of various families of probability density functions in order to obtain representations of practical situations that involve statistical models.

It has by many experts been argued that the information geometric approach may not add significantly to the understanding of the theory of parametric statistical models, and this we acknowledge. Nevertheless, we are of the opinion that there is benefit for those involved with practical modelling if essential qualitative features that are common across a wide range of applications can be presented in a way that allows geometrical tools to measure distances between and lengths along trajectories through perturbations of models of relevance. Historically, the richness of operations and structure in geometry has had a powerful influence on physics and those applications suggested new geometrical developments or methodologies; indeed, from molecular biology some years ago, the behaviour of certain enzymes in DNA manipulation led to the identification of useful geometrical operators. What we offer here is some elementary geometry to display the features common, and of most significance, to a wide range of typical statistical models for real processes. Many more geometrical tools are available to make further sophisticated studies, and we hope that these may attract the interest of those who model. For example, it would be interesting to explore the details of the role of curvature in a variety of applications, and to identify when the distinguished curves called geodesics, so important in fundamental physics, have particular significance in various real

K. Arwini, C.T.J. Dodson, *Information Geometry*.
Lecture Notes in Mathematics 1953,
© Springer-Verlag Berlin Heidelberg 2008

processes with essentially statistical features. Are there useful ways to compactify some parameter spaces of certain applications to benefit thereby from algebraic operations on the information geometry? Do universal connections on our information geometric spaces have a useful role in applications?

3.1 Fisher Information Metric

First we set up a smooth n-manifold, § 2.0.2, with a chart the n-dimensional parameter space of a smooth family of probability density functions, § 1.2. Let Θ be the parameter space of a parametric statistical model, that is an n-dimensional smooth family $\{p_\theta | \theta \in \Theta\}$ of probability density functions defined on some fixed event space Ω of unit measure,

$$\int_\Omega p_\theta = 1 \quad \text{for all } \theta \in \Theta.$$

Let Θ be the parameter space of an n-dimensional smooth such family defined on some fixed event space Ω

$$\{p_\theta | \theta \in \Theta\} \quad \text{with} \quad \int_\Omega p_\theta = 1 \quad \text{for all } \theta \in \Theta.$$

Then, the derivatives of the log-likelihood function, $l = \log p_\theta$, yield a matrix and the expectation of its entries is

$$g_{ij} = \int_\Omega p_\theta \left(\frac{\partial l}{\partial \theta^i} \frac{\partial l}{\partial \theta^j} \right) = -\int_\Omega p_\theta \left(\frac{\partial^2 l}{\partial \theta^i \partial \theta^j} \right), \tag{3.1}$$

for coordinates (θ^i) about $\theta \in \Theta \subseteq \mathbb{R}^n$.

This gives rise to a positive definite matrix, so inducing a Riemannian metric g, from §2.0.5, on Θ using for coordinates the parameters (θ^i). From the construction of (3.1), a smooth invertible transformation of random variables, that is of the labelling of the points in the event space Ω while keeping the same parameters (θ^i), will leave the Riemannian metric unaltered. Formally, it induces a smooth diffeomorphism of manifolds that preserves the metric, namely it induces a Riemannian isometry; in this situation the diffeomorphism is simply the identity map on parameters. We shall see this explicitly below for the case of the log-gamma distribution §3.6 and its associated Riemannian manifold.

The elements in the matrix (3.1) give the arc length function §2.1

$$ds^2 = \sum_{i,j} g_{ij} \, d\theta^i \, d\theta^j \tag{3.2}$$

which is commonly abbreviated to

$$ds^2 = g_{ij} \, d\theta^i \, d\theta^j \tag{3.3}$$

using the convention to sum over repeated indices.

The metric (3.1) is called the expected information metric or Fisher metric for the manifold obtained from the family of probability density functions; the original ideas are due to Fisher [86] and C.R. Rao [171]. Of course, the second equality in equation (3.1) depends on certain regularity conditions [188] but when it holds it can be particularly convenient to use. Amari [8, 9] and Amari and Nagaoka [11], Barndorff-Nielsen and Cox [20], Kass and Vos [113] and Murray and Rice [153] provide modern accounts of the differential geometry that arises from the information metric.

3.2 Exponential Family of Probability Density Functions

An n-dimensional parametric statistical model $\Theta \equiv \{p_\theta | \theta \in \Theta\}$ is said to be an exponential family or of exponential type, when the density function can be expressed in terms of functions $\{C, F_1, ..., F_n\}$ on Λ and a function φ on Θ as:

$$p(x; \theta) = e^{\{C(x) + \sum_i \theta_i F_i(x) - \varphi(\theta)\}} , \tag{3.4}$$

then we say that (θ_i) are its natural or canonical parameters, and φ is the potential function. From the normalization condition $\int p(x; \theta)\, dx = 1$ we obtain:

$$\psi(\theta) = \log \int e^{\{C(x) + \sum_i \theta_i F_i(x)\}}\, dx . \tag{3.5}$$

This potential function is therefore a distinguished function of the coordinates alone and in the sequel, § 3.4, we make use of it for the presentation of the manifold as an immersion in $\mathbb{R}^{n+1}$. From the normalization condition $\int_\Omega p_\theta(x)\, dx = 1$ we obtain:

$$\varphi(\theta) = \log \int_\Omega e^{\{C(x) + \sum_i (\theta^i F_i(x))\}}\, dx . \tag{3.6}$$

With $\partial_i = \frac{\partial}{\partial \theta^i}$, we use the log-likelihood function, § 1.4, $l(\theta, x) = \log(p_\theta(x))$ to obtain

$$\partial_i l(\theta, x) = F_i(x) - \partial_i \varphi(\theta)$$

and

$$\partial_i \partial_j l(\theta, x) = -\partial_i \partial_j \varphi(\theta) .$$

The information metric g on the n-dimensional space of parameters $\Theta \subset \mathbb{R}^n$, equivalently on the set $S = \{p_\theta | \theta \in \Theta \subset \mathbb{R}^n\}$, has coordinates:

$$[g_{ij}] = -\int_\Omega [\partial_i \partial_j l(\theta, x)]\, p_\theta(x)\, dx = \partial_i \partial_j \varphi(\theta) = \varphi_{ij}(\theta) . \tag{3.7}$$

Then, (S, g) is a Riemannian n-manifold, § 2.0.5, with Levi-Civita connection, § 2.0.6, given by:

$$\Gamma_{ij}^k(\theta) = \sum_{h=1}^{n} \frac{1}{2} g^{kh} \left(\partial_i g_{jh} + \partial_j g_{ih} - \partial_h g_{ij} \right)$$

$$= \sum_{h=1}^{n} \frac{1}{2} g^{kh} \, \partial_i \partial_j \partial_h \varphi(\theta) = \sum_{h=1}^{n} \frac{1}{2} \varphi^{kh}(\theta) \, \varphi_{ijh}(\theta)$$

where $[\varphi^{hk}(\theta)]$ represents the inverse to $[\varphi_{hk}(x)]$.

3.3 Statistical a-Connections

There is a family of symmetric linear connections, § 2.0.6, which includes the Levi-Civita case, § 2.0.6, and it has certain uniqueness properties and significance in mathematical statistics. See Amari [9] and Lauritzen [134] for more details and properties than we have space for here. In § 2.2 we discuss how families of linear connections have certain universal properties.

Consider for $\alpha \in \mathbb{R}$ the function $\Gamma_{ij,k}^{(\alpha)}$ which maps each point $\theta \in \Theta$ to the following value:

$$\Gamma_{ij,k}^{(\alpha)}(\theta) = \int_{\Omega} \left(\partial_i \partial_j l + \frac{1-\alpha}{2} \partial_i l \, \partial_j l \right) \partial_k l \, p_\theta$$

$$= \frac{1-\alpha}{2} \partial_i \partial_j \partial_k \varphi(\theta) = \frac{1-\alpha}{2} \varphi_{ijk}(\theta). \tag{3.8}$$

So we have an affine connection $\nabla^{(\alpha)}$ on the statistical manifold (S, g) defined by

$$g(\nabla_{\partial_i}^{(\alpha)} \partial_j, \partial_k) = \Gamma_{ij,k}^{(\alpha)},$$

where g is the Fisher information metric, §3.1. We call this $\nabla^{(\alpha)}$ the α-connection and it is clearly a symmetric connection, §2.0.6 and defines an α-curvature. We have also

$$\nabla^{(\alpha)} = (1-\alpha) \, \nabla^{(0)} + \alpha \, \nabla^{(1)},$$

$$= \frac{1+\alpha}{2} \, \nabla^{(1)} + \frac{1-\alpha}{2} \, \nabla^{(-1)}.$$

For a submanifold $M \subset S$, §2.0.5, the α-connection on M is simply the restriction with respect to g of the α-connection on S. Note that the 0-connection is the Riemannian or Levi-Civita, §2.0.6 connection with respect to the metric and its uniqueness implies that an α-connection is a metric connection if and only if $\alpha = 0$.

Proposition 3.1. *The 0-connection is the Riemannian connection, metric connection or Levi-Civita connection with respect to the Fisher metric.*

In general, when $\alpha \neq 0$, $\nabla^{(\alpha)}$ is not metric.

The notion of exponential family, §3.2 has a close relation to $\nabla^{(1)}$. From the definition of an exponential family given in Equation (3.4), with $\partial_i = \frac{\partial}{\partial \theta_i}$, we obtain

$$\partial_i \ell(x; \theta) = F_i(x) - \partial_i \varphi(\theta) \tag{3.9}$$

and

$$\partial_i \partial_j \ell(x; \theta) = -\partial_i \partial_j \varphi(\theta). \tag{3.10}$$

where $\ell(x; \theta) = \log f(x; \theta)$.

Hence we have $\Gamma^{(1)}_{ij,k} = -\partial_i \partial_j \varphi \, E_\theta[\partial_k \ell_\theta]$, which is 0. In other words, we see that (θ_i) is a 1-affine coordinate system, and Θ is 1-flat.

In particular, the 1-connection is said to be an exponential connection, and the (-1)-connection is said to be a mixture connection. We say that an α-connection and the $(-\alpha)$-connection are mutually dual with respect to the metric g since the following formula holds:

$$X g(Y, Z) = g(\nabla^{(\alpha)}_X Y, Z) + g(Y, \nabla^{(-\alpha)}_X Z),$$

where X, Y and Z are arbitrary vector fields on M.

Now, Θ is an exponential family, so a mixture coordinate system is given by a potential function, §3.2, that is,

$$\eta_i = \frac{\partial \varphi}{\partial \theta_i}. \tag{3.11}$$

Since (θ_i) is a 1-affine coordinate system, (η_i) is a (-1)-affine coordinate system, and they are mutually dual with respect to the metric. Therefore the statistical manifold has dually orthogonal foliations (Section 3.7 in [11]).

The coordinates in (η_i) admit a potential function given by:

$$\lambda = \theta_i \eta_i - \varphi(\theta). \tag{3.12}$$

3.4 Affine Immersions

Let M be an m-dimensional manifold, f an immersion from M to $\mathbb{R}^{m+1}$, and ξ a vector field along f. We can $\forall x \in R^{m+1}$, identify $T_x R^{m+1} \equiv \mathbb{R}^{m+1}$. The pair $\{f, \xi\}$ is said to be an affine immersion from M to $\mathbb{R}^{m+1}$ if, for each point $p \in M$, the following formula holds:

$$T_{f(p)} R^{m+1} = f_*(T_p M) \oplus Span\{\xi_p\}.$$

We call ξ a transversal vector field and it is a technical requirement to ensure that the differential structure is preserved into the immersion.

We denote by D the standard flat affine connection of $\mathbb{R}^{m+1}$. Identifying the covariant derivative along f with D, we have the following decompositions:

$$D_X f_* Y = f_*(\nabla_X Y) + h(X, Y)\xi,$$
$$D_X \xi = -f_*(Sh(X)) + \mu(X)\xi.$$

The induced objects ∇, h, Sh and μ are the induced connection, the affine fundamental form, the affine shape operator and the transversal connection form, respectively. If the affine fundamental form h is positive definite everywhere on M, the immersion f is said to be strictly convex. And if $\mu = 0$, the affine immersion $\{f, \xi\}$ is said to be equiaffine. It is known that a strictly convex equiaffine immersion induces a statistical manifold. Conversely, the condition when a statistical manifold can be realized in an affine space has been studied. We say that an affine immersion $\{f, \xi\} : \Theta \to \mathbb{R}^{m+1}$ is a graph immersion if the hypersurface is a graph of φ in $\mathbb{R}^{m+1}$:

$$f : M \to \mathbb{R}^{m+1} : \begin{bmatrix} \theta_1 \\ \cdot \\ \cdot \\ \cdot \\ \theta_m \end{bmatrix} \mapsto \begin{bmatrix} \theta_1 \\ \cdot \\ \cdot \\ \theta_m \\ \varphi(\theta) \end{bmatrix}, \quad \xi = \begin{bmatrix} 0 \\ 0 \\ 0 \\ 1 \end{bmatrix},$$

Set $\partial_i = \frac{\partial}{\partial \theta_i}$, $\varphi_{ij} = \frac{\partial^2 \varphi}{\partial \theta_i \partial \theta_j}$. Then we have

$$D_{\partial_i} f_* \partial_j = \varphi_{ij} \xi.$$

This implies that the induced connection ∇ is flat and (θ_i) is a ∇-affine coordinate system.

Proposition 3.2. *Let (M, h, ∇, ∇^*) be a simply connected dually flat space with a global coordinate system and (θ) an affine coordinate system of ∇. Suppose that φ is a θ-potential function. Then (M, h, ∇) can be realized in $\mathbb{R}^{m+1}$ by a graph immersion whose potential is φ.*

3.4.1 Weibull Distributions: Not of Exponential Type

We shall see that gamma density functions form an exponential family; however, some distributions do not and one such example is given by the Weibull family for nonnegative random variable x:

$$w(x; \kappa, \tau) = \kappa\tau(\kappa x)^{\tau - 1} e^{(\kappa x)^{\tau - 1}} \quad \kappa, \tau > 0. \tag{3.13}$$

Like the gamma family, the Weibull family (3.13) contains the exponential distribution as a special case and has wide application in models for reliability and lifetime statistics, but the lack of a natural affine immersion to present perturbations of the exponential distribution makes it unsuitable for our present purposes. We provide elsewhere [17] the $\alpha-$connection and $\alpha-$curvature for the Weibull family and illustrate the geometry of its information metric with examples of geodesics.

3.5 Gamma 2-Manifold $\mathcal{G}$

The family of gamma density functions has event space $\Omega = \mathbb{R}^+$ and probability density functions given by

$$\{f(x;\gamma,\kappa)|\gamma,\kappa \in \mathbb{R}^+\}$$

so here $M \equiv \mathbb{R}^+ \times \mathbb{R}^+$ and the random variable is $x \in \Omega = \mathbb{R}^+$ with

$$f(x;\gamma,\kappa) = \left(\frac{\kappa}{\gamma}\right)^\kappa \frac{x^{\kappa-1}}{\Gamma(\kappa)} e^{-x\kappa/\gamma}. \qquad (3.14)$$

Proposition 3.3. *Denote by $\mathcal{G}$ the gamma manifold based on the family of gamma density functions. Set $\nu = \frac{\kappa}{\gamma}$. Then the probability density functions have the form*

$$p(x;\nu,\kappa) = \nu^\kappa \frac{x^{\kappa-1} e^{-x\nu}}{\Gamma(\kappa)}. \qquad (3.15)$$

In this case (ν,κ) is a natural coordinate system of the 1-connection and

$$\varphi(\theta) = \log \Gamma(\kappa) - \kappa \log \nu \qquad (3.16)$$

is the corresponding potential function, §3.2.

Proof. Using $\nu = \frac{\kappa}{\gamma}$, the logarithm of gamma density functions can be written as

$$\log p(x;\nu,\kappa) = \log\left(\nu^\kappa \frac{x^{\kappa-1}}{\Gamma(\kappa)} e^{-\nu x}\right)$$
$$= -\log x + (\kappa \log x - \nu x) - (\log \Gamma(\kappa) - \kappa \log \nu) \quad (3.17)$$

Hence the set of all gamma density functions is an exponential family, §3.2. The coordinates $(\theta_1,\theta_2) = (\nu,\kappa)$ is a natural coordinate system, §3.3, and

$$\varphi(\nu,\kappa) = \log \Gamma(\kappa) - \kappa \log \nu$$

is its potential function. $\qquad \square$

Corollary 3.4. *Since $\varphi(\theta)$ is a potential function, the Fisher metric is given by the Hessian of φ, that is, with respect to natural coordinates:*

$$[g_{ij}](\nu,\kappa) = \left[\frac{\partial^2 \varphi(\theta)}{\partial \theta_i \partial \theta_j}\right] = \begin{bmatrix} \frac{\kappa}{\nu^2} & -\frac{1}{\nu} \\ -\frac{1}{\nu} & \psi''(\kappa) \end{bmatrix} = \begin{bmatrix} \frac{\kappa}{\nu^2} & -\frac{1}{\nu} \\ -\frac{1}{\nu} & \frac{d^2}{d\kappa^2} \log(\Gamma) \end{bmatrix}. \qquad (3.18)$$

In terms of the original coordinates (γ, κ) in equation (3.14), the gamma density functions are

$$f(x; \gamma, \kappa) = \left(\frac{\kappa}{\gamma}\right)^{\kappa} \frac{x^{\kappa-1}}{\Gamma(\kappa)} e^{-x\kappa/\gamma}$$

and then the metric components matrix takes a convenient diagonal form

$$[g_{ij}](\gamma, \kappa) = \begin{bmatrix} \frac{\kappa}{\gamma^2} & 0 \\ 0 & \frac{d^2}{d\kappa^2} \log(\Gamma) - \frac{1}{\kappa} \end{bmatrix}. \tag{3.19}$$

So the pair (γ, κ) yields an orthogonal basis of tangent vectors, which is useful in calculations because then the arc length function §2.1 is simply

$$ds^2 = \frac{\kappa}{\gamma^2} d\gamma^2 + \left(\left(\frac{\Gamma'(\kappa)}{\Gamma(\kappa)}\right)' - \frac{1}{\kappa}\right) d\kappa^2.$$

This orthogonality property of (γ, κ) coordinates is equivalent to asymptotic independence of the maximum likelihood estimates, cf. Barndorff-Nielsen and Cox [20], Kass and Vos [113] and Murray and Rice [153] [20, 113, 153].

3.5.1 Gamma a-Connection

For each $\alpha \in \mathbb{R}$, the α (or $\nabla^{(\alpha)}$)-connection is the torsion-free affine connection with components:

$$\Gamma_{ij,k}^{(\alpha)} = \frac{1-\alpha}{2} \partial_i \partial_j \partial_k \varphi(\theta),$$

where $\varphi(\theta)$ is the potential function, and $\partial_i = \frac{\partial}{\partial \theta_i}$.

Since the set of gamma density functions is an exponential family, §3.2, and the connection $\nabla^{(1)}$ is flat. In this case, (ν, κ) is a 1-affine coordinate system.

So the 1 and (-1)-connections on the gamma manifold are flat.

Proposition 3.5. *The functions $\Gamma_{ij,k}^{(\alpha)}$ are given by*

$$\Gamma_{11,1}^{(\alpha)} = -\frac{(1-\alpha)\,\kappa}{\nu^3},$$

$$\Gamma_{12,1}^{(\alpha)} = \Gamma_{12,2}^{(\alpha)} = \frac{1-\alpha}{2\,\nu^2},$$

$$\Gamma_{22,2}^{(\alpha)} = \frac{(1-\alpha)\,\psi''(\kappa)}{2} \tag{3.20}$$

while the other independent components are zero. □

We have an affine connection $\nabla^{(\alpha)}$ defined by

$$\langle \nabla^{(\alpha)}_{\partial_i} \partial_j, \partial_k \rangle = \Gamma^{(\alpha)}_{ij,k},$$

So by solving the equations

$$\Gamma^{(\alpha)}_{ij,k} = \sum_{h=1}^{2} g_{kh}\, \Gamma^{h(\alpha)}_{ij}, (k=1,2).$$

we obtain the components of $\nabla^{(\alpha)}$:

Proposition 3.6. *The components $\Gamma^{i(\alpha)}_{jk}$ of the $\nabla^{(\alpha)}$-connection are given by*

$$\Gamma^{(\alpha)1}_{11} = \frac{(\alpha-1)\,(-1+2\,\kappa\,\psi'(1,\kappa))}{2\,\nu\,(-1+\kappa\,\psi'(\kappa))},$$

$$\Gamma^{(\alpha)1}_{12} = -\frac{(\alpha-1)\,\psi'(1,\kappa)}{-2+2\,\kappa\,\psi'(\kappa)},$$

$$\Gamma^{(\alpha)1}_{22} = -\frac{(\alpha-1)\,\nu\,\psi''(\kappa)}{-2+2\,\kappa\,\psi'(\kappa)},$$

$$\Gamma^{(\alpha)2}_{11} = \frac{(\alpha-1)\,\kappa}{2\,\nu^2\,(-1+\kappa\,\psi'(\kappa))},$$

$$\Gamma^{(\alpha)2}_{12} = \frac{1-\alpha}{-2\,\nu+2\,\nu\,\kappa\,\psi'(\kappa)},$$

$$\Gamma^{(\alpha)2}_{22} = -\frac{(\alpha-1)\,\kappa\,\psi''(\kappa)}{-2+2\,\kappa\,\psi'(\kappa)}. \tag{3.21}$$

while the other independent components are zero. □

3.5.2 Gamma a-Curvatures

Proposition 3.7. *Direct calculation gives the α-curvature tensor of $\mathcal{G}$*

$$R^{\alpha}_{1212} = \frac{(\alpha^2-1)\,(\psi'(\kappa)+\kappa\,\psi''(\kappa))}{4\,\nu^2\,(-1+\kappa\,\psi'(\kappa))}, \tag{3.22}$$

while the other independent components are zero.
 By contraction we obtain:
 α-*Ricci tensor:*

$$[R^{(\alpha)}_{ij}] = (\alpha^2-1) \begin{bmatrix} \dfrac{-\kappa\,(\psi'(\kappa)+\kappa\,\psi''(\kappa))}{4\,\nu^2\,(-1+\kappa\,\psi'(\kappa))^2} & \dfrac{(\psi'(\kappa)+\kappa\,\psi''(\kappa))}{4\,\nu\,(-1+\kappa\,\psi'(\kappa))^2} \\[2ex] \dfrac{(\psi'(\kappa)+\kappa\,\psi''(\kappa))}{4\,\nu\,(-1+\kappa\,\psi'(\kappa))^2} & \dfrac{-\psi'(\kappa)\,(\psi'(\kappa)+\kappa\,\psi''(\kappa))}{4\,(-1+\kappa\,\psi'(\kappa))^2} \end{bmatrix} \tag{3.23}$$

Additionally, the eigenvalues and the eigenvectors for the α-Ricci tensor are given by

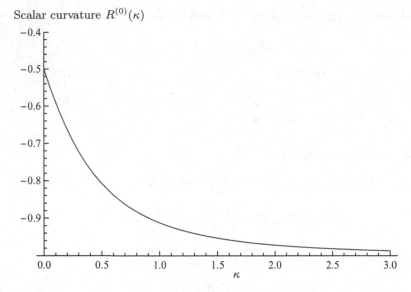

Scalar curvature $R^{(0)}(\kappa)$

Fig. 3.1. Scalar curvature for $\alpha = 0$ from equation (3.25) for the gamma manifold. The regime $\kappa < 1$ corresponds to clustering of events and $\kappa > 1$ corresponds to smoothing out of events, relative to an underlying Poisson process which corresponds to $\kappa = 1$.

$$(1-\alpha^2) \begin{pmatrix} \dfrac{\left(\kappa+\nu^2\,\psi'(\kappa)+\sqrt{4\,\nu^2+\kappa^2-2\,\nu^2\,\kappa\,\psi'(\kappa)+\nu^4\,\psi'(\kappa)^2}\right)\left(\psi'(\kappa)+\kappa\,\psi''(\kappa)\right)}{8\,\nu^2\,(-1+\kappa\,\psi'(\kappa))^2} \\[3mm] \dfrac{\left(\kappa+\nu^2\,\psi'(\kappa)-\sqrt{4\,\nu^2+\kappa^2-2\,\nu^2\,\kappa\,\psi'(\kappa)+\nu^4\,\psi'(\kappa)^2}\right)\left(\psi'(\kappa)+\kappa\,\psi''(\kappa)\right)}{8\,\nu^2\,(-1+\kappa\,\psi'(\kappa))^2} \end{pmatrix}$$

$$\begin{pmatrix} \dfrac{-\left(\kappa-\nu^2\,\psi'(\kappa)+\sqrt{4\,\nu^2+\kappa^2-2\,\nu^2\,\kappa\,\psi'(\kappa)+\nu^4\,\psi'(\kappa)^2}\right)}{2\,\nu} & 1 \\[3mm] \dfrac{-\kappa+\nu^2\,\psi'(\kappa)+\sqrt{4\,\nu^2+\kappa^2-2\,\nu^2\,\kappa\,\psi'(\kappa)+\nu^4\,\psi'(\kappa)^2}}{2\,\nu} & 1 \end{pmatrix} \qquad (3.24)$$

α-Scalar curvature:

$$R^{(\alpha)} = \frac{\left(1-\alpha^2\right)\left(\psi'(\kappa)+\kappa\,\psi''(\kappa)\right)}{2\left(-1+\kappa\,\psi'(\kappa)\right)^2}. \qquad (3.25)$$

This is shown in Figure 3.1 for the Levi-Civita case $\alpha = 0$. We note that $R^{(\alpha)} \to -\frac{(1-\alpha^2)}{2}$ *as* $\kappa \to 0$. $\qquad\Box$

3.5.3 Gamma Manifold Geodesics

The Fisher information metric for the gamma manifold is given in (γ, κ) coordinates by the arc length function §2.0.5 §3.2

$$ds^2 = \frac{\kappa}{\gamma^2}\, d\gamma^2 + \left(\left(\frac{\Gamma'(\kappa)}{\Gamma(\kappa)} \right)' - \frac{1}{\kappa} \right) d\kappa^2.$$

The Levi-Civita connection ∇ is that given by setting $\alpha = 0$ in the α-connections of the previous section. Geodesics for this case are curves satisfying

$$\nabla_{\dot{c}}\dot{c} = 0.$$

Background details can be found for example in [70]. In coordinate components $\dot{c} = \dot{c}^j \partial_j$ and we obtain from (2.5)

$$\ddot{c}^i + \Gamma^i_{jk}\, \dot{c}^j \dot{c}^k = 0 \quad \text{for each } i$$

and in our case with coordinates $(\gamma, \kappa) = (x, y)$ we have the nonlinear simultaneous equations

$$\ddot{x} = \frac{\dot{x}^2}{x} - \frac{\dot{x}\,\dot{y}}{y} \tag{3.26}$$

$$\ddot{y} = \frac{y\,\dot{x}^2}{2x^2\,(y\,\psi'(y) - 1)} - \frac{(\psi''(y)\,y^2 + 1)\,\dot{y}^2}{2y\,(y\,\psi'(y) - 1)} \tag{3.27}$$

$$\text{with} \quad \psi'(y) = \left(\frac{\Gamma'(y)}{\Gamma(y)} \right)'.$$

This system is difficult to solve analytically but we can find numerical solutions using the *Mathematica* programs of Gray [99]. Figure 3.2 shows a spray of some maximally extended geodesics emanating from the point $(\gamma, \kappa) = (1, 1)$.

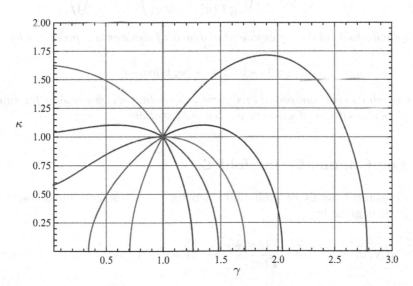

Fig. 3.2. Examples of maximally extended geodesics passing through $(\gamma, \kappa) = (1, 1)$ in the gamma manifold.

3.5.4 Mutually Dual Foliations

Now, $\mathcal{G}$ represents an exponential family of density functions, so a mixture coordinate system is given by a potential function. Since (ν, κ) is a 1-affine coordinate system, (η_1, η_2) given by

$$\eta_1 = \frac{\partial \varphi}{\partial \nu} = -\frac{\kappa}{\nu},$$

$$\eta_2 = \frac{\partial \varphi}{\partial \kappa} = \psi(\kappa) - \log \nu. \tag{3.28}$$

is a (-1)-affine coordinate system, and they are mutually dual with respect to the metric. Therefore the gamma manifold has dually orthogonal foliations and potential function, §3.2,

$$\lambda = -\kappa + \psi(\kappa) - \log \Gamma(\kappa). \tag{3.29}$$

3.5.5 Gamma Affine Immersion

The gamma manifold has an affine immersion in $\mathbb{R}^3$, §3.4

Proposition 3.8. *Let $\mathcal{G}$ be the gamma manifold with the Fisher metric g and the exponential connection $\nabla^{(1)}$. Denote by (ν, κ) a natural coordinate system. Then $\mathcal{G}$ can be realized in $\mathbb{R}^3$ by the graph of a potential function*

$$f : \mathcal{G} \to \mathbb{R}^3 \begin{pmatrix} \nu \\ \kappa \end{pmatrix} \mapsto \begin{pmatrix} \nu \\ \kappa \\ \log \Gamma(\kappa) - \kappa \log \nu \end{pmatrix}, \quad \xi = \begin{pmatrix} 0 \\ 0 \\ 1 \end{pmatrix}.$$

The submanifold, §2.0.5, of exponential density functions is represented by the curve

$$(0, \infty) \to \mathbb{R}^3 : \nu \mapsto \{\nu, 1, \log \frac{1}{\nu}\}$$

and a tubular neighbourhood of this curve will contain all immersions for small enough perturbations of exponential density functions. $\qquad \square$

3.6 Log-Gamma 2-Manifold $\mathcal{L}$

The log-gamma family of probability density functions for random variable $N \in (0, 1]$ is given by

$$q(N; \nu, \tau) = \frac{\nu^\tau N^{\nu-1} \left(\log \frac{1}{N}\right)^{\tau-1}}{\Gamma(\tau)} \quad \text{for } \nu > 0 \text{ and } \tau > 0. \tag{3.30}$$

Some of these density functions with central mean $\overline{N} = \frac{1}{2}$ are shown in Figure 3.3.

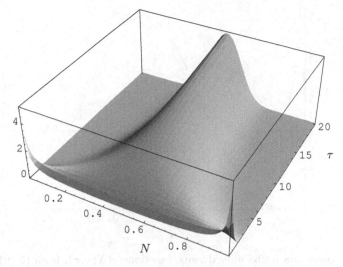

Fig. 3.3. The log-gamma family of probability densities (3.30) with central mean $\overline{N} = \frac{1}{2}$ as a surface. The surface tends to the delta function as $\tau \to \infty$ and coincides with the constant 1 at $\tau = 1$.

Proposition 3.9. *The log-gamma family (3.30) with information metric determines a Riemannian 2-manifold $\mathcal{L}$ with the following properties*
- *it contains the uniform distribution*
- *it contains approximations to truncated Gaussian density functions*
- *it is an isometric isomorph of the manifold $\mathcal{G}$ of gamma density functions.*

Proof. By integration, it is easily checked that the family given by equation (3.30) consists of probability density functions for the random variable $N \in (0,1]$. The limiting densities are given by

$$\lim_{\tau \to 1^+} q(N, \nu, \tau) = q(N, \nu, 1) = \frac{1}{\nu} \left(\frac{1}{N} \right)^{1-\frac{1}{\nu}} \qquad (3.31)$$

$$\lim_{\kappa \to 1} q(N, \nu, 1) = q(N, 1, 1) = 1. \qquad (3.32)$$

The mean, $\overline{N}$, standard deviation σ_N, and coefficient of variation, §1.2, cv_N, of N are given by

$$\overline{N} = \left(\frac{\nu}{1+\nu} \right)^{\tau} \qquad (3.33)$$

$$\sigma_N = \sqrt{\left(\frac{\nu}{\nu+2} \right)^{\tau} - \left(\frac{\nu}{1+\nu} \right)^{2\tau}} \qquad (3.34)$$

$$cv_N = \frac{\sigma_N}{\overline{N}} = \sqrt{\frac{(1+\nu)^{2\tau}}{\nu^{\tau}(2+\nu)^{\tau}} - 1}. \qquad (3.35)$$

$q(N; \nu, \tau)$

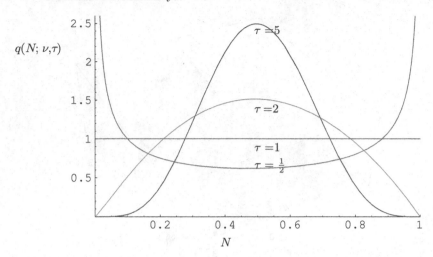

Fig. 3.4. Log-gamma probability density functions $q(N; \nu, \tau)$, from (3.30) for $N \in (0, 1]$, with central mean $\overline{N} = \frac{1}{2}$, and $\tau = \frac{1}{2}, 1, 2, 5$. The case $\tau = 1$ corresponds to a uniform distribution, $\tau < 1$ corresponds to clustering and conversely, $\tau > 1$ corresponds to dispersion.

We can obtain the family of densities having central mean in $(0, 1]$, by solving $\overline{N} = \frac{1}{2}$, which corresponds to the locus $(2^{1/\tau} - 1)\nu = 1$; some of these are shown in Figure 3.3 and Figure 3.4. Evidently, the density functions with central mean and large τ provide approximations to Gaussian density functions truncated on $(0, 1]$.

For the log-gamma densities, the Fisher information metric on the parameter space $\Theta = \{(\nu, \tau) \in (0, \infty) \times (0, \infty)\}$ is given by

$$[g_{ij}](\nu, \tau) = \begin{bmatrix} \frac{\tau}{\nu^2} & -\frac{1}{\nu} \\ -\frac{1}{\nu} & \frac{d^2}{d\tau^2} \log(\Gamma) \end{bmatrix} \tag{3.36}$$

In fact, (3.30) arises for the non-negative random variable $x = \log \frac{1}{N}$ from the gamma family equation (3.14) above with change of parameters to $\nu = \frac{\kappa}{\gamma}$ and $\tau = \kappa$, namely

$$f(x, \nu, \tau) = \frac{x^{\tau - 1} \nu^\tau}{\Gamma(\tau)} e^{-x\nu}. \tag{3.37}$$

It is known that the gamma family (3.37) has also the information metric (3.36) so the identity map on the space of coordinates (ν, τ) is an isometry of Riemannian manifolds $\mathcal{L}$ and $\mathcal{G}$.   □

In terms of the coordinates $(\gamma = \tau/\nu, \tau)$ the log-gamma densities (3.30) become

$$q(N; \frac{\tau}{\gamma}, \tau) = \frac{1}{\Gamma(\tau)} \left(\frac{\tau}{\gamma}\right)^\tau N^{\frac{\tau}{\gamma} - 1} \left(\log \frac{1}{N}\right)^{\tau - 1} \quad \text{for } \gamma > 0 \text{ and } \tau > 0. \tag{3.38}$$

Then the metric components form a diagonal matrix

$$[g_{ij}](\gamma, \tau) = \begin{bmatrix} \frac{\tau}{\gamma^2} & 0 \\ 0 & \frac{d^2}{d\tau^2}\log(\Gamma) - \frac{1}{\tau} \end{bmatrix}. \qquad (3.39)$$

Hence the coordinates (γ, τ) yield orthogonal tangent vectors, which can be very convenient in applications.

3.6.1 Log-Gamma Random Walks

We can illustrate the sensitivity of the parameter τ in the vicinity of the uniform distribution at $\tau = 1$ by constructing random walks using unit steps in the plane with log-gamma distributed directions. This is done by wrapping the log-gamma random variable $N \in (0, 1]$ from Equation 3.30 around the unit circle $\mathbb{S}^1$ to give a random angle $\theta \in (0, 2\pi]$. Then, starting at the origin $r(0) = (0, 0)$ in the x, y plane we take a random sequence $\{N_i \in (0, 1]\}_{i=1,n}$ drawn from the log-gamma distribution Equation 3.30 to generate a sequence of angles $\{\theta_i = 2\pi N_i \in (0, 2\pi]\}_{i=1,n}$ and hence define the random walk

$$r : \{1, 2, \ldots, n\} \to \mathbb{R}^2 : k \mapsto \sum_{i=1}^{k}(\cos\theta_i, \sin\theta_i). \qquad (3.40)$$

Figure 3.5 shows typical such random walks with 10,000 steps for the cases from Equation 3.30 with $\nu = 1$ and $\tau = 0.9, 1, 1.1$. So, the $\tau = 1$ case is a standard or isotropic random walk with uniformly distributed directions for each successive unit step.

3.7 Gaussian 2-Manifold

The family of univariate normal or Gaussian density functions has event space $\Omega = \mathbb{R}$ and probability density functions given by

$$N \equiv \{N(\mu, \sigma^2)\} = \{n(x; \mu, \sigma) \mid \mu \in \mathbb{R},\ \sigma \in \mathbb{R}^+\} \qquad (3.41)$$

with mean μ and variance σ^2. So here $N = \mathbb{R} \times \mathbb{R}^+$ is the upper half -plane, and the random variable is $x \in \Omega = \mathbb{R}$ with

$$n(x; \mu, \sigma) = \frac{1}{\sqrt{2\pi}\,\sigma} e^{-\frac{(x-\mu)^2}{2\sigma^2}}. \qquad (3.42)$$

The mean μ and standard deviation σ are frequently used as a local coordinate system $\xi = (\xi_1, \xi_2) = (\mu, \sigma)$. Lauritzen [134] gave a detailed discussion of the information geometry of the manifold N of Gaussian density functions, including its geodesic curves.

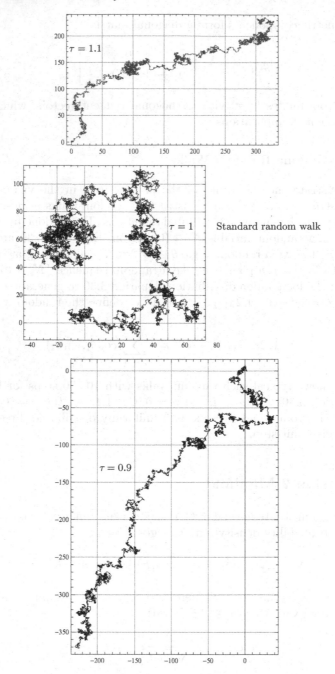

Fig. 3.5. Log-gamma random walks from Equation 3.40. Each has 10,000 unit length steps starting from the origin for the cases in Equation 3.30 with $\nu = 1$ for $\tau = 0.9,\ 1,\ 1.1$. The central graph shows the standard random walk with uniformly distributed directions in the plane.

Shannon's information theoretic entropy is given by:

$$S_N(\mu, \sigma) = -\int_{-\infty}^{\infty} \log(n(t; \mu, \sigma))\, n(t; \mu, \sigma)\, dt = \frac{1}{2}\left(1 + \log(2\pi)\right) + \log(\sigma)$$

$$(3.43)$$

At unit variance the entropy is $S_N = \frac{1}{2}\left(1 + \log(2\pi)\right)$.

3.7.1 Gaussian Natural Coordinates

Proposition 3.10. *In the manifold of Gaussian or normal densities N, set $\theta_1 = \frac{\mu}{\sigma^2}$ and $\theta_2 = -\frac{1}{2\sigma^2}$. Then $(\frac{\mu}{\sigma^2}, -\frac{1}{2\sigma^2})$ is a natural coordinate system and*

$$\varphi = -\frac{\theta_1^2}{4\theta_2} + \frac{1}{2}\log\left(-\frac{\pi}{\theta_2}\right) = \frac{\mu^2}{2\sigma^2} + \log(\sqrt{2\pi}\,\sigma) \qquad (3.44)$$

is the corresponding potential function, §3.2.

Proof. Set $\theta_1 = \frac{\mu}{\sigma^2}$ and $\theta_2 = -\frac{1}{2\sigma^2}$. Then the logarithm of the univariate Gaussian density can be written as

$$\log n(x; \theta_1, \theta_2) = \log c^{\left(\frac{\mu}{\sigma^2}\right)x + \left(\frac{-1}{2\sigma^2}\right)x^2 - \left(\frac{\mu^2}{2\sigma^2} + \log(\sqrt{2\pi}\,\sigma)\right)}$$

$$= \theta_1 x + \theta_2 x^2 - \left(-\frac{\theta_1^2}{4\theta_2} + \frac{1}{2}\log\left(-\frac{\pi}{\theta_2}\right)\right) \qquad (3.45)$$

Hence the set of all univariate Gaussian density functions is an exponential family, §3.2. The coordinates (θ_1, θ_2) is a natural coordinate system, §3.3 and $\varphi = -\frac{\theta_1^2}{4\theta_2} + \frac{1}{2}\log\left(-\frac{\pi}{\theta_2}\right) = \frac{\mu^2}{2\sigma^2} + \log(\sqrt{2\pi}\,\sigma)$ is its potential function.   □

3.7.2 Gaussian Information Metric

Proposition 3.11. *The Fisher metric, §3.1 with respect to natural coordinates (θ_1, θ_2) is given by:*

$$[g_{ij}] = \begin{bmatrix} \frac{-1}{2\theta_2} & \frac{\theta_1}{2\theta_2^2} \\ \frac{\theta_1}{2\theta_2^2} & \frac{\theta_2 - \theta_1^2}{2\theta_2^3} \end{bmatrix} = \begin{bmatrix} \sigma^2 & 2\mu\sigma^2 \\ 2\mu\sigma^2 & 2\sigma^2\left(2\mu^2 + \sigma^2\right) \end{bmatrix} \qquad (3.46)$$

Proof. Since φ is a potential function, the Fisher metric is given by the Hessian of φ, that is,

$$g_{ij} = \frac{\partial^2 \varphi}{\partial\theta^i \partial\theta^j}. \qquad (3.47)$$

Then, we have the metric by a straightforward calculation.   □

3.7.3 Gaussian Mutually Dual Foliations

Since N represents an exponential family, a mixture coordinate system is given by a potential function. We have

$$\eta_1 = \frac{\partial \varphi}{\partial \theta^1} = \frac{-\theta_1}{2\,\theta_2} = \mu,$$

$$\eta_2 = \frac{\partial \varphi}{\partial \theta^2} = \frac{\theta_1{}^2 - 2\,\theta_2}{4\,\theta_2{}^2} = \mu^2 + \sigma^2. \tag{3.48}$$

Since (θ_1, θ_2) is a 1-affine coordinate system, (η_1, η_2) is a (-1)-affine coordinate system, and they are mutually dual with respect to the metric. Therefore the Gaussian manifold has dually orthogonal foliations. The coordinates in (3.48) admit the potential function

$$\lambda = -\frac{1}{2}\left(1 + \log(-\frac{\pi}{\theta_2})\right) = \frac{-1}{2}\left(1 + \log(2\,\pi) + 2\,\log(\sigma)\right). \tag{3.49}$$

3.7.4 Gaussian Affine Immersions

We show that the Gaussian manifold can be realized in Euclidean $\mathbb{R}^3$ by an affine immersion, §3.4.

Proposition 3.12. *Let N be the Gaussian manifold with the Fisher metric g and the exponential connection $\nabla^{(1)}$. Denote by (θ_1, θ_2) a natural coordinate system. Then M can be realized in $\mathbb{R}^3$ by the graph of a potential function, namely, $\mathcal{G}$ can be realized by the affine immersion $\{f, \xi\}$:*

$$f : N \to \mathbb{R}^3 : \begin{pmatrix} \theta_1 \\ \theta_2 \end{pmatrix} \mapsto \begin{pmatrix} \theta_1 \\ \theta_2 \\ \varphi \end{pmatrix}, \quad \xi = \begin{pmatrix} 0 \\ 0 \\ 1 \end{pmatrix}.$$

where ψ is the potential function $\varphi = -\frac{\theta_1{}^2}{4\,\theta_2} + \frac{1}{2}\log(-\frac{\pi}{\theta_2})$.
The submanifold, §2.0.5, of univariate Gaussian density functions with zero mean (i.e. $\theta_1 = 0$) is represented by the curve

$$(-\infty, 0) \to \mathbb{R}^3 : \theta_2 \mapsto \{0, \theta_2, \frac{1}{2}\log(-\frac{\pi}{\theta_2})\},$$

In addition, the submanifold of univariate Gaussian density functions with unit variance (i.e. $\theta_2 = -\frac{1}{2}$) is represented by the curve

$$\mathbb{R} \to \mathbb{R}^3 : \theta_1 \mapsto \{\theta_1, -\frac{1}{2}, \frac{\theta_1{}^2}{2} + \frac{1}{2}\log(2\,\pi)\}.$$

$\square$

3.8 Gaussian a-Geometry

By direct calculation we provide the α-connections, §3.3, and various α-curvature objects of the Gaussian 2-manifold N: the α-curvature tensor, the α-Ricci curvature with its eigenvalues and eigenvectors, the α-sectional curvature, and the α-Gaussian curvature. Henceforth, we use the coordinate system (μ, σ).

3.8.1 Gaussian a-Connection

For each $\alpha \in \mathbb{R}$, the α (or $\nabla^{(\alpha)}$)-connection is the torsion-free affine connection with components, §2.0.6:

$$\Gamma_{ij,k}^{(\alpha)} = \int_{-\infty}^{\infty} \left(\frac{\partial^2 \log p}{\partial \xi^i \partial \xi^j} \frac{\partial \log p}{\partial \xi^k} + \frac{1-\alpha}{2} \frac{\partial \log p}{\partial \xi^i} \frac{\partial \log p}{\partial \xi^j} \frac{\partial \log p}{\partial \xi^k} \right) p \, dx$$

Proposition 3.13. *The functions $\Gamma_{ij,k}^{(\alpha)}$ are given by:*

$$\Gamma_{11,2}^{(\alpha)} = \frac{1-\alpha}{\sigma^3},$$

$$\Gamma_{12,1}^{(\alpha)} = -\frac{1+\alpha}{\sigma^3},$$

$$\Gamma_{22,2}^{(\alpha)} = -\frac{2+4\alpha}{\sigma^3} \tag{3.50}$$

while the other independent components are zero. $\square$

We have an affine connection $\nabla^{(\alpha)}$ defined by:

$$\langle \nabla_{\partial_i}^{(\alpha)} \partial_j, \partial_k \rangle = \Gamma_{ij,k}^{(\alpha)},$$

So by solving the equations

$$\Gamma_{ij,k}^{(\alpha)} = \sum_{h=1}^{3} g_{kh} \, \Gamma_{ij}^{h(\alpha)}, (k = 1, 2).$$

we obtain the components of $\nabla^{(\alpha)}$:

Proposition 3.14. *The components $\Gamma_{jk}^{i(\alpha)}$ of the $\nabla^{(\alpha)}$-connections are given by:*

$$\Gamma_{12}^{(\alpha)1} = -\frac{\alpha+1}{\sigma},$$

$$\Gamma_{11}^{(\alpha)2} = \frac{1-\alpha}{2\sigma},$$

$$\Gamma_{22}^{(\alpha)2} = -\frac{2\alpha+1}{\sigma}. \tag{3.51}$$

while the other independent components are zero. $\square$

3.8.2 Gaussian α-Curvatures

Proposition 3.15. *By direct calculation we have; the α-curvature tensor of N, §2.0.6, is given by*

$$R^{(\alpha)}_{1212} = \frac{1 - \alpha^2}{\sigma^4},$$
(3.52)

while the other independent components are zero. □

Proposition 3.16. *The components of the α-Ricci tensor are given by the symmetric matrix $R^{(\alpha)} = [R^{(\alpha)}_{ij}]$:*

$$R^{(\alpha)}_{11} = \frac{\alpha^2 - 1}{2\,\sigma^2}$$
(3.53)

while the other components are zero.
    We note that $R^{(\alpha)}_{11} \to \frac{\alpha^2 - 1}{2}$ as $\sigma \to 1$.

In addition, the eigenvalues and the eigenvectors for the α-Ricci tensor are given by:

$$\frac{(\alpha^2 - 1)}{\sigma^2} \begin{pmatrix} \frac{1}{2} \\ 1 \end{pmatrix}$$
(3.54)

$$\begin{pmatrix} 1 & 0 \\ 0 & 1 \end{pmatrix}. \quad \square$$
(3.55)

Proposition 3.17. *The α-Gaussian curvature $K^{(\alpha)}$ of N is given by:*

$$K^{(\alpha)} = \frac{\alpha^2 - 1}{2}.$$
(3.56)

So geometrically Gaussian manifold constitutes part of a pseudosphere when $\alpha^2 < 1$. □

3.9 Gaussian Mutually Dual Foliations

Since the family of Gaussian density functions N is an exponential family, a mixture coordinate system is given by the potential function $\varphi = -\frac{\theta_1^2}{4\theta_2} + \frac{1}{2}\log(-\frac{\pi}{\theta_2})$, that is

$$\eta_1 = \frac{\partial \varphi(\theta)}{\partial \theta_1} = -\frac{\theta_1}{2\,\theta_2} = \mu\,,$$

$$\eta_2 = \frac{\partial \varphi(\theta)}{\partial \theta_2} = \frac{\theta_1^2 - 2\,\theta_2^2}{4\,\theta_2^2} = \mu^2 + \sigma^2\,.$$
(3.57)

Since (θ_1, θ_2) is a 1-affine coordinate system, (η_1, η_2) is a (-1)-affine coordinate system, and they are mutually dual with respect to the Fisher metric. The coordinates in (η_i) have a potential function given by:

$$\lambda = \frac{-1}{2}\left(1 + \log(-\frac{\pi}{\theta_2})\right) = \frac{-1}{2}\left(1 + \log(2\pi) + 2\log(\sigma)\right). \qquad (3.58)$$

The coordinates (θ_i) and (η_i) are mutually dual. Therefore the Gaussian manifold N has dually orthogonal foliations, for example:

Take $(\eta_1, \theta_2) = (\mu, -\frac{1}{2\sigma^2})$ as a coordinate system for N, then the Gaussian density functions take the form:

$$p(x; \eta_1, \theta_2) = \sqrt{-\frac{\theta_2}{\pi}}\, e^{(x-\eta_1)^2 \theta_2} \quad \text{for } \eta_1 \in \mathbb{R},\ \theta_2 \in \mathbb{R}^-, \qquad (3.59)$$

and the Fisher metric is

$$ds_g^2 = -2\,\theta_2\, d\eta_1{}^2 + \frac{1}{2\theta_2{}^2}\, d\theta_2{}^2 \quad \text{for } \eta_1 \in \mathbb{R},\ \theta_2 \in \mathbb{R}^-. \qquad (3.60)$$

We remark that (θ_i) is a geodesic coordinate system of $\nabla^{(1)}$, and (η_i) is a geodesic coordinate system of $\nabla^{(-1)}$.

3.10 Gaussian Submanifolds

We consider three submanifolds N_1, N_2 and N_3 of the Gaussian manifold N (3.41). These submanifolds have dimension 1 and so all the curvatures are zero.

3.10.1 Central Mean Submanifold

This is defined by $N_1 \subset N$: $\mu = 0$. The Gaussian density functions with zero mean are of form:

$$p(x; \sigma) = \frac{1}{\sqrt{2\pi}\,\sigma}\, e^{-\frac{x^2}{2\sigma^2}} \quad \text{for } x \in \mathbb{R},\ \sigma \in \mathbb{R}^+. \qquad (3.61)$$

Proposition 3.18. *The information metric $[g_{ij}]$ is as follows:*

$$ds_g^2 = G\, d\sigma^2 = \frac{2}{\sigma^2}\, d\sigma^2 \quad \text{for } \sigma \in \mathbb{R}^+. \qquad \Box \qquad (3.62)$$

Proposition 3.19. *By direct calculation the α-connections of N_1 are*

$$\Gamma_{11,1}^{(\alpha)} = -\frac{2 + 4\alpha}{\sigma^3},$$

$$\Gamma_{11}^{(\alpha)1} = -\frac{1 + 2\alpha}{\sigma}. \qquad \Box \qquad (3.63)$$

3.10.2 Unit Variance Submanifold

This is defined as $N_2 \subset N$: $\sigma = 1$. The Gaussian density functions with unit variance are of form:

$$p(x; \mu) = \frac{1}{\sqrt{2\pi}} e^{-\frac{(x-\mu)^2}{2}} \quad \text{for } x \in \mathbb{R}, \mu \in \mathbb{R}. \tag{3.64}$$

Proposition 3.20. *The information metric $[g_{ij}]$ is as follows:*

$$ds_g^2 = G \, d\sigma^2 = d\mu^2 \quad \text{for } \mu \in \mathbb{R} . \quad \square \tag{3.65}$$

Proposition 3.21. *The components of α-connections of the submanifold N_2 are zero.* $\square$

3.10.3 Unit Coefficient of Variation Submanifold

This is defined as $N_3 \subset N$: $\mu = \sigma$. The Gaussian density functions with identical mean and standard deviation are of form:

$$p(x; \sigma) = \frac{1}{\sqrt{2\pi}\,\sigma} e^{-\frac{(x-\sigma)^2}{2\sigma^2}} \quad \text{for } x \in \mathbb{R}, \sigma \in \mathbb{R}^+. \tag{3.66}$$

Proposition 3.22. *The information metric $[g_{ij}]$ is as follows:*

$$ds_g^2 = G \, d\sigma^2 = \frac{3}{\sigma^2} \, d\sigma^2 \quad \text{for } \sigma \in \mathbb{R}^+. \quad \square \tag{3.67}$$

Proposition 3.23. *By direct calculation the α-connections of N_3 are*

$$\Gamma_{11,1}^{(\alpha)} = -\frac{3 + 7\alpha}{\sigma^3},$$

$$\Gamma_{11}^{(\alpha)1} = -\frac{3 + 7\alpha}{3\sigma} . \quad \square \tag{3.68}$$

3.11 Gaussian Affine Immersions

Proposition 3.24. *Let N be the Gaussian manifold with the Fisher metric g and the exponential connection $\nabla^{(1)}$. Denote by $(\theta_1, \theta_2) = (\frac{\mu}{\sigma^2}, -\frac{1}{2\sigma^2})$ the natural coordinate system. Then N can be realized in $\mathbb{R}^3$ by the graph of a potential function, namely, $\mathcal{G}$ can be realized by the affine immersion, §3.4, $\{f, \xi\}$:*

$$f : N \to \mathbb{R}^3 : \begin{pmatrix} \theta_1 \\ \theta_2 \end{pmatrix} \mapsto \begin{pmatrix} \theta_1 \\ \theta_2 \\ \varphi \end{pmatrix}, \quad \xi = \begin{pmatrix} 0 \\ 0 \\ 1 \end{pmatrix} . \tag{3.69}$$

where φ is the potential function

$$\varphi = -\frac{\theta_1{}^2}{4\,\theta_2} + \frac{1}{2}\log(-\frac{\pi}{\theta_2}) = \frac{\mu^2}{2\,\sigma^2} + \log(\sqrt{2\,\pi}\,\sigma)\,.$$

We consider particular submanifolds; central mean submanifold, unit variance submanifold and the submanifold with identical mean and standard deviation. We represent them as curves in $\mathbb{R}^3$ as follows:

1. *The central mean submanifold N_1:*
   The Gaussian density functions with zero mean (i.e. $\theta_1 = 0$ and $\theta_2 = -\frac{1}{2\,\sigma^2}$) are represented by the curve

$$(-\infty, 0) \to \mathbb{R}^3 : \theta_2 \mapsto \{0, \theta_2, \frac{1}{2}\log(-\frac{\pi}{\theta_2})\}\,,$$

2. *The unit variance submanifold N_2:*
   The Gaussian density functions with unit variance (i.e. $\theta_1 = \mu$ and $\theta_2 = -\frac{1}{2}$) are represented by the curve

$$\mathbb{R} \to \mathbb{R}^3 : \theta_1 \mapsto \{\theta_1, -\frac{1}{2}, \frac{\theta_1{}^2}{2} + \frac{1}{2}\log(2\,\pi)\}\,.$$

3. *The submanifold N_3 with identical mean and standard deviation:*
   The Gaussian density functions with identical mean and standard deviation $\mu = \sigma$ (i.e. $\theta_1 = \frac{1}{\sigma}$ and $\theta_2 = -\frac{1}{2\,\sigma^2}$) are represented by the curve

$$(0, \infty) \to \mathbb{R}^3 : \theta_1 \mapsto \{\theta_1, -2\,\theta_1{}^2, \frac{1}{2} + \log(\frac{\sqrt{2\,\pi}}{\theta_1})\}\,.$$

□

3.12 Log-Gaussian Manifold

The log-Gaussian density functions arise from the Gaussian density functions (3.42) for the non negative random variable $x = \log\frac{1}{m}$, or equivalently, $m = e^{-x}$:

$$g(m) = \frac{1}{m\,\sqrt{2\,\pi}\,\sigma}\,e^{-\frac{(\log(m)+\mu)^2}{2\,\sigma^2}} \quad \text{for } \mu \in \mathbb{R},\ \sigma \in \mathbb{R}^+, \qquad (3.70)$$

The case where $\mu = 0$ and $\sigma = 1$ is called the standard log-Gaussian distribution. Figure 3.6 shows a plot of log-Gaussian density functions $g(m; \mu, \sigma)$ for the range $n \in [0, 3]$; with $\mu = 0$, and $\sigma = 1, 1.5, 2$. The case $\sigma = 1$ corresponds to the standard Log-Gaussian distribution $g(m; 0, 1)$.

Corollary 3.25. *The mean $\overline{m}$, standard deviation σ_m, and coefficient of variation, §1.2, cv_m, for log-Gaussian density functions are given by:*

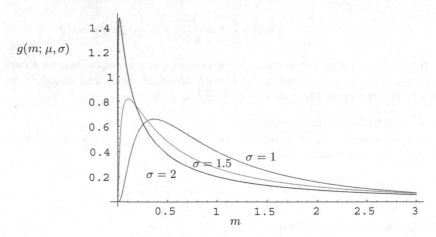

$g(m; \mu, \sigma)$

Fig. 3.6. Log-Gaussian probability density functions $g(m; \mu, \sigma)$; with $\mu = 0$, and $\sigma = 1, 1.5, 2$. The case $\sigma = 1$ corresponds to the standard log-Gaussian density $g(m; 0, 1)$.

$$\overline{m} = e^{\frac{\sigma^2}{2} - \mu},$$
$$\sigma_m = \sqrt{\left(e^{\sigma^2 - 2\mu}\right)\left(e^{\sigma^2} - 1\right)},$$
$$cv_m = e^{\mu - \frac{\sigma^2}{2}}\sqrt{\left(e^{\sigma^2 - 2\mu}\right)\left(e^{\sigma^2} - 1\right)}. \quad \Box \qquad (3.71)$$

Directly from the definition of the Fisher metric we deduce:

Proposition 3.26. *The family of log-Gaussian density functions for random variable n determines a Riemannian 2-manifold which is an isometric isomorph of the Gaussian 2-manifold.*

Proof. We show that the log-Gaussian and Gaussian families have the same Fisher metric. From

$$g(m) = p(x) \frac{dx}{dm}$$

Hence

$$\log g(m) = \log p(x) + \log\left(\frac{dx}{dm}\right)$$

then differentiation of this relation with respect to θ_i and θ_j, with $(\theta_1, \theta_2) = (\frac{\mu}{\sigma^2}, -\frac{1}{2\sigma^2})$, yields

$$\frac{\partial^2 \log g(m)}{\partial \theta_i \partial \theta_j} = \frac{\partial^2 \log p(x)}{\partial \theta_i \partial \theta_j}$$

from (3.1) we can see that $p(x)$ and $g(m)$ have the same Fisher metric. So the identity map on the parameter space $\mathbb{R} \times \mathbb{R}^+$ determines an isometry, §2.0.5, of Riemannian manifolds. $\qquad \Box$

4

Information Geometry of Bivariate Families

From the study by Arwini [13], we provide information geometry, including the α-geometry, of several important families of bivariate probability density functions. They have marginal density functions that are gamma density functions, exponential density functions and Gaussian density functions. These are used for applications in the sequel, when we have two random variables that have non-zero covariance—such as will arise for a coupled pair of random processes.

The multivariate Gaussian is well-known and its information geometry has been reported before [183, 189]; our recent work has contributed the bivariate Gaussian α-geometry. Surprisingly, it is very difficult to construct a bivariate exponential distribution, or for that matter a bivariate Poisson distribution that has tractable information geometry. However we have calculated the case of the Freund bivariate mixture exponential distribution [89]. The only bivariate gamma distribution for which we have found the information geometry tractable is the McKay case [146] which is restricted to positive covariance, and we begin with this.

4.1 McKay Bivariate Gamma 3-Manifold M

The results in this section were computed in [13] and first reported in [14]. The McKay bivariate gamma distribution is one of the bivariate density functions constructed by statisticians using the so-called conditional method. McKay [146] derived it as follows:

Let $(X_1, X_2, ..., X_N)$ be a random sample from a normal population. Suppose s_N^2 is the sample variance, and let s_n^2 be the variance in a sub-sample of size n. Then s_N^2 and s_n^2 jointly have the McKay bivariate gamma distribution.

The information geometry of the 3-manifold of McKay bivariate gamma density functions can provide a metrization of departures from Poisson randomness and departures from independence for bivariate processes. The curvature objects are derived, including those on three submanifolds. As in the case

K. Arwini, C.T.J. Dodson, *Information Geometry.*
Lecture Notes in Mathematics 1953,
© Springer-Verlag Berlin Heidelberg 2008

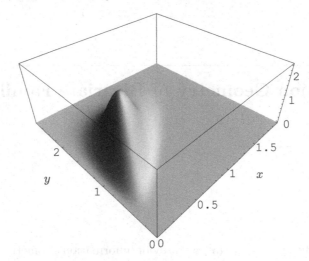

Fig. 4.1. Part of the family of McKay bivariate gamma probability density functions; observe that these are zero outside the octant $0 < x < y < \infty$. Here the correlation coefficient has been set to $\rho_{xy} = 0.6$ and $\alpha_1 = 5$.

of bivariate Gaussian manifolds, we have negative scalar curvature but here it is not constant and we show how it depends on correlation. The multiplicative group action of the real numbers as a scale-changing process influences the Riemannian geometric properties. These results have potential applications, for example, in the characterization of stochastic materials.

Fisher Information Metric

The classical family of McKay bivariate gamma density functions M, is defined on $0 < x < y < \infty$ with parameters $\alpha_1, \sigma_{12}, \alpha_2 > 0$ and probability density functions

$$f^*(x,y;\alpha_1,\sigma_{12},\alpha_2) = \frac{\left(\frac{\alpha_1}{\sigma_{12}}\right)^{\frac{(\alpha_1+\alpha_2)}{2}} x^{\alpha_1-1}(y-x)^{\alpha_2-1}e^{-\sqrt{\frac{\alpha_1}{\sigma_{12}}}y}}{\Gamma(\alpha_1)\Gamma(\alpha_2)} \,. \qquad (4.1)$$

Here σ_{12} is the covariance, §1.3, of X and Y. One way to view this is that $f^*(x,y)$ is the probability density for the two random variables X and $Y = X + Z$ where X and Z both have gamma density functions.

The correlation coefficient, §1.3, and marginal functions, §1.3 of X and Y are given by

$$\rho(X,Y) = \sqrt{\frac{\alpha_1}{\alpha_1+\alpha_2}} > 0 \qquad (4.2)$$

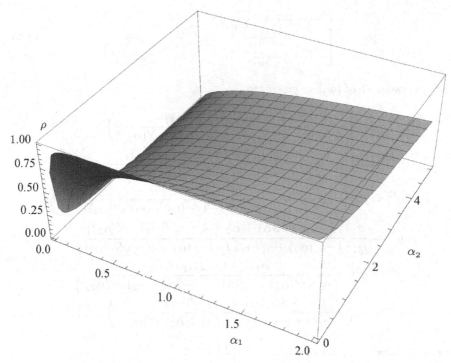

Fig. 4.2. Correlation coefficient ρ from equation (4.2) for McKay probablity density functions, in terms of α_1, α_2.

$$f_X^*(x) = \frac{\left(\frac{\alpha_1}{\sigma_{12}}\right)^{\frac{\alpha_1}{2}} x^{\alpha_1 - 1} e^{-\sqrt{\frac{\alpha_1}{\sigma_{12}}} x}}{\Gamma(\alpha_1)}, \quad x > 0 \tag{4.3}$$

$$f_Y^*(y) = \frac{\left(\frac{\alpha_1}{\sigma_{12}}\right)^{\frac{(\alpha_1 + \alpha_2)}{2}} y^{(\alpha_1 + \alpha_2) - 1} e^{-\sqrt{\frac{\alpha_1}{\sigma_{12}}} y}}{\Gamma(\alpha_1 + \alpha_2)}, \quad y > 0 \tag{4.4}$$

Figure 4.2 shows a plot of the correlation coefficient from equation (4.2). The marginal density functions of X and Y are gamma with shape parameters α_1 and $\alpha_1 + \alpha_2$, respectively; note that it is not possible to choose parameters such that both marginal functions are exponential, §1.2.2.

Examples of the outputs of two McKay distribution simulators are shown in Figure 9.16 and Figure 10.23. In Figure 9.16, each plot shows 5000 points with coordinates (x, y) and $x < y$, for three values of the correlation coefficient: $\rho = 0.5, 0.7, 0.9$.

Proposition 4.1. *Let M be the family of McKay bivariate gamma density functions (4.1), then $(\alpha_1, \sigma_{12}, \alpha_2)$ is a local coordinate system, and M becomes a 3-manifold, §2.0.2, with Fisher information metric, §2.0.5, §3.1,*

$$[g_{ij}] = \begin{bmatrix} \frac{-3\,\alpha_1+\alpha_2}{4\,\alpha_1{}^2} + (\frac{\Gamma'(\alpha_1)}{\Gamma(\alpha_1)})' & \frac{\alpha_1-\alpha_2}{4\,\alpha_1\,\sigma_{12}} & -\frac{1}{2\,\alpha_1} \\ \frac{\alpha_1-\alpha_2}{4\,\alpha_1\,\sigma_{12}} & \frac{\alpha_1+\alpha_2}{4\,\sigma_{12}{}^2} & \frac{1}{2\,\sigma_{12}} \\ -\frac{1}{2\,\alpha_1} & \frac{1}{2\,\sigma_{12}} & (\frac{\Gamma'(\alpha_2)}{\Gamma(\alpha_2)})' \end{bmatrix} \tag{4.5}$$

The inverse $[g^{ij}]$ of $[g_{ij}]$ is given by:

$$g^{11} = -\left(\frac{-1+(\alpha_1+\alpha_2)\,\psi'(\alpha_2)}{\psi'(\alpha_2)+\psi'(\alpha_1)\,(1-(\alpha_1+\alpha_2)\,\psi'(\alpha_2))} \right),$$

$$g^{12} = g^{21} = \frac{\sigma_{12}\,(1+(\alpha_1-\alpha_2)\,\psi'(\alpha_2))}{\alpha_1\,(\psi'(\alpha_2)+\psi'(\alpha_1)\,(1-(\alpha_1+\alpha_2)\,\psi'(\alpha_2)))},$$

$$g^{13} = g^{31} = \frac{1}{-\psi'(\alpha_2)+\psi'(\alpha_1)\,(-1+(\alpha_1+\alpha_2)\,\psi'(\alpha_2))},$$

$$g^{22} = \frac{\sigma_{12}{}^2\,(-1+(-3\,\alpha_1+\alpha_2+4\,\alpha_1{}^2\,\psi'(\alpha_1))\,\psi'(\alpha_2))}{\alpha_1{}^2\,(-\psi'(\alpha_2)+\psi'(\alpha_1)\,(-1+(\alpha_1+\alpha_2)\,\psi'(\alpha_2)))},$$

$$g^{23} = g^{32} = \frac{\sigma_{12}\,(-1+2\,\alpha_1\,\psi'(\alpha_1))}{\alpha_1\,(\psi'(\alpha_2)+\psi'(\alpha_1)\,(1-(\alpha_1+\alpha_2)\,\psi'(\alpha_2)))},$$

$$g^{33} = -\left(\frac{-1+(\alpha_1+\alpha_2)\,\psi'(\alpha_1)}{\psi'(\alpha_2)+\psi'(\alpha_1)\,(1-(\alpha_1+\alpha_2)\,\psi'(\alpha_2))} \right), \tag{4.6}$$

where we have abbreviated $\psi'(\alpha_1) = (\frac{\Gamma'(\alpha_1)}{\Gamma(\alpha_1)})'$. □

4.2 McKay Manifold Geometry in Natural Coordinates

The original presentation of the McKay distribution was in the form:

$$f(x,y) = \frac{c^{(\alpha_1+\alpha_2)}x^{\alpha_1-1}(y-x)^{\alpha_2-1}e^{-cy}}{\Gamma(\alpha_1)\Gamma(\alpha_2)} \tag{4.7}$$

defined on $0 < x < y < \infty$ with parameters $\alpha_1, c, \alpha_2 > 0$. The marginal functions, §1.3, of X and Y are given by:

$$f_X(x) = \frac{c^{\alpha_1}x^{\alpha_1-1}e^{-cx}}{\Gamma(\alpha_1)}, \quad x > 0 \tag{4.8}$$

$$f_Y(y) = \frac{c^{(\alpha_1+\alpha_2)}y^{(\alpha_1+\alpha_2)-1}e^{-cy}}{\Gamma(\alpha_1+\alpha_2)}, \quad y > 0 \tag{4.9}$$

The covariance and correlation coefficient, §1.3, of X and Y are given by:

$$\sigma_{12} = \frac{\alpha_1}{c^2}$$

$$\rho(X,Y) = \sqrt{\frac{\alpha_1}{\alpha_1+\alpha_2}}$$

4.3 McKay Densities Have Exponential Type

The McKay family (4.7) forms an exponential family, §3.2.

Proposition 4.2. *Let M be the set of McKay bivariate gamma density functions, that is*

$$M = \{f | f(x, y; \alpha_1, c, \alpha_2) = \frac{c^{(\alpha_1 + \alpha_2)} x^{\alpha_1 - 1} (y - x)^{\alpha_2 - 1} e^{-cy}}{\Gamma(\alpha_1)\Gamma(\alpha_2)}, \quad (4.10)$$

$$\text{for } y > x > 0, \ \alpha_1, c, \alpha_2 > 0\}.$$

Then (α_1, c, α_2) is a natural coordinate system, §3.3, and

$$\varphi(\theta) = \log \Gamma(\alpha_1) + \log \Gamma(\alpha_2) - (\alpha_1 + \alpha_2) \log c \qquad (4.11)$$

is the corresponding potential function, §3.2.

Proof.

$$\log f(x, y; \alpha_1, c, \alpha_2) = \log \left(\frac{c^{(\alpha_1 + \alpha_2)} x^{\alpha_1 - 1} (y - x)^{\alpha_2 - 1} e^{-cy}}{\Gamma(\alpha_1)\Gamma(\alpha_2)} \right)$$

$$= -\log x - \log(y - x) \qquad (4.12)$$

$$+ \alpha_1 (\log x) + c(-y) + \alpha_2 (\log(y - x)) \qquad (4.13)$$

$$- (\log \Gamma(\alpha_1) + \log \Gamma(\alpha_2) - (\alpha_1 + \alpha_2) \log c). \qquad (4.14)$$

Hence the set of all McKay bivariate gamma density functions is an exponential family. The terms in the line (4.13) imply that $(\theta_1, \theta_2, \theta_3) = (\alpha_1, c, \alpha_2)$ is a natural coordinate system and $(x_1, x_2, x_3) = (F_1(x), F_2(x), F_3(x)) = (\log x, -y, \log(y - x))$ is a random variable space (the random variables x_1, x_2, x_3 are not independent, but related by $x_3 = \log(-x_2 - e^{x_1})$, and (4.14) implies that $\varphi(\theta) = \log \Gamma(\alpha_1) + \log \Gamma(\alpha_2) - (\alpha_1 + \alpha_2) \log c = \log \Gamma(\theta_1) + \log \Gamma(\theta_3) - (\theta_1 + \theta_3) \log \theta_3$ is its potential function.
We remark that (4.12) implies that $C(X, Y) = -\log x - \log(y - x) = -x_1 - x_3$ is the normalization function since f is a probability density function. $\square$

4.3.1 McKay Information Metric

Proposition 4.3. *Let M be the set of McKay bivariate gamma density functions (4.7), then using coordinates (α_1, c, α_2), M is a 3-manifold with Fisher information metric $[g_{ij}]$, §3.1, given by:*

$$[g_{ij}] = \begin{bmatrix} \psi'(\alpha_1) & -\frac{1}{c} & 0 \\ -\frac{1}{c} & \frac{\alpha_1 + \alpha_2}{c^2} & -\frac{1}{c} \\ 0 & -\frac{1}{c} & \psi'(\alpha_2) \end{bmatrix} \qquad (4.15)$$

Proof. Since (α_1, c, α_2) is the natural coordinate system, and $\varphi(\theta)$ its potential function, §3.2 (4.11), the Fisher metric is given by the Hessian of $\varphi(\theta)$, that is,

$$g_{ij} = \frac{\partial^2 \varphi(\theta)}{\partial \theta_i \partial \theta_j}.$$

Then, we have the Fisher metric by a straightforward calculation.   □

4.4 McKay a-Geometry

By direct calculation, using §3.3, we provide the α-connections , and various α-curvature objects of the McKay 3-manifold M: the α-curvature tensor, the α-Ricci curvature, the α-scalar curvature, the α-sectional curvature, and the α-mean curvature.

4.4.1 McKay a-Connection

For each $\alpha \in \mathbb{R}$, the $\alpha-$ (or $\nabla^{(\alpha)}$)-connection is the torsion-free affine connection with components:

$$\Gamma_{ij,k}^{(\alpha)} = \frac{1 - \alpha}{2} \, \partial_i \, \partial_j \, \partial_k \varphi(\theta) \,,$$

where $\varphi(\theta)$ is the potential function, and $\partial_i = \frac{\partial}{\partial \theta_i}$.

Since the set of McKay bivariate gamma density functions is an exponential family, the connection $\nabla^{(1)}$ is flat. In this case, (α_1, c, α_2) is a 1-affine coordinate system. So the 1 and (-1)-connections on the McKay manifold are flat.

Proposition 4.4. *The functions $\Gamma_{ij,k}^{(\alpha)}$ are given by:*

$$\Gamma_{11,1}^{(\alpha)} = \frac{(1 - \alpha) \, \psi''(\alpha_1)}{2} \,,$$

$$\Gamma_{22,1}^{(\alpha)} = \Gamma_{12,2}^{(\alpha)} = \Gamma_{23,2}^{(\alpha)} = \Gamma_{22,3}^{(\alpha)} = \frac{(1 - \alpha)}{2 \, c^2} \,,$$

$$\Gamma_{22,2}^{(\alpha)} = -\frac{(1 - \alpha) \, (\alpha_1 + \alpha_2)}{c^3} \,,$$

$$\Gamma_{33,3}^{(\alpha)} = \frac{(1 - \alpha) \, \psi''(\alpha_2)}{2} \tag{4.16}$$

while the other independent components are zero.   □

We have an affine connection $\nabla^{(\alpha)}$ defined by:

$$\langle \nabla_{\partial_i}^{(\alpha)} \partial_j, \partial_k \rangle = \Gamma_{ij,k}^{(\alpha)} \,,$$

So by solving the equations

$$\Gamma_{ij,k}^{(\alpha)} = \sum_{h=1}^{3} g_{kh} \, \Gamma_{ij}^{h(\alpha)}, (k = 1, 2, 3).$$

we obtain the components of $\nabla^{(\alpha)}$:

Proposition 4.5. *The components* $\Gamma_{jk}^{i(\alpha)}$ *of the* $\nabla^{(\alpha)}$-*connections are given from §2.0.6 and §3.3 by:*

$$\Gamma_{11}^{(\alpha)1} = \frac{(1 - \alpha) \, \psi''(\alpha_1) \, (-1 + \psi'(\alpha_2) \, (\alpha_1 + \alpha_2))}{2 \, (\psi'(\alpha_1) + \psi'(\alpha_2) - \psi'(\alpha_1) \, \psi'(\alpha_2) \, (\alpha_1 + \alpha_2))},$$

$$\Gamma_{11}^{(\alpha)2} = \frac{c \, (1 - \alpha) \, \psi'(\alpha_2) \, \psi''(\alpha_1)}{2 \, (\psi'(\alpha_1) + \psi'(\alpha_2) - \psi'(\alpha_1) \, \psi'(\alpha_2) \, (\alpha_1 + \alpha_2))},$$

$$\Gamma_{11}^{(\alpha)3} = \frac{(1 - \alpha) \, \psi''(\alpha_1)}{2 \, (\psi'(\alpha_1) + \psi'(\alpha_2) - \psi'(\alpha_1) \, \psi'(\alpha_2) \, (\alpha_1 + \alpha_2))},$$

$$\Gamma_{12}^{(\alpha)1} = \Gamma_{23}^{(\alpha)1} = \frac{(1 - \alpha) \, \psi'(\alpha_2)}{2 \, c \, (\psi'(\alpha_1) + \psi'(\alpha_2) - \psi'(\alpha_1) \, \psi'(\alpha_2) \, (\alpha_1 + \alpha_2))},$$

$$\Gamma_{12}^{(\alpha)2} = \Gamma_{23}^{(\alpha)2} = \frac{(1 - \alpha) \, \psi'(\alpha_1) \, \psi'(\alpha_2)}{2 \, (\psi'(\alpha_1) + \psi'(\alpha_2) - \psi'(\alpha_1) \, \psi'(\alpha_2) \, (\alpha_1 + \alpha_2))},$$

$$\Gamma_{12}^{(\alpha)3} = \Gamma_{23}^{(\alpha)3} = \frac{(1 - \alpha) \, \psi'(\alpha_1)}{2 \, c \, (\psi'(\alpha_1) + \psi'(\alpha_2) - \psi'(\alpha_1) \, \psi'(\alpha_2) \, (\alpha_1 + \alpha_2))},$$

$$\Gamma_{22}^{(\alpha)1} = \frac{-(1 - \alpha) \, \psi'(\alpha_2) \, (\alpha_1 + \alpha_2)}{2 \, c^2 \, (\psi'(\alpha_1) + \psi'(\alpha_2) - \psi'(\alpha_1) \, \psi'(\alpha_2) \, (\alpha_1 + \alpha_2))},$$

$$\Gamma_{22}^{(\alpha)2} = \frac{-(1 - \alpha) \, (-\psi'(\alpha_1) - \psi'(\alpha_2) + 2 \, \psi'(\alpha_1) \, \psi'(\alpha_2) \, (\alpha_1 + \alpha_2))}{2 \, c \, (\psi'(\alpha_1) + \psi'(\alpha_2) - \psi'(\alpha_1) \, \psi'(\alpha_2) \, (\alpha_1 + \alpha_2))},$$

$$\Gamma_{22}^{(\alpha)3} = \frac{-(1 - \alpha) \, \psi'(\alpha_1) \, (\alpha_1 + \alpha_2)}{2 \, c^2 \, (\psi'(\alpha_1) + \psi'(\alpha_2) - \psi'(\alpha_1) \, \psi'(\alpha_2) \, (\alpha_1 + \alpha_2))},$$

$$\Gamma_{33}^{(\alpha)1} = \frac{(1 - \alpha) \, \psi''(\alpha_2)}{2 \, (\psi'(\alpha_1) + \psi'(\alpha_2) - \psi'(\alpha_1) \, \psi'(\alpha_2) \, (\alpha_1 + \alpha_2))},$$

$$\Gamma_{33}^{(\alpha)2} = \frac{c \, (1 - \alpha) \, \psi'(\alpha_1) \, \psi''(\alpha_2)}{2 \, (\psi'(\alpha_1) + \psi'(\alpha_2) - \psi'(\alpha_1) \, \psi'(\alpha_2) \, (\alpha_1 + \alpha_2))},$$

$$\Gamma_{33}^{(\alpha)3} = \frac{(1 - \alpha) \, \psi''(\alpha_2) \, (-1 + \psi'(\alpha_1) \, (\alpha_1 + \alpha_2))}{2 \, (\psi'(\alpha_1) + \psi'(\alpha_2) - \psi'(\alpha_1) \, \psi'(\alpha_2) \, (\alpha_1 + \alpha_2))}. \tag{4.17}$$

while the other independent components are zero. ☐

4.4.2 McKay a-Curvatures

Proposition 4.6. *The components* $R_{ijkl}^{(\alpha)}$ *of the* α-*curvature tensor, §3.3, are given by:*

$$R^{(\alpha)}_{1212} = \frac{-\left(\alpha^2 - 1\right) \psi'(\alpha_2) \left(\psi'(\alpha_1) + \psi''(\alpha_1) (\alpha_1 + \alpha_2)\right)}{4 c^2 \left(\psi'(\alpha_1) + \psi'(\alpha_2) - \psi'(\alpha_1) \psi'(\alpha_2) (\alpha_1 + \alpha_2)\right)},$$

$$R^{(\alpha)}_{1213} = \frac{\left(\alpha^2 - 1\right) \psi'(\alpha_2) \psi''(\alpha_1)}{4 c \left(\psi'(\alpha_1) + \psi'(\alpha_2) - \psi'(\alpha_1) \psi'(\alpha_2) (\alpha_1 + \alpha_2)\right)},$$

$$R^{(\alpha)}_{1223} = \frac{\left(\alpha^2 - 1\right) \psi'(\alpha_1) \psi'(\alpha_2)}{4 c^2 \left(\psi'(\alpha_1) + \psi'(\alpha_2) - \psi'(\alpha_1) \psi'(\alpha_2) (\alpha_1 + \alpha_2)\right)},$$

$$R^{(\alpha)}_{1313} = \frac{\left(\alpha^2 - 1\right) \psi''(\alpha_1) \psi''(\alpha_2)}{4 \left(\psi'(\alpha_1) + \psi'(\alpha_2) - \psi'(\alpha_1) \psi'(\alpha_2) (\alpha_1 + \alpha_2)\right)},$$

$$R^{(\alpha)}_{1323} = \frac{\left(\alpha^2 - 1\right) \psi'(\alpha_1) \psi''(\alpha_2)}{4 c \left(\psi'(\alpha_1) + \psi'(\alpha_2) - \psi'(\alpha_1) \psi'(\alpha_2) (\alpha_1 + \alpha_2)\right)},$$

$$R^{(\alpha)}_{2323} = \frac{-\left(\alpha^2 - 1\right) \psi'(\alpha_1) \left(\psi'(\alpha_2) + \psi''(\alpha_2) (\alpha_1 + \alpha_2)\right)}{4 c^2 \left(\psi'(\alpha_1) + \psi'(\alpha_2) - \psi'(\alpha_1) \psi'(\alpha_2) (\alpha_1 + \alpha_2)\right)} \qquad (4.18)$$

while the other independent components are zero. □

Proposition 4.7. *The components of the α-Ricci tensor are given by the symmetric matrix $R^{(\alpha)} = [R^{(\alpha)}_{ij}]$:*

$$R^{(\alpha)}_{11} = \left(\alpha^2 - 1\right) \left(\frac{-\psi'(\alpha_1) \psi''(\alpha_1) \left(\psi'(\alpha_2)^2 - \psi''(\alpha_2)\right) (\alpha_1 + \alpha_2)}{4 \left(\psi'(\alpha_1) + \psi'(\alpha_2) - \psi'(\alpha_1) \psi'(\alpha_2) (\alpha_1 + \alpha_2)\right)^2} + \right.$$
$$\left. \frac{-\left(\psi'(\alpha_1) \psi'(\alpha_2) (\psi'(\alpha_1) \psi'(\alpha_2) - 2 \psi''(\alpha_1))\right) - \psi''(\alpha_1) \psi''(\alpha_2)}{4 \left(\psi'(\alpha_1) + \psi'(\alpha_2) - \psi'(\alpha_1) \psi'(\alpha_2) (\alpha_1 + \alpha_2)\right)^2} \right),$$

$$R^{(\alpha)}_{12} = \left(\alpha^2 - 1\right) \left(\frac{\left(\psi'(\alpha_2)^2 \psi''(\alpha_1) + \psi'(\alpha_1)^2 \psi''(\alpha_2)\right) (\alpha_1 + \alpha_2)}{4 c \left(\psi'(\alpha_1) + \psi'(\alpha_2) - \psi'(\alpha_1) \psi'(\alpha_2) (\alpha_1 + \alpha_2)\right)^2} + \right.$$
$$\left. \frac{\psi'(\alpha_2) \left(\psi'(\alpha_1) (\psi'(\alpha_1) + \psi'(\alpha_2)) - \psi''(\alpha_1)\right) - \psi'(\alpha_1) \psi''(\alpha_2)}{4 c \left(\psi'(\alpha_1) + \psi'(\alpha_2) - \psi'(\alpha_1) \psi'(\alpha_2) (\alpha_1 + \alpha_2)\right)^2} \right),$$

$$R^{(\alpha)}_{13} = -\left(\alpha^2 - 1\right) \frac{\left(\psi'(\alpha_1)^2 + \psi''(\alpha_1)\right) \left(\psi'(\alpha_2)^2 + \psi''(\alpha_2)\right)}{4 \left(\psi'(\alpha_1) + \psi'(\alpha_2) - \psi'(\alpha_1) \psi'(\alpha_2) (\alpha_1 + \alpha_2)\right)^2},$$

$$R^{(\alpha)}_{22} = -\left(\alpha^2 - 1\right) (\alpha_1 + \alpha_2) \left(\frac{\psi'(\alpha_2) \left(\psi'(\alpha_1) (\psi'(\alpha_1) + \psi'(\alpha_2)) - \psi''(\alpha_1)\right)}{4 c^2 \left(\psi'(\alpha_1) + \psi'(\alpha_2) - \psi'(\alpha_1) \psi'(\alpha_2) (\alpha_1 + \alpha_2)\right)^2} \right.$$
$$\left. + \frac{-\psi'(\alpha_1) \psi''(\alpha_2) + \left(\psi'(\alpha_2)^2 \psi''(\alpha_1) + \psi'(\alpha_1)^2 \psi''(\alpha_2)\right) (\alpha_1 + \alpha_2)}{4 c^2 \left(\psi'(\alpha_1) + \psi'(\alpha_2) - \psi'(\alpha_1) \psi'(\alpha_2) (\alpha_1 + \alpha_2)\right)^2} \right),$$

$$R_{23}^{(\alpha)} = (\alpha^2 - 1) \, \Big(\frac{\psi'(\alpha_2) \, (\psi'(\alpha_1) \, (\psi'(\alpha_1) + \psi'(\alpha_2)) - \psi''(\alpha_1))}{4 \, c \, (\psi'(\alpha_1) + \psi'(\alpha_2) - \psi'(\alpha_1) \, \psi'(\alpha_2) \, (\alpha_1 + \alpha_2))^2}$$

$$+ \frac{-\psi'(\alpha_1) \, \psi''(\alpha_2) + \Big(\psi'(\alpha_2)^2 \, \psi''(\alpha_1) + \psi'(\alpha_1)^2 \, \psi''(\alpha_2) \Big) \, (\alpha_1 + \alpha_2)}{4 \, c \, (\psi'(\alpha_1) + \psi'(\alpha_2) - \psi'(\alpha_1) \, \psi'(\alpha_2) \, (\alpha_1 + \alpha_2))^2} \Big),$$

$$R_{33}^{(\alpha)} = (\alpha^2 - 1) \, \Big(\frac{-\Big(\psi'(\alpha_1)^2 \, \psi'(\alpha_2)^2 \Big) + (2 \, \psi'(\alpha_1) \, \psi'(\alpha_2) - \psi''(\alpha_1)) \, \psi''(\alpha_2)}{4 \, (\psi'(\alpha_1) + \psi'(\alpha_2) - \psi'(\alpha_1) \, \psi'(\alpha_2) \, (\alpha_1 + \alpha_2))^2}$$

$$+ \frac{-\psi'(\alpha_2) \, \Big(\psi'(\alpha_1)^2 - \psi''(\alpha_1) \Big) \, \psi''(\alpha_2) \, (\alpha_1 + \alpha_2)}{4 \, (\psi'(\alpha_1) + \psi'(\alpha_2) - \psi'(\alpha_1) \, \psi'(\alpha_2) \, (\alpha_1 + \alpha_2))^2} \Big) . \quad \square \quad (4.19)$$

Proposition 4.8. *The α-scalar curvature $R^{(\alpha)}$ of M is given by:*

$$R^{(\alpha)} = - (\alpha^2 - 1) \, \Big(\frac{\psi'(\alpha_2) \, (\psi'(\alpha_1) \, (\psi'(\alpha_1) + \psi'(\alpha_2)) - 2 \, \psi''(\alpha_1)) - 2 \, \psi'(\alpha_1) \, \psi''(\alpha_2)}{2 \, (\psi'(\alpha_1) + \psi'(\alpha_2) - \psi'(\alpha_1) \, \psi'(\alpha_2) \, (\alpha_1 + \alpha_2))^2}$$

$$+ \frac{\Big(\psi'(\alpha_2)^2 \, \psi''(\alpha_1) + \Big(\psi'(\alpha_1)^2 - \psi''(\alpha_1) \Big) \, \psi''(\alpha_2) \Big) \, (\alpha_1 + \alpha_2)}{2 \, (\psi'(\alpha_1) + \psi'(\alpha_2) - \psi'(\alpha_1) \, \psi'(\alpha_2) \, (\alpha_1 + \alpha_2))^2} \Big) . \quad (4.20)$$

The α-scalar curvature $R^{(\alpha)}$ has limiting value $\frac{(\alpha^2 - 1)}{2}$ as $\alpha_1, \alpha_2 \to 0$. So M has a negative scalar curvature $R^{(0)}$, and this has limiting value $-\frac{1}{2}$ as $\alpha_1, \alpha_2 \to 0$. Figure 4.3 shows a plot of $R^{(0)}$ for the range $\alpha_1, \alpha_2 \in [0, 4]$. $\quad \square$

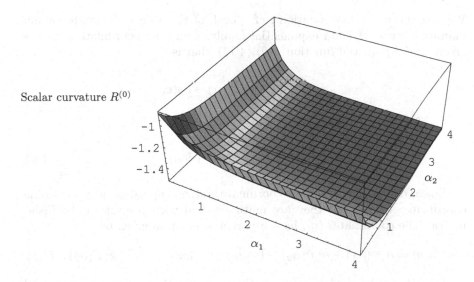

Fig. 4.3. The scalar curvature $R^{(0)}$ for the McKay bivariate gamma 3-manifold M; the limiting value at the origin is $-\frac{1}{2}$.

Proposition 4.9. *The α-sectional curvatures of M are given by:*

$$\varrho^{(\alpha)}(1,2) = \frac{(\alpha^2 - 1)\,\psi'(\alpha_2)\,(\psi'(\alpha_1) + \psi''(\alpha_1)\,(\alpha_1 + \alpha_2))}{4\,(-1 + \psi'(\alpha_1)(\alpha_1 + \alpha_2))\,(\psi'(\alpha_1) + \psi'(\alpha_2) - \psi'(\alpha_1)\,\psi'(\alpha_2)(\alpha_1 + \alpha_2))},$$

$$\varrho^{(\alpha)}(1,3) = \frac{-(\alpha^2 - 1)\,\psi''(\alpha_1)\,\psi''(\alpha_2)}{4\,\psi'(\alpha_1)\,\psi'(\alpha_2)\,(\psi'(\alpha_1) + \psi'(\alpha_2) - \psi'(\alpha_1)\,\psi'(\alpha_2)\,(\alpha_1 + \alpha_2))},$$

$$\varrho^{(\alpha)}(2,3) = \frac{(\alpha^2 - 1)\,\psi'(\alpha_1)\,(\psi'(\alpha_2) + \psi''(\alpha_2)\,(\alpha_1 + \alpha_2))}{4\,(\psi'(\alpha_2)\,(\alpha_1 + \alpha_2) - 1)\,(\psi'(\alpha_1) + \psi'(\alpha_2) - \psi'(\alpha_1)\psi'(\alpha_2)\,(\alpha_1 + \alpha_2))}.$$

$$(4.21)$$

Proposition 4.10. *The α-mean curvatures $\varrho^{(\alpha)}(\lambda)$ $(\lambda = 1, 2, 3)$ are given by:*

$$\varrho^{(\alpha)}(1) = \frac{(\alpha^2 - 1)\left(-\frac{\psi''(\alpha_1)\,\psi''(\alpha_2)}{\psi'(\alpha_1)} + \frac{\psi'(\alpha_2)^2\left(\psi'(\alpha_1) + \psi''(\alpha_1)\,(\alpha_1 + \alpha_2)\right)}{-1 + \psi'(\alpha_1)\,(\alpha_1 + \alpha_2)}\right)}{8\,\psi'(\alpha_2)\,(\psi'(\alpha_1) + \psi'(\alpha_2) - \psi'(\alpha_1)\,\psi'(\alpha_2)\,(\alpha_1 + \alpha_2))},$$

$$\varrho^{(\alpha)}(2) = \frac{(\alpha^2 - 1)\left(\frac{\psi'(\alpha_2)\left(\psi'(\alpha_1) + \psi''(\alpha_1)\,(\alpha_1 + \alpha_2)\right)}{-1 + \psi'(\alpha_1)\,(\alpha_1 + \alpha_2)} + \frac{\psi'(\alpha_1)\left(\psi'(\alpha_2) + \psi''(\alpha_2)\,(\alpha_1 + \alpha_2)\right)}{-1 + \psi'(\alpha_2)\,(\alpha_1 + \alpha_2)}\right)}{8\,(\psi'(\alpha_1) + \psi'(\alpha_2) - \psi'(\alpha_1)\,\psi'(\alpha_2)\,(\alpha_1 + \alpha_2))},$$

$$\varrho^{(\alpha)}(3) = \frac{(\alpha^2 - 1)\left(-\frac{\psi''(\alpha_1)\,\psi''(\alpha_2)}{\psi'(\alpha_2)} + \frac{\psi'(\alpha_1)^2\left(\psi'(\alpha_2) + \psi''(\alpha_2)\,(\alpha_1 + \alpha_2)\right)}{-1 + \psi'(\alpha_2)\,(\alpha_1 + \alpha_2)}\right)}{8\,\psi'(\alpha_1)\,(\psi'(\alpha_1) + \psi'(\alpha_2) - \psi'(\alpha_1)\,\psi'(\alpha_2)\,(\alpha_1 + \alpha_2))}. \quad \Box \ (4.22)$$

4.5 McKay Mutually Dual Foliations

We give mutually dual foliations, cf. §3.5.4, of the McKay bivariate gamma manifold. Since M is an exponential family, a mixture coordinate system is given by the potential function $\varphi(\theta)$ (4.14), that is

$$\eta_1 = \frac{\partial \varphi(\theta)}{\partial \alpha_1} = \psi(\alpha_1) - \log c,$$

$$\eta_2 = \frac{\partial \varphi(\theta)}{\partial c} = -\frac{\alpha_1 + \alpha_2}{c},$$

$$\eta_3 = \frac{\partial \varphi(\theta)}{\partial \alpha_2} = \psi(\alpha_2) - \log c. \quad (4.23)$$

Since (α_1, c, α_2) is a 1-affine coordinate system, (η_1, η_2, η_3) is a (-1)-affine coordinate system, and they are mutually dual with respect to the Fisher metric. The coordinates (η_i) have a potential function given by:

$$\lambda(\eta) = \alpha_1 \psi(\alpha_1) + \alpha_2 \psi(\alpha_2) - (\alpha_1 + \alpha_2) - \log \Gamma(\alpha_1) - \log \Gamma(\alpha_2). \quad (4.24)$$

In fact we have also dually orthogonal foliations. For example, with $(\alpha_1, \eta_2, \alpha_2)$ as a coordinate system for M the McKay densities take the form:

$$f(x, y; \alpha_1, \eta_2, \alpha_2) = \left(-\frac{\alpha_1 + \alpha_2}{\eta_2} \right)^{\alpha_1 + \alpha_2} \frac{x^{\alpha_1 - 1} (y - x)^{\alpha_2 - 1}}{\Gamma(\alpha_1) \Gamma(\alpha_2)} e^{\frac{\alpha_1 + \alpha_2}{\eta_2} y}. \quad (4.25)$$

and the Fisher metric is

$$\begin{bmatrix} \psi'(\alpha_1) - \frac{1}{\alpha_1 + \alpha_2} & 0 & -\frac{1}{\alpha_1 + \alpha_2} \\ 0 & \frac{\alpha_1 + \alpha_2}{(\eta_2)^2} & 0 \\ -\frac{1}{\alpha_1 + \alpha_2} & 0 & \psi'(\alpha_2) - \frac{1}{\alpha_1 + \alpha_2} \end{bmatrix}. \quad (4.26)$$

We remark that (α_1, c, α_2) is a geodesic coordinate system of $\nabla^{(1)}$, and (η_1, η_2, η_3) is a geodesic coordinate system of $\nabla^{(-1)}$.

4.6 McKay Submanifolds

We consider three submanifolds, §2.0.5, M_1, M_2 and M_3 of the 3-manifold M of McKay bivariate gamma density functions (4.11) $f(x, y; \alpha_1, c, \alpha_2)$, where we use the coordinate system (α_1, c, α_2). These submanifolds have dimension 2 and so it follows that the scalar curvature is twice the Gaussian curvature, $R = 2K$. Recall from §1.3 that the correlation is given by

$$\rho = \sqrt{\frac{\alpha_1}{\alpha_1 + \alpha_2}}.$$

In the cases of M_1 and M_2 the scalar curvature can be shown as a function of ρ only.

4.6.1 Submanifold M_1

This is defined as $M_1 \subset M$: $\alpha_1 = 1$. The density functions are of form:

$$f(x, y; 1, c, \alpha_2) = \frac{c^{1 + \alpha_2} (y - x)^{\alpha_2 - 1} e^{-cy}}{\Gamma(\alpha_2)}, \quad (4.27)$$

defined on $0 < x < y < \infty$ with parameters $c, \alpha_2 > 0$. The correlation coefficient and marginal functions, §1.3, of X and Y are given by:

$$\rho(X, Y) = \frac{1}{\sqrt{1 + \alpha_2}} \quad (4.28)$$

$$f_X(x) = c e^{-cx}, \quad x > 0 \quad (4.29)$$

$$f_Y(y) = \frac{c^{(1 + \alpha_2)} y^{\alpha_2} e^{-cy}}{\alpha_2 \, \Gamma(\alpha_2)}, \quad y > 0 \quad (4.30)$$

So here we have $\alpha_2 = \frac{1 - \rho^2}{\rho^2}$, which in practice would give a measure of the variability not due to the correlation.

Proposition 4.11. *The metric tensor $[g_{ij}]$ has component matrix*

$$G = [g_{ij}] = \begin{bmatrix} \frac{1+\alpha_2}{c^2} & -\frac{1}{c} \\ -\frac{1}{c} & \psi'(\alpha_2) \end{bmatrix} . \quad \square \qquad (4.31)$$

Proposition 4.12. *The α-connections of M_1 are*

$$\Gamma_{11,1}^{(\alpha)} = \frac{(\alpha-1)\,(1+\alpha_2)}{c^3},$$

$$\Gamma_{22,1}^{(\alpha)} = \frac{-(\alpha-1)}{2\,c^2},$$

$$\Gamma_{22,2}^{(\alpha)} = \frac{-(\alpha-1)\,\psi''(\alpha_2)}{2},$$

$$\Gamma_{11}^1 = \frac{(\alpha-1)}{2\,c}\left(2 + \frac{1}{-1+\psi'(\alpha_2)\,(1+\alpha_2)}\right),$$

$$\Gamma_{12}^1 = \frac{-(\alpha-1)\,\psi'(\alpha_2)}{2\,(-1+\psi'(\alpha_2)\,(1+\alpha_2))},$$

$$\Gamma_{22}^1 = \frac{-c\,(\alpha-1)\,\psi''(\alpha_2)}{2\,(-1+\psi'(\alpha_2)\,(1+\alpha_2))},$$

$$\Gamma_{11}^2 = \frac{(\alpha-1)\,(1+\alpha_2)}{2\,c^2\,(-1+\psi'(\alpha_2)\,(1+\alpha_2))},$$

$$\Gamma_{12}^2 = \frac{-(\alpha-1)}{2\,c\,(-1+\psi'(\alpha_2)\,(1+\alpha_2))},$$

$$\Gamma_{22}^2 = \frac{-(\alpha-1)\,\psi''(\alpha_2)\,(1+\alpha_2)}{2\,(-1+\psi'(\alpha_2)\,(1+\alpha_2))} . \quad \square \qquad (4.32)$$

The Levi-Civita connection, §2.0.6, ∇ is that given by setting $\alpha = 0$ in the α-connections above and geodesics, §2.1, for this case are curves $h : (a, b) \to M_1$ satisfying

$$\nabla_{\dot{h}}\dot{h} = 0$$

which expands using (2.5). This equation is difficult to solve analytically but we can find numerical solutions using the *Mathematica* programs of Gray [99]. Figure 4.4 shows some geodesics passing through $(c, \alpha_2) = (1, 1)$ in the gamma submanifold M_1.

Proposition 4.13. *The curvature tensor, §2.0.6, of M_1 is given by*

$$R_{1212}^{(\alpha)} = \frac{(\alpha^2-1)\,(\psi'(\alpha_2)+\psi''(\alpha_2)\,(1+\alpha_2))}{4\,c^2\,(-1+\psi'(\alpha_2)\,(1+\alpha_2))}, \qquad (4.33)$$

while the other independent components are zero.

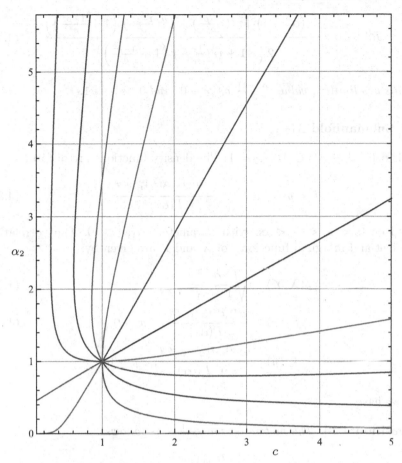

Fig. 4.4. Geodesics passing through $(c, \alpha_2) = (1, 1)$ in the McKay submanifold M_1.

By contraction we obtain:
Ricci tensor $[R_{ij}^{(\alpha)}] =$

$$\begin{bmatrix} \dfrac{-(\alpha^2-1)\,(1+\alpha_2)\,(\psi'(\alpha_2)+\psi''(\alpha_2)\,(1+\alpha_2))}{4\,c^2\,(-1+\psi'(\alpha_2)\,(1+\alpha_2))^2} & \dfrac{(\alpha^2-1)\,(\psi'(\alpha_2)+\psi''(\alpha_2)\,(1+\alpha_2))}{4\,c\,(-1+\psi'(\alpha_2)\,(1+\alpha_2))^2} \\[4mm] \dfrac{(\alpha^2-1)\,(\psi'(\alpha_2)+\psi''(\alpha_2)\,(1+\alpha_2))}{4\,c\,(-1+\psi'(\alpha_2)\,(1+\alpha_2))^2} & \dfrac{-(\alpha^2-1)\,\psi'(\alpha_2)\,(\psi'(\alpha_2)+\psi''(\alpha_2)\,(1+\alpha_2))}{4\,(-1+\psi'(\alpha_2)\,(1+\alpha_2))^2} \end{bmatrix}$$

$$(4.34)$$

Scalar curvature:

$$R^{(\alpha)} = \frac{-\left(\alpha^2-1\right)\,(\psi'(\alpha_2)+\psi''(\alpha_2)\,(1+\alpha_2))}{2\left(-1+\psi'(\alpha_2)\,(1+\alpha_2)\right)^2}\,. \qquad (4.35)$$

The α-scalar curvature $R^{(\alpha)}$ for M_1 can be written as a function of ρ only as follows:

$$R^{(\alpha)}(\rho) = \frac{-(\alpha^2 - 1)\left(\psi'(\frac{1-\rho^2}{\rho^2}) + \psi''(\frac{1-\rho^2}{\rho^2})\left(1 + \frac{1-\rho^2}{\rho^2}\right)\right)}{2\left(-1 + \psi'(\frac{1-\rho^2}{\rho^2})\left(1 + \frac{1-\rho^2}{\rho^2}\right)\right)^2}. \quad (4.36)$$

and this has limiting value $\frac{(\alpha^2-1)}{3}$ *as* $\rho \to 0$*, and* 0 *as* $\rho \to 1$. □

4.6.2 Submanifold M_2

This is defined as $M_2 \subset M$: $\alpha_2 = 1$. The density functions are of form:

$$f(x, y; \alpha_1, c, 1) = \frac{c^{(\alpha_1+1)} x^{\alpha_1-1} e^{-cy}}{\Gamma(\alpha_1)}, \quad (4.37)$$

defined on $0 < x < y < \infty$ with parameters $\alpha_1, c > 0$. The correlation coefficient and marginal functions, of X and Y are given by:

$$\rho(X, Y) = \sqrt{\frac{\alpha_1}{1 + \alpha_1}}, \quad (4.38)$$

$$f_X(x) = \frac{c^{\alpha_1} x^{\alpha_1-1} e^{-cx}}{\Gamma(\alpha_1)}, \quad x > 0 \quad (4.39)$$

$$f_Y(y) = \frac{c^{(\alpha_1+1)} y^{\alpha_1} e^{-cy}}{\alpha_1 \Gamma(\alpha_1)}, \quad y > 0 \quad (4.40)$$

Here we have $\alpha_1 = \frac{\rho^2}{1-\rho^2}$.

Proposition 4.14. *The metric tensor* $[g_{ij}]$ *is as follows:*

$$G = [g_{ij}] = \begin{bmatrix} \psi'(\alpha_1) & -\frac{1}{c} \\ -\frac{1}{c} & \frac{1+\alpha_1}{c^2} \end{bmatrix}. \quad □ \quad (4.41)$$

It follows that the geodesic curves in M_2 are essentially the same as in M_1, but with the order of coordinates interchanged. Hence they look like those in Figure 4.4 but with coordinates (α_1, c) instead of (c, α_2).

Proposition 4.15. *The* α-*connections of* M_2 *are*

$$\Gamma_{11,1}^{(\alpha)} = \frac{-(\alpha - 1)\, \psi''(\alpha_1)}{2},$$

$$\Gamma_{12,2}^{(\alpha)} = \frac{-(\alpha - 1)}{2\, c^2},$$

$$\Gamma_{22,2}^{(\alpha)} = \frac{(\alpha - 1)\, (1 + \alpha_1)}{c^3},$$

$$\Gamma_{11}^1 = \frac{-(\alpha - 1)\, \psi''(\alpha_1)\, (1 + \alpha_1)}{2\, (-1 + \psi'(\alpha_1)\, (1 + \alpha_1))},$$

$$\Gamma_{12}^1 = \frac{-(\alpha - 1)}{2\,c\,(-1 + \psi'(\alpha_1)\,(1 + \alpha_1))},$$

$$\Gamma_{22}^1 = \frac{(\alpha - 1)\,(1 + \alpha_1)}{2\,c^2\,(-1 + \psi'(\alpha_1)\,(1 + \alpha_1))},$$

$$\Gamma_{11}^2 = \frac{-c\,(\alpha - 1)\,\psi''(\alpha_1)}{2\,(-1 + \psi'(\alpha_1)\,(1 + \alpha_1))},$$

$$\Gamma_{12}^2 = \frac{-(\alpha - 1)\,\psi'(\alpha_1)}{2\,(-1 + \psi'(\alpha_1)\,(1 + \alpha_1))},$$

$$\Gamma_{22}^2 = \frac{(\alpha - 1)}{2\,c}\left(2 + \frac{1}{-1 + \psi'(\alpha_1)\,(1 + \alpha_1)}\right). \qquad \square \qquad (4.42)$$

Proposition 4.16. *The α-curvature tensor is given by*

$$R_{1212}^{(\alpha)} = \frac{(\alpha^2 - 1)\,(\psi'(\alpha_1) + \psi''(\alpha_1)\,(1 + \alpha_1))}{4\,c^2\,(-1 + \psi'(\alpha_1)\,(1 + \alpha_1))}. \qquad (4.43)$$

while the other independent components are zero.

By contraction we obtain:

α-*Ricci tensor* $[R_{ij}^{(\alpha)}] =$

$$\begin{bmatrix} \frac{-(\alpha^2-1)\,\psi'(\alpha_1)\,(\psi'(\alpha_1)+\psi''(\alpha_1)\,(1+\alpha_1))}{4\,(-1+\psi'(\alpha_1)\,(1+\alpha_1))^2} & \frac{(\alpha^2-1)\,(\psi'(\alpha_1)+\psi''(\alpha_1)\,(1+\alpha_1))}{4\,c\,(-1+\psi'(\alpha_1)\,(1+\alpha_1))^2} \\ \frac{(\alpha^2-1)\,(\psi'(\alpha_1)+\psi''(\alpha_1)\,(1+\alpha_1))}{4\,c\,(-1+\psi'(\alpha_1)\,(1+\alpha_1))^2} & \frac{-(\alpha^2-1)\,(1+\alpha_1)\,(\psi'(\alpha_1)+\psi''(\alpha_1)\,(1+\alpha_1))}{4\,c^2\,(-1+\psi'(\alpha_1)\,(1+\alpha_1))^2} \end{bmatrix}. $$

$$(4.44)$$

α-*Scalar curvature:*

$$R^{(\alpha)}(\alpha_1, \alpha_2) = \frac{-(\alpha^2 - 1)\,(\psi'(\alpha_1) + \psi''(\alpha_1)\,(1 + \alpha_1))}{2\,(-1 + \psi'(\alpha_1)\,(1 + \alpha_1))^2}. \qquad (4.45)$$

Note that, the α-scalar curvature $R^{(\alpha)}$ for M_2 can be written as a function ρ only:

$$R^{(\alpha)}(\rho) = \frac{-(\alpha^2 - 1)\,\left(\psi'(\frac{\rho^2}{1-\rho^2}) + \psi''(\frac{\rho^2}{1-\rho^2})\,\left(1 + \frac{\rho^2}{1-\rho^2}\right)\right)}{2\,\left(-1 + \psi'(\frac{\rho^2}{1-\rho^2})\,\left(1 + \frac{\rho^2}{1-\rho^2}\right)\right)^2}. \qquad (4.46)$$

and this has limiting value 0 as $\rho \to 0$, and $\frac{(\alpha^2-1)}{3}$ as $\rho \to 1$. Figure 4.5 shows a plot of $R^{(0)}$ as a function of correlation ρ for M_1 and M_2, for the range $\rho \in [0,1]$. $\qquad \square$

4.6.3 Submanifold M_3

This is defined as $M_3 \subset M$: $\alpha_1 + \alpha_2 = 1$. The density functions are of form:

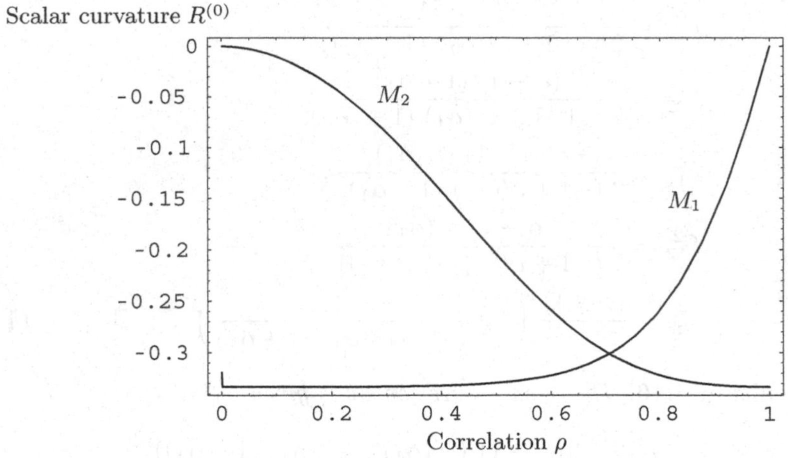

Scalar curvature $R^{(0)}$

Fig. 4.5. The scalar curvature $R^{(0)}$ as a function of correlation ρ for McKay submanifolds: M_1 (M with $\alpha_1 = 1$) where $R^{(0)}$ increases from $-\frac{1}{3}$ to 0, and M_2 (M with $\alpha_2 = 1$) where $R^{(0)}$ decreases from 0 to $-\frac{1}{3}$.

$$f(x, y; \alpha_1, c) = \frac{c\, x^{\alpha_1 - 1} (y - x)^{-\alpha_1}}{\Gamma(1 - \alpha_1)\, \Gamma(\alpha_1)}\, e^{-cy}\,, \tag{4.47}$$

defined on $0 < x < y < \infty$ with parameters $\alpha_1 < 1$, $c > 0$. The correlation coefficient and marginal functions, of X and Y are given by:

$$\rho(X, Y) = \sqrt{\alpha_1} \tag{4.48}$$

$$f_X(x) = \frac{c^{\alpha_1} x^{\alpha_1 - 1} e^{-cx}}{\Gamma(\alpha_1)}, \quad x > 0 \tag{4.49}$$

$$f_Y(y) = c\, e^{-cy}, \quad y > 0 \tag{4.50}$$

So here we have $\alpha_1 = \rho^2$.

Proposition 4.17. *The metric tensor $[g_{ij}]$ is as follows:*

$$G = [g_{ij}] = \begin{bmatrix} \pi^2 \csc(\pi\,\alpha_1)^2 & 0 \\ 0 & \frac{1}{c^2} \end{bmatrix} . \quad \Box \tag{4.51}$$

Proposition 4.18. *The α-connections of M_3 are*

$$\Gamma_{11,1}^{(\alpha)} = \pi^3\, (\alpha - 1)\, \cot(\pi\,\alpha_1)\, \csc(\pi\,\alpha_1)^2\,,$$

$$\Gamma_{22,2}^{(\alpha)} = \frac{\alpha - 1}{c^3},$$

$$\Gamma_{11}^1 = \pi\, (\alpha - 1)\, \cot(\pi\,\alpha_1)\,,$$

$$\Gamma_{22}^2 = \frac{\alpha - 1}{c} . \quad \Box \tag{4.52}$$

Proposition 4.19. *The α-curvature tensor, the α-Ricci tensor and the α-scalar curvature of M_3 are zero.* $\quad \Box$

4.7 McKay Bivariate Log-Gamma Manifold $\widetilde{M}$

We introduce a McKay bivariate log-gamma distribution, which has log-gamma marginal functions, §1.3, §3.6. This family of densities determines a Riemannian 3-manifold which is isometric to the McKay manifold, §3.1.

McKay bivariate log-gamma density functions arise from the McKay bivariate gamma density functions (4.7) for the non-negative random variables $x = log\frac{1}{n}$ and $y = log\frac{1}{m}$, or equivalently, $n = e^{-x}$ and $m = e^{-y}$. The McKay bivariate log-gamma density functions

$$g(n,m) = \frac{c^{\alpha_1+\alpha_2}\, m^{c-1}\,(-\log n)^{\alpha_1-1}\,(\log(n)-\log(m))^{\alpha_2-1}}{n\,\Gamma(\alpha_1)\,\Gamma(\alpha_2)} \quad (4.53)$$

are defined on $0 < m < n < 1$ with parameters $\alpha_i, c > 0$ $(i = 1, 2)$.

Corollary 4.20. *The covariance, §1.3, and marginal density functions, §1.3, of n and m arc given by:*

$$Cov(n,m) = c^{\alpha_1+\alpha_2}\left(\frac{(-1)^{2\alpha_2+1}\,c^{\alpha_1}}{(1+c)^{2\,\alpha_1+\alpha_2}} + \frac{(-1)^{\alpha_2}}{(-1-c)^{\alpha_2}\,(2+c)^{\alpha_1}}\right), \quad (4.54)$$

$$g_n(n) = \frac{c^{\alpha_1}\,n^{c-1}\,(-\log n)^{\alpha_1-1}}{\Gamma(\alpha_1)}, \quad (4.55)$$

$$g_m(m) = \frac{c^{\alpha_1+\alpha_2}\,m^{c-1}\,(-\log m)^{\alpha_1+\alpha_2-1}}{\Gamma(\alpha_1+\alpha_2)}. \quad (4.56)$$

Note that the marginal functions are log-gamma density functions. □

Proposition 4.21. *The family of McKay bivariate log-gamma density functions for random variables n and m where $0 < m < n < 1$, determines a Riemannian 3-manifold $\widetilde{M}$ with the following properties*
- *it contains the uniform-log gamma distribution when $c = 1$, as $\alpha_1, \alpha_2 \to 1$.*
- *the structure is given by the information-theoretic metric*
- *$\widetilde{M}$ is an isometric isomorph §3.1 of a the McKay 3-manifold M.*

Proof. Firstly, when $c = 1$ and $\alpha_1 \to 1$, we have the limiting joint density function is given by

$$\lim_{\alpha_1 \to 1} g(n,m;\alpha_1,1,\alpha_2) = g(n,m;1,1,\alpha_2) = \frac{(\log(n)-\log(m))^{\alpha_2-1}}{n\,\Gamma(\alpha_2)}$$

$$\lim_{\alpha_2 \to 1} g(n,m;1,1,\alpha_2) = g(n,m;1,1,1) = \frac{1}{n}.$$

So $g(n,m;1,1,1)$ determines the uniform-log gamma distribution, which has marginal functions: uniform distribution $g_n(n) = 1$ and log-gamma distribution $g_m(m) = -\frac{\log m}{\Gamma(2)}$.

We show that the McKay bivariate log-gamma and McKay bivariate gamma families have the same Fisher information metric.

$$g(n, m) = f(x, y) \frac{dx\, dy}{dn\, dm}$$

Hence

$$\log g(n, m) = \log f(x, y) + \log(\frac{dx\, dy}{dn\, dm})$$

then double differentiation of this relation with respect to θ_i and θ_j; when $(\theta_1, \theta_2, \theta_3) = (\alpha_1, c, \alpha_2)$ yields

$$\frac{\partial^2 \log g(n, m)}{\partial \theta_i \partial \theta_j} = \frac{\partial^2 \log f(x, y)}{\partial \theta_i \partial \theta_j}$$

we can see that $f(x, y)$ and $g(n, m)$ have the same Fisher metric.

Finally, the identity map on the parameter space

$$\{(\alpha_1, c, \alpha_2) \in \mathbb{R}^+ \times \mathbb{R}^+ \times \mathbb{R}^+\}$$

determines an isometry of Riemannian manifolds. $\square$

4.8 Generalized McKay 5-Manifold

Here we introduce a bivariate gamma distribution which is a slight generalization of that due to McKay, by substituting $(x - \gamma_1)$ for x and $(y - \gamma_2)$ for y in equation (4.7). We call this a bivariate 3-parameter gamma distribution, because the marginal functions are univariate 3-parameter gamma density functions, the extra parameters γ_i simply shifting the density function in the space of random variables. Then we consider the bivariate 3-parameter gamma family as a Riemannian 5-manifold. The geometrical objects have been calculated [13] but are not listed here because their expressions are lengthy.

4.8.1 Bivariate 3-Parameter Gamma Densities

Proposition 4.22. *Let X and Y be continuous random variables, then*

$$f(x, y) = \frac{c^{(\alpha_1 + \alpha_2)}(x - \gamma_1)^{\alpha_1 - 1}(y - \gamma_2 - x + \gamma_1)^{\alpha_2 - 1} e^{-c(y - \gamma_2)}}{\Gamma(\alpha_1)\Gamma(\alpha_2)} \quad (4.57)$$

defined on $(y - \gamma_2) > (x - \gamma_1) > 0$, $\alpha_1, c, \alpha_2 > 0$, $\gamma_1, \gamma_2 \geq 0$, is a density function. The covariance and marginal density functions, of X and Y are given by:

$$\sigma_{12} = \frac{\alpha_1}{c^2} \tag{4.58}$$

$$f_X(x) = \frac{c^{\alpha_1}(x - \gamma_1)^{\alpha_1 - 1}e^{-c(x - \gamma_1)}}{\Gamma(\alpha_1)}, \quad x > \gamma_1 \geq 0 \tag{4.59}$$

$$f_Y(y) = \frac{c^{(\alpha_1 + \alpha_2)}(y - \gamma_2)^{(\alpha_1 + \alpha_2) - 1}e^{-c(y - \gamma_2)}}{\Gamma(\alpha_1 + \alpha_2)}, \quad y > \gamma_2 \geq 0 . \ \square \tag{4.60}$$

Note that the marginal functions f_X and f_Y are univariate 3-parameter gamma density functions with parameters (c, α_1, γ_1) and $(c, \alpha_1 + \alpha_2, \gamma_2)$, where γ_1 and γ_2 are location parameters. When $\gamma_1, \gamma_2 = 0$ we recover the McKay distribution (4.7), and when $\alpha_1 = 1$ the marginal function f_X is exponential distribution with two parameters (c, γ_1), while we obtain exponential marginal function f_Y with two parameters (c, γ_2) when $\alpha_1 + \alpha_2 = 1$.

4.8.2 Generalized McKay Information Metric

Proposition 4.23. *We define the set of bivariate 3-parameter gamma density functions as*

$$M^* = \{f | f(x,y) = \frac{c^{(\alpha_1 + \alpha_2)}(x - \gamma_1)^{\alpha_1 - 1}(y - \gamma_2 - x + \gamma_1)^{\alpha_2 - 1}e^{-c(y - \gamma_2)}}{\Gamma(\alpha_1)\Gamma(\alpha_2)},$$

$$(y - \gamma_2) > (x - \gamma_1) > 0, \ \alpha_1, \alpha_2 > 2, \ c > 0, \ \gamma_1, \gamma_2 \geq 0.\} \tag{4.61}$$

Then we have:

1. *Identifying $(\alpha_1, c, \alpha_2, \gamma_1, \gamma_2)$ as a local coordinate system, M^* can be regarded as a 5-manifold.*
2. *M^* is a Riemannian manifold with Fisher information matrix $G = [g_{ij}]$ where*

$$g_{ij} = \int_{\gamma_1}^{\infty} \int_{x - \gamma_1 \mid \gamma_2}^{\infty} \frac{\partial^2 \log f(x,y)}{\partial \xi^i \partial \xi^j} f(x,y) \ dy \ dx$$

and $(\xi^1, \xi^2, \xi^3, \xi^4, \xi^5) = (\alpha_1, c, \alpha_2, \gamma_1, \gamma_2)$.
is given by:

$$[g_{ij}] = \begin{bmatrix} \psi'(\alpha_1) & -\frac{1}{c} & 0 & \frac{c}{\alpha_1 - 1} & 0 \\ -\frac{1}{c} & \frac{\alpha_1 + \alpha_2}{c^2} & -\frac{1}{c} & 0 & -1 \\ 0 & -\frac{1}{c} & \psi'(\alpha_2) & \frac{c}{1 - \alpha_2} & \frac{c}{\alpha_2 - 1} \\ \frac{c}{-1 + \alpha_1} & 0 & \frac{c}{1 - \alpha_2} & c^2\left(\frac{1}{\alpha_1 - 2} + \frac{1}{\alpha_2 - 2}\right) & \frac{c^2}{2 - \alpha_2} \\ 0 & -1 & \frac{c}{\alpha_2 - 2} & \frac{c^2}{\alpha_2 - 2} & \frac{c^2}{\alpha_2 - 2} \end{bmatrix} \tag{4.62}$$

where $\psi(\alpha_i) = \frac{\Gamma'(\alpha_i)}{\Gamma(\alpha_i)}$ $(i = 1, 2)$. $\square$

4.9 Freund Bivariate Exponential 4-Manifold $\mathcal{F}$

The results in this section were computed in [13] and first reported in [15]. The Freund [89] bivariate exponential mixture distribution arises from reliability models. Consider an instrument that has two components A, B with lifetimes X, Y respectively having density functions (when both components are in operation)

$$f_X(x) = \alpha_1 e^{-\alpha_1 x}$$
$$f_Y(y) = \alpha_2 e^{-\alpha_2 y}$$

for $(\alpha_1, \alpha_2 > 0; x, y > 0)$. Then X and Y are dependent by a failure of either component changing the parameter of the life distribution of the other component. Explicitly, when A fails, the parameter for Y becomes β_2; when B fails, the parameter for X becomes β_1. There is no other dependence. Hence the joint density function of X and Y is

$$f(x,y) = \begin{cases} \alpha_1 \beta_2 e^{-\beta_2 y - (\alpha_1 + \alpha_2 - \beta_2)x} & \text{for } 0 < x < y, \\ \alpha_2 \beta_1 e^{-\beta_1 x - (\alpha_1 + \alpha_2 - \beta_1)y} & \text{for } 0 < y < x \end{cases} \tag{4.63}$$

where $\alpha_i, \beta_i > 0 \quad (i = 1, 2)$.

For $\alpha_1 + \alpha_2 \neq \beta_1$, the marginal density function of X is

$$f_X(x) = \left(\frac{\alpha_2}{\alpha_1 + \alpha_2 - \beta_1} \right) \beta_1 e^{-\beta_1 x}$$
$$+ \left(\frac{\alpha_1 - \beta_1}{\alpha_1 + \alpha_2 - \beta_1} \right) (\alpha_1 + \alpha_2) e^{-(\alpha_1 + \alpha_2)x}, \ x \geq 0 \tag{4.64}$$

and for $\alpha_1 + \alpha_2 \neq \beta_2$, The marginal density function of Y is

$$f_Y(y) = \left(\frac{\alpha_1}{\alpha_1 + \alpha_2 - \beta_2} \right) \beta_2 e^{-\beta_2 y}$$
$$+ \left(\frac{\alpha_2 - \beta_2}{\alpha_1 + \alpha_2 - \beta_2} \right) (\alpha_1 + \alpha_2) e^{-(\alpha_1 + \alpha_2)y}, \ y \geq 0 \tag{4.65}$$

We can see that the marginal density functions are not actually exponential but mixtures of exponential density functions if $\alpha_i > \beta_i$, that is, they are weighted averages. So these are bivariate mixture exponential density functions. The marginal density functions $f_X(x)$ and $f_Y(y)$ are exponential density functions only in the special case $\alpha_i = \beta_i \quad (i = 1, 2)$.

For the special case when $\alpha_1 + \alpha_2 = \beta_1 = \beta_2$, Freund obtained the joint density function as:

$$f(x,y) = \begin{cases} \alpha_1 (\alpha_1 + \alpha_2) e^{-(\alpha_1 + \alpha_2)y} & \text{for } 0 < x < y, \\ \alpha_2 (\alpha_1 + \alpha_2) e^{-(\alpha_1 + \alpha_2)x} & \text{for } 0 < y < x \end{cases} \tag{4.66}$$

with marginal density functions, §1.3:

$$f_X(x) = (\alpha_1 + \alpha_2(\alpha_1 + \alpha_2)x)\, e^{-(\alpha_1+\alpha_2)x}\, x \geq 0, \tag{4.67}$$
$$f_Y(y) = (\alpha_2 + \alpha_1(\alpha_1 + \alpha_2)y)\, e^{-(\alpha_1+\alpha_2)y}\, y \geq 0 \tag{4.68}$$

The covariance, §1.3 and correlation coefficient, §1.3 of X and Y are

$$Cov(X,Y) = \frac{\beta_1\beta_2 - \alpha_1\alpha_2}{\beta_1\beta_2(\alpha_1+\alpha_2)^2}, \tag{4.69}$$

$$\rho(X,Y) = \frac{\beta_1\beta_2 - \alpha_1\alpha_2}{\sqrt{\alpha_2{}^2 + 2\alpha_1\alpha_2 + \beta_1{}^2}\sqrt{\alpha_1{}^2 + 2\alpha_1\alpha_2 + \beta_2{}^2}} \tag{4.70}$$

Note that $-\frac{1}{3} < \rho(X,Y) < 1$. The correlation coefficient $\rho(X,Y) \to 1$ when $\beta_1, \beta_2 \to \infty$, and $\rho(X,Y) \to -\frac{1}{3}$ when $\alpha_1 = \alpha_2$ and $\beta_1, \beta_2 \to 0$. In many applications, $\beta_i > \alpha_i$ $(i = 1, 2)$ (i.e., lifetime tends to be shorter when the other component is out of action); in such cases the correlation is positive.

4.9.1 Freund Fisher Metric

As in the case of the McKay manifold, the multiplicative group action of the real numbers as a scale-changing process influences the Riemannian geometric properties.

Proposition 4.24. $\mathcal{F}$, the set of Freund bivariate mixture exponential density functions, (4.63) becomes a 4-manifold with Fisher information metric, §2.0.5 §3.1,

$$g_{ij} = \int_0^\infty \int_0^\infty \frac{\partial^2 \log f(x,y)}{\partial x_i \partial x_j}\, f(x,y)\, dx\, dy$$

and $(x_1, x_2, x_3, x_4) = (\alpha_1, \beta_1, \alpha_2, \beta_2)$.
is given by

$$
[g_{ij}] = \int_0^\infty \int_0^\infty \frac{\partial^2 \log f(x,y)}{\partial x_i \partial x_j}\, f(x,y)\, dx\, dy
$$
$$
= \begin{bmatrix}
\frac{1}{\alpha_1{}^2 + \alpha_1\alpha_2} & 0 & 0 & 0 \\
0 & \frac{\alpha_2}{\beta_1{}^2(\alpha_1+\alpha_2)} & 0 & 0 \\
0 & 0 & \frac{1}{\alpha_2{}^2 + \alpha_1\alpha_2} & 0 \\
0 & 0 & 0 & \frac{\alpha_1}{\beta_2{}^2(\alpha_1+\alpha_2)}
\end{bmatrix} \tag{4.71}
$$

The inverse matrix $[g^{ij}] = [g_{ij}]^{-1}$ *is given by*

$$
[g^{ij}] = \begin{bmatrix}
\alpha_1{}^2 + \alpha_1\alpha_2 & 0 & 0 & 0 \\
0 & \frac{\beta_1{}^2(\alpha_1+\alpha_2)}{\alpha_2} & 0 & 0 \\
0 & 0 & \alpha_2{}^2 + \alpha_1\alpha_2 & 0 \\
0 & 0 & 0 & \frac{\beta_2{}^2(\alpha_1+\alpha_2)}{\alpha_1}
\end{bmatrix}. \tag{4.72}
$$

$\square$

The orthogonality property of $(\alpha_1, \beta_1, \alpha_2, \beta_2)$ coordinates is equivalent to asymptotic independence of the maximum likelihood estimates[20, 113, 153].

4.10 Freund Natural Coordinates

Leurgans, Tsai, and Crowley [135] pointed out that the Freund density functions form an exponential family, §3.2, with natural parameters

$$(\theta_1, \theta_2, \theta_3, \theta_4) = \left(\alpha_1 + \beta_1, \alpha_2, \log\left(\frac{\alpha_1\,\beta_2}{\alpha_2\,\beta_1}\right), \beta_2\right) \tag{4.73}$$

and potential function

$$\varphi(\theta) = -\log\left(\frac{\theta_1\,\theta_2\,\theta_4}{e^{\theta_3}\,\theta_2 + \theta_4}\right) = -\log(\alpha_2\,\beta_1). \tag{4.74}$$

Hence

$$\theta_1 = \alpha_1 + \beta_1, \ \ \theta_2 = \alpha_2, \ \ \theta_3 = \log\left(\frac{\alpha_1\,\beta_2}{\alpha_2\,\beta_1}\right), \ \ \theta_4 = \beta_2$$

yield:

$$\alpha_1 = \frac{\theta_1\,\theta_2}{e^{\theta_3}\,\theta_2 + \theta_4}\,e^{\theta_3}, \ \ \beta_1 = \frac{\theta_1\,\theta_4}{e^{\theta_3}\,\theta_2 + \theta_4}, \ \ \alpha_2 = \theta_2, \ \ \beta_2 = \theta_4.$$

so (4.63) in natural coordinates is

$$f(x,y) = \begin{cases} \frac{\theta_1\,\theta_2\,\theta_4}{e^{\theta_3}\,\theta_2+\theta_4}\,e^{\theta_3}e^{-\theta_4 y - (\theta_1-\theta_4)x} & \text{for } 0 < x < y \\ \frac{\theta_1\,\theta_2\,\theta_4}{e^{\theta_3}\,\theta_2+\theta_4}\,e^{-\theta_2 x - (\theta_1-\theta_2)y} & \text{for } 0 < y < x \end{cases}$$

$$= \begin{cases} e^{\theta_1(-x)+\theta_3+\theta_4(x-y)+\log\left(\frac{\theta_1\,\theta_2\,\theta_4}{e^{\theta_3}\,\theta_2+\theta_4}\right)} & \text{for } 0 < x < y \\ e^{\theta_1(y)+\theta_2(y-x)+\log\left(\frac{\theta_1\,\theta_2\,\theta_4}{e^{\theta_3}\,\theta_2+\theta_4}\right)} & \text{for } 0 < y < x \end{cases}. \tag{4.75}$$

In natural coordinates (θ_i) (4.73) the Fisher metric is given by

$$\left[\frac{\partial^2 \varphi(\theta)}{\partial\theta_i \partial\theta_j}\right] = \begin{bmatrix} \frac{1}{\theta_1^2} & 0 & 0 & 0 \\ 0 & \frac{\theta_4\left(2e^{\theta_3}\theta_2+\theta_4\right)}{\theta_2^2\left(e^{\theta_3}\theta_2+\theta_4\right)^2} & \frac{e^{\theta_3}\theta_4}{\left(e^{\theta_3}\theta_2+\theta_4\right)^2} & -\frac{e^{\theta_3}}{\left(e^{\theta_3}\theta_2+\theta_4\right)^2} \\ 0 & \frac{e^{\theta_3}\theta_4}{\left(e^{\theta_3}\theta_2+\theta_4\right)^2} & \frac{e^{\theta_3}\theta_2\theta_4}{\left(e^{\theta_3}\theta_2+\theta_4\right)^2} & -\frac{e^{\theta_3}\theta_2}{\left(e^{\theta_3}\theta_2+\theta_4\right)^2} \\ 0 & -\frac{e^{\theta_3}}{\left(e^{\theta_3}\theta_2+\theta_4\right)^2} & -\frac{e^{\theta_3}\theta_2}{\left(e^{\theta_3}\theta_2+\theta_4\right)^2} & \frac{1}{\theta_4^2} - \frac{1}{\left(e^{\theta_3}\theta_2+\theta_4\right)^2} \end{bmatrix},$$

$$= \begin{bmatrix} \frac{1}{(\alpha_1+\beta_1)^2} & 0 & 0 & 0 \\ 0 & \frac{\beta_1\left(2\alpha_1+\beta_1\right)}{\alpha_2^2\left(\alpha_1+\beta_1\right)^2} & \frac{\alpha_1\beta_1}{\alpha_2\left(\alpha_1+\beta_1\right)^2} & -\frac{\alpha_1\beta_1}{\alpha_2\left(\alpha_1+\beta_1\right)^2\beta_2} \\ 0 & \frac{\alpha_1\beta_1}{\alpha_2\left(\alpha_1+\beta_1\right)^2} & \frac{\alpha_1\beta_1}{\left(\alpha_1+\beta_1\right)^2} & -\frac{\alpha_1\beta_1}{\left(\alpha_1+\beta_1\right)^2\beta_2} \\ 0 & -\frac{\alpha_1\beta_1}{\alpha_2\left(\alpha_1+\beta_1\right)^2\beta_2} & \frac{\alpha_1\beta_1}{\left(\alpha_1+\beta_1\right)^2\beta_2} & \frac{\alpha_1\left(\alpha_1+2\beta_1\right)}{\left(\alpha_1+\beta_1\right)^2\beta_2^2} \end{bmatrix}. \tag{4.76}$$

4.11 Freund a-Geometry

The α-connection §3.3, α-curvature tensor, α-Ricci tensor with its eigenvalues and eigenvectors, α-scalar curvature, α-sectional curvatures and the α-mean curvatures are more simply reported with respect to the coordinate system $(\alpha_1, \beta_1, \alpha_2, \beta_2)$.

4.11.1 Freund a-Connection

For each $\alpha \in \mathbb{R}$, the α *(or $\nabla^{(\alpha)}$)-connection* is the torsion-free affine connection with components

$$\Gamma_{ij,k}^{(\alpha)} = \int_0^\infty \int_x^\infty \left(\frac{\partial^2 \log f_1}{\partial \xi^i \partial \xi^j} \frac{\partial \log f_1}{\partial \xi^k} + \frac{1-\alpha}{2} \frac{\partial \log f_1}{\partial \xi^i} \frac{\partial \log f_1}{\partial \xi^j} \frac{\partial \log f_1}{\partial \xi^k} \right) f_1 \, dy \, dx$$

$$+ \int_0^\infty \int_x^\infty \left(\frac{\partial^2 \log f_2}{\partial \xi^i \partial \xi^j} \frac{\partial \log f_2}{\partial \xi^k} + \frac{1-\alpha}{2} \frac{\partial \log f_2}{\partial \xi^i} \frac{\partial \log f_2}{\partial \xi^j} \frac{\partial \log f_2}{\partial \xi^k} \right) f_2 \, dy \, dx$$

Proposition 4.25. *The nonzero independent components* $\Gamma_{ij,k}^{(\alpha)}$ *are*

$$\Gamma_{11,1}^{(\alpha)} = \frac{2(\alpha-1)\alpha_1 - (1+\alpha)\alpha_2}{2\alpha_1^2 (\alpha_1 + \alpha_2)^2},$$

$$\Gamma_{11,3}^{(\alpha)} = \frac{1+\alpha}{2\alpha_1 (\alpha_1 + \alpha_2)^2},$$

$$\Gamma_{12,2}^{(\alpha)} = \frac{(\alpha-1)\alpha_2}{2(\alpha_1 + \alpha_2)^2 \beta_1^2},$$

$$\Gamma_{13,3}^{(\alpha)} = \frac{\alpha-1}{2\alpha_2 (\alpha_1 + \alpha_2)^2},$$

$$\Gamma_{14,4}^{(\alpha)} = \frac{-(\alpha-1)\alpha_2}{2(\alpha_1 + \alpha_2)^2 \beta_2^2},$$

$$\Gamma_{22,2}^{(\alpha)} = \frac{(\alpha-1)\alpha_2}{(\alpha_1 + \alpha_2) \beta_1^3},$$

$$\Gamma_{22,3}^{(\alpha)} = \frac{-(1+\alpha)\alpha_1}{2(\alpha_1 + \alpha_2)^2 \beta_1^2},$$

$$\Gamma_{33,3}^{(\alpha)} = \frac{-(1+\alpha)\alpha_1 + 2(\alpha-1)\alpha_2}{2\alpha_2^2 (\alpha_1 + \alpha_2)^2},$$

$$\Gamma_{34,4}^{(\alpha)} = \frac{(\alpha-1)\alpha_1}{2(\alpha_1 + \alpha_2)^2 \beta_2^2},$$

$$\Gamma_{44,4}^{(\alpha)} = \frac{(\alpha-1)\alpha_1}{(\alpha_1 + \alpha_2) \beta_2^3}. \qquad \square \tag{4.77}$$

We have a symmetric linear connection $\nabla^{(\alpha)}$ defined by:

$$\langle \nabla^{(\alpha)}_{\partial_i} \partial_j, \partial_k \rangle = \Gamma^{(\alpha)}_{ij,k},$$

Hence from

$$\Gamma^{(\alpha)}_{ij,k} = \sum_{h=1}^{4} g_{kh} \, \Gamma^{h(\alpha)}_{ij}, (k = 1, 2, 3, 4).$$

we obtain the components of $\nabla^{(\alpha)}$.

Proposition 4.26. *The nonzero components* $\Gamma^{i(\alpha)}_{jk}$ *of the* $\nabla^{(\alpha)}$-*connections are given by:*

$$\Gamma^{(\alpha)1}_{11} = -\frac{1+\alpha}{2\,\alpha_1} + \frac{-1+3\,\alpha}{2\,(\alpha_1 + \alpha_2)},$$

$$\Gamma^{(\alpha)1}_{13} = \Gamma^{(\alpha)2}_{12} = \Gamma^{(\alpha)3}_{13} = \Gamma^{(\alpha)4}_{34} = \frac{\alpha - 1}{2\,(\alpha_1 + \alpha_2)},$$

$$\Gamma^{(\alpha)1}_{22} = -\Gamma^{(\alpha)3}_{22} = \frac{(1+\alpha)\,\alpha_1\,\alpha_2}{2\,(\alpha_1 + \alpha_2)\,\beta_1^2},$$

$$\Gamma^{(\alpha)1}_{33} = \Gamma^{(\alpha)2}_{23} = \frac{(1+\alpha)\,\alpha_1}{2\,\alpha_2\,(\alpha_1 + \alpha_2)},$$

$$\Gamma^{(\alpha)1}_{44} = -\Gamma^{(\alpha)3}_{44} = \frac{-(1+\alpha)\,\alpha_1\,\alpha_2}{2\,(\alpha_1 + \alpha_2)\,\beta_2^2},$$

$$\Gamma^{(\alpha)3}_{11} = \Gamma^{(\alpha)4}_{14} = \frac{(1+\alpha)\,\alpha_2}{2\,\alpha_1\,(\alpha_1 + \alpha_2)},$$

$$\Gamma^{(\alpha)3}_{33} = -\frac{1+\alpha}{2\,\alpha_2} + \frac{-1+3\,\alpha}{2\,(\alpha_1 + \alpha_2)},$$

$$\Gamma^{(\alpha)4}_{44} = \frac{\alpha - 1}{\beta_2}. \quad \square \tag{4.78}$$

4.11.2 Freund α-Curvatures

Proposition 4.27. *The nonzero independent components* $R^{(\alpha)}_{ijkl}$ *of the* α-*curvature tensor are given by:*

$$R^{(\alpha)}_{1212} = \frac{(\alpha^2 - 1)\,\alpha_2^2}{4\,\alpha_1\,(\alpha_1 + \alpha_2)^3\,\beta_1^2},$$

$$R^{(\alpha)}_{1223} = \frac{(\alpha^2 - 1)\,\alpha_2}{4\,(\alpha_1 + \alpha_2)^3\,\beta_1^2},$$

$$R^{(\alpha)}_{1414} = \frac{(\alpha^2 - 1)\,\alpha_2}{4\,(\alpha_1 + \alpha_2)^3\,\beta_2^2},$$

$$R^{(\alpha)}_{1434} = \frac{-\left(\alpha^2 - 1\right)\alpha_1}{4\left(\alpha_1 + \alpha_2\right)^3 \beta_2{}^2},$$

$$R^{(\alpha)}_{2323} = \frac{\left(\alpha^2 - 1\right)\alpha_1}{4\left(\alpha_1 + \alpha_2\right)^3 \beta_1{}^2},$$

$$R^{(\alpha)}_{2424} = \frac{\left(\alpha^2 - 1\right)\alpha_1\alpha_2}{4\left(\alpha_1 + \alpha_2\right)^2 \beta_1{}^2 \beta_2{}^2},$$

$$R^{(\alpha)}_{3434} = \frac{\left(\alpha^2 - 1\right)\alpha_1{}^2}{4\,\alpha_2\left(\alpha_1 + \alpha_2\right)^3 \beta_2{}^2}. \qquad \Box \tag{4.79}$$

Contracting $R^{(\alpha)}_{ijkl}$ with g^{il} we obtain the components $R^{(\alpha)}_{jk}$ of the α-Ricci tensor.

Proposition 4.28. *The α-Ricci tensor $R^{(\alpha)} = [R^{(\alpha)}_{jk}]$ is given by:*

$$R^{(\alpha)} = [R^{(\alpha)}_{jk}] = \begin{bmatrix} \frac{-\left(\alpha^2-1\right)\alpha_2}{2\,\alpha_1\,(\alpha_1+\alpha_2)^2} & 0 & \frac{\alpha^2-1}{2\,(\alpha_1+\alpha_2)^2} & 0 \\ 0 & \frac{-\left(\alpha^2-1\right)\alpha_2}{2\,(\alpha_1+\alpha_2)\,\beta_1{}^2} & 0 & 0 \\ \frac{\alpha^2-1}{2\,(\alpha_1+\alpha_2)^2} & 0 & \frac{-\left(\alpha^2-1\right)\alpha_1}{2\,\alpha_2\,(\alpha_1+\alpha_2)^2} & 0 \\ 0 & 0 & 0 & \frac{-\left(\alpha^2-1\right)\alpha_1}{2\,(\alpha_1+\alpha_2)\,\beta_2{}^2} \end{bmatrix}$$

$$\tag{4.80}$$

The α-eigenvalues and the α-eigenvectors of the α-Ricci tensor are:

$$(\alpha^2 - 1)\begin{pmatrix} 0 \\ \frac{1}{(\alpha_1+\alpha_2)^2} - \frac{1}{2\,\alpha_1\,\alpha_2} \\ \frac{-\alpha_2}{2\,(\alpha_1+\alpha_2)\,\beta_1{}^2} \\ \frac{-\alpha_1}{2\,(\alpha_1+\alpha_2)\,\beta_2{}^2} \end{pmatrix} \tag{4.81}$$

$$\begin{pmatrix} \frac{\alpha_1}{\alpha_2} & 0 & 1 & 0 \\ -\frac{\alpha_2}{\alpha_1} & 0 & 1 & 0 \\ 0 & 1 & 0 & 0 \\ 0 & 0 & 0 & 1 \end{pmatrix} \qquad \Box \tag{4.82}$$

Proposition 4.29. *The Freund manifold F has a constant α-scalar curvature*

$$R^{(\alpha)} = \frac{3}{2}\left(1 - \alpha^2\right). \tag{4.83}$$

The Freund manifold has a positive scalar curvature $R^{(0)} = \frac{3}{2}$ when $\alpha = 0$, so geometrically it constitutes part of a sphere. $\qquad \Box$

Proposition 4.30. *The α-sectional curvatures $\varrho^{(\alpha)}(\lambda,\mu)$ $(\lambda,\mu=1,2,3,4)$ are given by:*

$$\varrho^{(\alpha)}(1,2) = \varrho^{(\alpha)}(1,4) = \frac{\left(1-\alpha^2\right)\alpha_2}{4\left(\alpha_1+\alpha_2\right)},$$

$$\varrho^{(\alpha)}(1,3) = 0,$$

$$\varrho^{(\alpha)}(2,3) = \frac{\left(1-\alpha^2\right)\alpha_1}{4\left(\alpha_1+\alpha_2\right)},$$

$$\varrho^{(\alpha)}(2,4) = \frac{1-\alpha^2}{4},$$

$$\varrho^{(\alpha)}(3,4) = \varrho(2,3) . \quad \square \tag{4.84}$$

Proposition 4.31. *The α-mean curvatures $\varrho^{(\alpha)}(\lambda)$ $(\lambda = 1,2,3,4)$ are given by:*

$$\varrho^{(\alpha)}(1) = \frac{\left(1-\alpha^2\right)\alpha_2}{6\left(\alpha_1+\alpha_2\right)},$$

$$\varrho^{(\alpha)}(2) = \varrho(4) = \frac{1-\alpha^2}{6},$$

$$\varrho^{(\alpha)}(3) = \frac{\left(1-\alpha^2\right)\alpha_1}{6\left(\alpha_1+\alpha_2\right)} . \quad \square \tag{4.85}$$

4.12 Freund Foliations

Since F is an exponential family, §3.2 a mutually dual coordinate system is given by the potential function $\varphi(\theta)$ (4.74), §3.2, that is

$$\eta_1 = \frac{\partial\varphi(\theta)}{\partial\theta_1} = -\frac{1}{\theta_1} = -\frac{1}{\alpha_1+\beta_1},$$

$$\eta_2 = \frac{\partial\varphi(\theta)}{\partial\theta_2} = -\frac{\theta_4}{\theta_2\left(e^{\theta_3}\theta_2+\theta_4\right)} = -\frac{\beta_1}{\alpha_2\left(\alpha_1+\beta_1\right)},$$

$$\eta_3 = \frac{\partial\varphi(\theta)}{\partial\theta_3} = 1 - \frac{\theta_4}{e^{\theta_3}\theta_2+\theta_4} = \frac{\alpha_1}{\alpha_1+\beta_1},$$

$$\eta_4 = \frac{\partial\varphi(\theta)}{\partial\theta_4} = -\frac{1}{\theta_4} + \frac{1}{e^{\theta_3}\theta_2+\theta_4} = -\frac{\alpha_1}{\left(\alpha_1+\beta_1\right)\beta_2} . \tag{4.86}$$

Then $(\theta_1,\theta_2,\theta_3,\theta_4)$ is a 1-affine coordinate system, $(\eta_1,\eta_2,\eta_3,\eta_4)$ is a (-1)-affine coordinate system, and they are mutually dual with respect to the Fisher information metric. The (η_i) have a potential function given by:

$$\lambda = \log\left(\frac{\theta_1\theta_2\theta_4}{e^{\theta_3}\theta_2+\theta_4}\right) + \frac{e^{\theta_3}\theta_2\theta_3}{e^{\theta_3}\theta_2+\theta_4} - 2$$

$$= \log(\alpha_2\beta_1) + \frac{\alpha_1}{\alpha_1+\beta_1}\log\left(\frac{\alpha_1\beta_2}{\alpha_2\beta_1}\right) - 2. \tag{4.87}$$

Coordinates (θ_i) and (η_i) are mutually dual so F has dually orthogonal foliations. With coordinates

$$(\eta_1, \theta_2, \theta_3, \theta_4) = (-\frac{1}{\theta_1}, \theta_2, \theta_3, \theta_4)$$

the Freund density functions are:

$$f(x, y; \eta_1, \theta_2, \theta_3, \theta_4) = \begin{cases} -\frac{\theta_2\,\theta_4\,e^{\theta_3}}{(\theta_2\,e^{\theta_3}+\theta_4)\,\eta_1}\,e^{\theta_4\,(x-y)+\frac{x}{\eta_1}} & \text{for } 0 < x < y \\ -\frac{\theta_2\,\theta_4}{(\theta_2\,e^{\theta_3}+\theta_4)\,\eta_1}\,e^{\theta_2\,(y-x)+\frac{y}{\eta_1}} & \text{for } 0 < y < x \end{cases} \qquad (4.88)$$

where $\eta_1 < 0$ and $\theta_i > 0 \quad (i = 2, 3, 4)$.
The Fisher metric is

$$[g_{ij}] = \begin{bmatrix} \frac{1}{\eta_1{}^2} & 0 & 0 & 0 \\ 0 & \frac{\theta_4\left(2\,e^{\theta_3}\,\theta_2+\theta_4\right)}{\theta_2{}^2\left(e^{\theta_3}\,\theta_2+\theta_4\right)^2} & \frac{e^{\theta_3}\,\theta_4}{\left(e^{\theta_3}\,\theta_2+\theta_4\right)^2} & -\frac{e^{\theta_3}}{\left(e^{\theta_3}\,\theta_2\mid\theta_4\right)^2} \\ 0 & \frac{e^{\theta_3}\,\theta_4}{\left(e^{\theta_3}\,\theta_2+\theta_4\right)^2} & \frac{e^{\theta_3}\,\theta_2\,\theta_4}{\left(e^{\theta_3}\,\theta_2+\theta_4\right)^2} & -\frac{e^{\theta_3}\,\theta_2}{\left(e^{\theta_3}\,\theta_2+\theta_4\right)^2} \\ 0 & -\frac{e^{\theta_3}}{\left(e^{\theta_3}\,\theta_2+\theta_4\right)^2} & -\frac{e^{\theta_3}\,\theta_2}{\left(e^{\theta_3}\,\theta_2+\theta_4\right)^2} & \frac{1}{\theta_4{}^2}-\frac{1}{\left(e^{\theta_3}\,\theta_2+\theta_4\right)^2} \end{bmatrix}$$

$$= \begin{bmatrix} (\alpha_1+\beta_1)^2 & 0 & 0 & 0 \\ 0 & \frac{\beta_1\,(2\,\alpha_1+\beta_1)}{\alpha_2{}^2\,(\alpha_1+\beta_1)^2} & \frac{\alpha_1\,\beta_1}{\alpha_2\,(\alpha_1+\beta_1)^2} & -\frac{\alpha_1\,\beta_1}{\alpha_2\,(\alpha_1+\beta_1)^2\,\beta_2} \\ 0 & \frac{\alpha_1\,\beta_1}{\alpha_2\,(\alpha_1+\beta_1)^2} & \frac{\alpha_1\,\beta_1}{(\alpha_1+\beta_1)^2} & -\frac{\alpha_1\,\beta_1}{(\alpha_1+\beta_1)^2\,\beta_2} \\ 0 & -\frac{\alpha_1\,\beta_1}{\alpha_2\,(\alpha_1+\beta_1)^2\,\beta_2} & -\frac{\alpha_1\,\beta_1}{(\alpha_1+\beta_1)^2\,\beta_2} & \frac{\alpha_1\,(\alpha_1+2\,\beta_1)}{(\alpha_1+\beta_1)^2\,\beta_2{}^2} \end{bmatrix}.$$

$$(4.89)$$

It follows that (θ_i) is a geodesic coordinate system of $\nabla^{(1)}$, and (η_i) is a geodesic coordinate system of $\nabla^{(-1)}$.

4.13 Freund Submanifolds

Four submanifolds F_i $(i = 1, 2, 3, 4)$, §2.0.5, of the 4-manifold F are of interest, including the case of statistically independent random variables, §1.3. Also one is the special case of an Absolutely Continuous Bivariate Exponential Distribution called ACBED (or ACBVE) by Block and Basu (cf. Hutchinson and Lai [105]). We use the coordinate system $(\alpha_1, \beta_1, \alpha_2, \beta_2)$ for the submanifolds F_i $(i \neq 3)$, and the coordinate system $(\lambda_1, \lambda_{12}, \lambda_2)$ for ACBED of the Block and Basu case.

4.13.1 Independence Submanifold F_1

This is defined as

$$F_1 \subset F : \beta_1 = \alpha_1, \; \beta_2 = \alpha_2.$$

The density functions are of form:

$$f(x, y; \alpha_1, \alpha_2) = f_1(x; \alpha_1) f_2(y; \alpha_2) \tag{4.90}$$

where f_i are the density functions of the univariate exponential density functions with the parameters $\alpha_i > 0$ $(i = 1, 2)$. This is the case for statistical independence, §1.3, of X and Y, so the space F_1 is the direct product of the two corresponding Riemannian spaces $\{f_1(x; \alpha_1) : f_1(x; \alpha_1) = \alpha_1 e^{-\alpha_1 x}, \alpha_1 > 0\}$ and $\{f_2(y; \alpha_2) : f_2(y; \alpha_2) = -\alpha_2 e^{-\alpha_2 y}, \alpha_2 > 0\}$.

Proposition 4.32. *The metric tensor $[g_{ij}]$ is as follows:*

$$[g_{ij}] = \begin{bmatrix} \frac{1}{\alpha_1^2} & 0 \\ 0 & \frac{1}{\alpha_2^2} \end{bmatrix} . \quad \square \tag{4.91}$$

Proposition 4.33. *The nonzero independent components of the α-connection are*

$$\Gamma_{11,1}^{(\alpha)} = \frac{\alpha - 1}{\alpha_1^{3}} ,$$

$$\Gamma_{22,2}^{(\alpha)} = \frac{\alpha - 1}{\alpha_2^{3}} ,$$

$$\Gamma_{11}^{(\alpha)1} = \frac{\alpha - 1}{\alpha_1} ,$$

$$\Gamma_{22}^{(\alpha)2} = \frac{\alpha - 1}{\alpha_2} . \quad \square \tag{4.92}$$

Proposition 4.34. *The α-curvature tensor, α-Ricci tensor, and α-scalar curvature of F_1 are zero.* $\qquad\qquad\square$

4.13.2 Submanifold F_2

This is defined as

$$F_2 \subset F : \alpha_1 = \alpha_2, \ \beta_1 = \beta_2.$$

The density functions are of form:

$$f(x, y; \alpha_1, \beta_1) = \begin{cases} \alpha_1 \beta_1 \, e^{-\beta_1 y - (2\alpha_1 - \beta_1)x} & \text{for } 0 < x < y \\ \alpha_1 \beta_1 \, e^{-\beta_1 x - (2\alpha_1 - \beta_1)y} & \text{for } 0 < y < x \end{cases} \tag{4.93}$$

with parameters $\alpha_1, \beta_1 > 0$. The covariance, correlation coefficient and marginal density functions, of X and Y are given by:

$$Cov(X, Y) = \frac{1}{4} \left(\frac{1}{\alpha_1^2} - \frac{1}{\beta_1^2} \right) , \tag{4.94}$$

$$\rho(X, Y) = 1 - \frac{4\,\alpha_1{}^2}{3\,\alpha_1{}^2 + \beta_1{}^2}, \tag{4.95}$$

$$f_X(x) = \left(\frac{\alpha_1}{2\,\alpha_1 - \beta_1}\right) \beta_1\, e^{-\beta_1 x} + \left(\frac{\alpha_1 - \beta_1}{2\,\alpha_1 - \beta_1}\right)(2\,\alpha_1)\, e^{-2\,\alpha_1 x}, \; x \geq 0, \tag{4.96}$$

$$f_Y(y) = \left(\frac{\alpha_1}{2\,\alpha_1 - \beta_1}\right) \beta_1\, e^{-\beta_1 y} + \left(\frac{\alpha_1 - \beta_1}{2\,\alpha_1 - \beta_1}\right)(2\,\alpha_1)\, e^{-2\,\alpha_1 y}, \; y \geq 0. \tag{4.97}$$

When $\alpha_1 = \beta_1$, $\rho(X, Y) = 0$. Submanifold F_2 forms an exponential family, with natural parameters (α_1, β_1) and potential function

$$\varphi = -\log(\alpha_1\,\beta_1). \tag{4.98}$$

Proposition 4.35. *The submanifold F_2 is an isometric isomorph of the manifold F_1 §3.1.*

Proof. Since $\varphi = -\log(\alpha_1\,\beta_1)$ is a potential function, the Fisher metric is the Hessian of φ, that is,

$$[g_{ij}] = [\frac{\partial^2 \varphi}{\partial \theta_i \partial \theta_j}] = \begin{bmatrix} \frac{1}{\alpha_1^2} & 0 \\ 0 & \frac{1}{\beta_1^2} \end{bmatrix} \tag{4.99}$$

where $(\theta_1, \theta_2) = (\alpha_1, \beta_1)$. $\qquad \square$

4.13.3 Submanifold F_3

This is defined as

$$F_3 \subset F : \beta_1 = \beta_2 = \alpha_1 + \alpha_2.$$

The density functions are of form:

$$f(x, y; \alpha_1, \alpha_2, \beta_2) = \begin{cases} \alpha_1\,(\alpha_1 + \alpha_2)\, e^{-(\alpha_1 + \alpha_2)y} & \text{for } 0 < x < y \\ \alpha_2\,(\alpha_1 + \alpha_2)\, e^{-(\alpha_1 + \alpha_2)x} & \text{for } 0 < y < x \end{cases} \tag{4.100}$$

with parameters $\alpha_1, \alpha_2 > 0$. The covariance, correlation coefficient and marginal functions, of X and Y are given by:

$$Cov(X, Y) = \frac{\alpha_1{}^2 + \alpha_1\,\alpha_2 + \alpha_2{}^2}{(\alpha_1 + \alpha_2)^4}, \tag{4.101}$$

$$\rho(X, Y) = \frac{\alpha_1{}^2 + \alpha_1\,\alpha_2 + \alpha_2{}^2}{\sqrt{2\,(\alpha_1 + \alpha_2)^2 - \alpha_1{}^2}\,\sqrt{2\,\alpha_1{}^2 + 4\,\alpha_1\alpha_2 + \alpha_2{}^2}}, \tag{4.102}$$

$$f_X(x) = (\alpha_2\,(\alpha_1 + \alpha_2)x + \alpha_1)\, e^{-(\alpha_1 + \alpha_2)x}, \; x \geq 0 \tag{4.103}$$

$$f_Y(y) = (\alpha_1\,(\alpha_1 + \alpha_2)y + \alpha_2)\, e^{-(\alpha_1 + \alpha_2)y}, \; y \geq 0 \tag{4.104}$$

Note that the correlation coefficient is positive.

Proposition 4.36. *The metric tensor on F_3 is*

$$[g_{ij}] = \begin{bmatrix} \dfrac{\alpha_2+2\,\alpha_1}{\alpha_1\,(\alpha_1+\alpha_2)^2} & \dfrac{1}{(\alpha_1+\alpha_2)^2} \\[2ex] \dfrac{1}{(\alpha_1+\alpha_2)^2} & \dfrac{\alpha_1+2\,\alpha_2}{\alpha_2\,(\alpha_1+\alpha_2)^2} \end{bmatrix} . \qquad \Box \qquad (4.105)$$

Proposition 4.37. *The nonzero independent components of the α-connection of F_3 are*

$$\Gamma_{11}^{(\alpha)1} = -\frac{1+\alpha}{2\,\alpha_1} + \frac{-1+3\,\alpha}{2\,(\alpha_1+\alpha_2)},$$

$$\Gamma_{12}^{(\alpha)1} = \frac{-1+\alpha}{2\,(\alpha_1+\alpha_2)},$$

$$\Gamma_{22}^{(\alpha)1} = \frac{(1+\alpha)\,\alpha_1}{2\,\alpha_2\,(\alpha_1+\alpha_2)},$$

$$\Gamma_{11}^{(\alpha)2} = \frac{(1+\alpha)\,\alpha_2}{2\,\alpha_1\,(\alpha_1+\alpha_2)},$$

$$\Gamma_{22}^{(\alpha)2} = -\frac{1+\alpha}{2\,\alpha_2} + \frac{-1+3\,\alpha}{2\,(\alpha_1+\alpha_2)} . \qquad \Box \qquad (4.106)$$

Proposition 4.38. *The α-curvature tensor, α-Ricci curvature, and α-scalar curvature of F_3 are zero.* $\qquad \Box$

4.13.4 Submanifold F_4

This is $F_4 \subset F$, ACBED of Block and Basu with the density functions:

$$f(x,y;\lambda_1,\lambda_{12},\lambda_2) = \begin{cases} \dfrac{\lambda_1\,\lambda\,(\lambda_2+\lambda_{12})}{\lambda_1+\lambda_2}\,e^{-\lambda_1\,x-(\lambda_2+\lambda_{12})\,y} & \text{for } 0 < x < y \\[2ex] \dfrac{\lambda_2\,\lambda\,(\lambda_1+\lambda_{12})}{\lambda_1+\lambda_2}\,e^{-(\lambda_1+\lambda_{12})\,x-\lambda_2\,y} & \text{for } 0 < y < x \end{cases} \qquad (4.107)$$

where the parameters $\lambda_1, \lambda_{12}, \lambda_2$ are positive, and $\lambda = \lambda_1 + \lambda_2 + \lambda_{12}$.

This distribution originated by omitting the singular part of the Marshall and Olkin distribution (cf. [122], page [139]); Block and Basu called it the ACBED to emphasize that they are the absolutely continuous bivariate exponential density functions. Alternatively, it can be derived by Freund's method (4.63), with

$$\alpha_1 = \lambda_1 + \frac{\lambda_1\,\lambda_{12}}{(\lambda_1+\lambda_2)},$$

$$\beta_1 = \lambda_1 + \lambda_{12},$$

$$\alpha_2 = \lambda_2 + \frac{\lambda_2\,\lambda_{12}}{(\lambda_1+\lambda_2)},$$

$$\beta_2 = \lambda_2 + \lambda_{12} .$$

By substitution we obtain the covariance, correlation coefficient and marginal functions:

$$Cov(X,Y) = \frac{(\lambda_1 + \lambda_2)^2 (\lambda_1 + \lambda_{12})(\lambda_2 + \lambda_{12}) - \lambda^2 \lambda_1 \lambda_2}{\lambda^2 (\lambda_1 + \lambda_2)^2 (\lambda_1 + \lambda_{12})(\lambda_2 + \lambda_{12})}, \qquad (4.108)$$

$$\rho(X,Y) = \frac{(\lambda_1 + \lambda_2)^2 (\lambda_1 + \lambda_{12})(\lambda_2 + \lambda_{12}) - \lambda^2 \lambda_1 \lambda_2}{\sqrt{\prod_{i=1,\,j\neq i}^{2} \left((\lambda_1 + \lambda_2)^2 (\lambda_i + \lambda_{12})^2 + \lambda_j \lambda^2 (\lambda_j + 2\lambda_i) \right)}}, \qquad (4.109)$$

$$f_X(x) = \left(\frac{-\lambda_{12}}{\lambda_1 + \lambda_2} \right) \lambda e^{-\lambda x}$$

$$+ \left(\frac{\lambda}{\lambda_1 + \lambda_2} \right) (\lambda_1 + \lambda_{12}) \, e^{-(\lambda_1 + \lambda_{12}) x}, \; x \geq 0 \qquad (4.110)$$

$$f_Y(y) = \left(\frac{-\lambda_{12}}{\lambda_1 + \lambda_2} \right) \lambda e^{-\lambda y}$$

$$+ \left(\frac{\lambda}{\lambda_1 + \lambda_2} \right) (\lambda_2 + \lambda_{12}) \, e^{-(\lambda_2 + \lambda_{12}) y}, \; y \geq 0 \qquad (4.111)$$

The correlation coefficient is positive, and the marginal functions are a negative mixture of two exponentials.

Proposition 4.39. *The metric tensor in the coordinate system* $(\lambda_1, \lambda_{12}, \lambda_2)$ *is* $[g_{ij}] =$

$$\begin{bmatrix} \frac{\lambda_2 \left(\frac{1}{\lambda_1} + \frac{\lambda_1 + \lambda_2}{(\lambda_1 + \lambda_{12})^2} \right)}{(\lambda_1 + \lambda_2)^2} + \frac{1}{\lambda^2} & \frac{\lambda_2}{(\lambda_1 + \lambda_2)(\lambda_1 + \lambda_{12})^2} + \frac{1}{\lambda^2} & \frac{-1}{(\lambda_1 + \lambda_2)^2} + \frac{1}{\lambda^2} \\[2ex] \frac{\lambda_2}{(\lambda_1 + \lambda_2)(\lambda_1 + \lambda_{12})^2} + \frac{1}{\lambda^2} & \frac{\frac{\lambda_2}{(\lambda_1 + \lambda_{12})^2} + \frac{\lambda_1}{(\lambda_2 + \lambda_{12})^2}}{\lambda_1 + \lambda_2} + \frac{1}{\lambda^2} & \frac{\lambda_1}{(\lambda_1 + \lambda_2)(\lambda_2 + \lambda_{12})^2} + \frac{1}{\lambda^2} \\[2ex] \frac{-1}{(\lambda_1 + \lambda_2)^2} + \frac{1}{\lambda^2} & \frac{\lambda_1}{(\lambda_1 + \lambda_2)(\lambda_2 + \lambda_{12})^2} + \frac{1}{\lambda^2} & \frac{\lambda_1 \left(\frac{1}{\lambda_2} + \frac{\lambda_1 + \lambda_2}{(\lambda_2 + \lambda_{12})^2} \right)}{(\lambda_1 + \lambda_2)^2} + \frac{1}{\lambda^2} \end{bmatrix}. \qquad (4.112)$$

The α-connections and the α-curvatures were computed [13] but are not listed because they have lengthy expressions. When $\lambda_1 = \lambda_2$, this family of density functions becomes

$$f(x,y;\lambda_1,\lambda_{12}) = \begin{cases} \frac{(2\lambda_1 + \lambda_{12})(\lambda_1 + \lambda_{12})}{2} \, e^{-\lambda_1 x - (\lambda_1 + \lambda_{12}) y}, \; 0 < x < y \\[2ex] \frac{(2\lambda_1 + \lambda_{12})(\lambda_1 + \lambda_{12})}{2} \, e^{-\lambda_1 y - (\lambda_1 + \lambda_{12}) x} \; 0 < y < x \end{cases} \qquad (4.113)$$

That is an exponential family with natural parameters $(\theta_1, \theta_2) = (\lambda_1, \lambda_{12})$ and potential function

$$\varphi(\theta) = \log(2) - \log(\lambda_1 + \lambda_{12}) - \log(2\lambda_1 + \lambda_{12}).$$

From equations (4.110, 4.111), this family of density functions has identical marginal density functions.

The metric tensor $[g_{ij}] = [\frac{\partial^2 \varphi}{\partial \theta_i \partial \theta_j}]$ is

$$[g_{ij}] = \begin{bmatrix} \frac{1}{(\lambda_1+\lambda_{12})^2} + \frac{4}{(2\,\lambda_1+\lambda_{12})^2} & \frac{1}{(\lambda_1+\lambda_{12})^2} + \frac{2}{(2\,\lambda_1+\lambda_{12})^2} \\ \frac{1}{(\lambda_1+\lambda_{12})^2} + \frac{2}{(2\,\lambda_1+\lambda_{12})^2} & \frac{1}{(\lambda_1+\lambda_{12})^2} + \frac{1}{(2\,\lambda_1+\lambda_{12})^2} \end{bmatrix}$$

(4.114)

Then using

$$\Gamma_{ij,k}^{(\alpha)} = \frac{1-\alpha}{2} \frac{\partial^3 \varphi}{\partial \theta_i \partial \theta_j \partial \theta_k}$$

$$\Gamma_{11,1}^{(\alpha)} = (1-\alpha) \left(\frac{-1}{(\lambda_1+\lambda_{12})^3} - \frac{8}{(2\,\lambda_1+\lambda_{12})^3} \right),$$

$$\Gamma_{11,2}^{(\alpha)} = (1-\alpha) \left(\frac{-1}{(\lambda_1+\lambda_{12})^3} - \frac{4}{(2\,\lambda_1+\lambda_{12})^3} \right),$$

$$\Gamma_{12,2}^{(\alpha)} = (1-\alpha) \left(\frac{-1}{(\lambda_1+\lambda_{12})^3} - \frac{2}{(2\,\lambda_1+\lambda_{12})^3} \right),$$

$$\Gamma_{22,2}^{(\alpha)} = (1-\alpha) \left(\frac{-1}{(\lambda_1+\lambda_{12})^3} - \frac{1}{(2\,\lambda_1+\lambda_{12})^3} \right).$$

(4.115)

From

$$\Gamma_{ij,k}^{(\alpha)} = \sum_{h=1}^{3} g_{kh} \, \Gamma_{ij}^{h(\alpha)}, (k = 1,2)$$

the components of $\nabla^{(\alpha)}$ are

$$\Gamma^{(\alpha)1} = [\Gamma_{ij}^{(\alpha)1}] = \begin{bmatrix} \frac{1-\alpha}{\lambda_1+\lambda_{12}} + \frac{4\,(\alpha-1)}{2\,\lambda_1+\lambda_{12}} & \frac{(\alpha-1)\,\lambda_{12}}{(\lambda_1+\lambda_{12})\,(2\,\lambda_1+\lambda_{12})} \\ \frac{(\alpha-1)\,\lambda_{12}}{(\lambda_1+\lambda_{12})\,(2\,\lambda_1+\lambda_{12})} & \frac{-(\alpha-1)\,\lambda_1}{(\lambda_1+\lambda_{12})\,(2\,\lambda_1+\lambda_{12})} \end{bmatrix},$$

$$\Gamma^{(\alpha)2} = [\Gamma_{ij}^{(\alpha)2}] = \begin{bmatrix} \frac{-2\,(\alpha-1)\,\lambda_{12}}{(\lambda_1+\lambda_{12})\,(2\,\lambda_1+\lambda_{12})} & \frac{2\,(\alpha-1)\,\lambda_1}{(\lambda_1+\lambda_{12})\,(2\,\lambda_1+\lambda_{12})} \\ \frac{2\,(\alpha-1)\,\lambda_1}{(\lambda_1+\lambda_{12})\,(2\,\lambda_1+\lambda_{12})} & \frac{2\,(\alpha-1)}{\lambda_1+\lambda_{12}} + \frac{1-\alpha}{2\,\lambda_1+\lambda_{12}} \end{bmatrix}.$$

(4.116)

The α-curvature tensor, α-Ricci curvature, and α-scalar curvature are zero. Since $(\lambda_1, \lambda_{12})$ is a 1-affine coordinate system, then there is a (-1)-affine coordinate system

$$(\eta_1, \eta_2) = (-\frac{1}{\lambda_1 + \lambda_{12}} - \frac{1}{\lambda_1 + 2\,\lambda_{12}}, -\frac{1}{\lambda_1 + \lambda_{12}} - \frac{1}{2\,\lambda_1 + \lambda_{12}})$$

with potential function

$$\lambda = -2 - \log(2) + \log(2\,\lambda_1 + \lambda_{12}) + \log(\lambda_1 + \lambda_{12}).$$

4.14 Freund Affine Immersion

Proposition 4.40. *Let F be the Freund 4-manifold with the Fisher metric g and the exponential connection $\nabla^{(1)}$. Denote by (θ_i) the natural coordinate system (4.73). Then, by §3.4, F can be realized in $\mathbb{R}^5$ by the graph of a potential function, the affine immersion $\{f, \xi\}$:*

$$f : F \to \mathbb{R}^5 : [\theta_i] \mapsto \begin{bmatrix} \theta_i \\ \varphi(\theta) \end{bmatrix}, \quad \xi = \begin{bmatrix} 0 \\ 0 \\ 0 \\ 0 \\ 1 \end{bmatrix}, \tag{4.117}$$

where $\varphi(\theta)$ is the potential function

$$\varphi(\theta) = -\log\left(\frac{\theta_1 \theta_2 \theta_4}{e^{\theta_3} \theta_2 + \theta_4}\right) = -\log(\alpha_2 \beta_1).$$

$\square$

4.15 Freund Bivariate Log-Exponential Manifold

The Freund bivariate mixture log-exponential density functions arise from the Freund density functions (4.63) for the non-negative random variables $x = \log\frac{1}{n}$ and $y = \log\frac{1}{m}$, or equivalently, $n = e^{-x}$ and $m = e^{-y}$. So the Freund log-exponential density functions are given by:

$$g(n,m) = \begin{cases} \alpha_1 \beta_2 \, m^{(\beta_2-1)} \, n^{(\alpha_1+\alpha_2-\beta_2-1)} & \text{for } 0 < m < n < 1, \\ \alpha_2 \beta_1 \, n^{(\beta_1-1)} \, m^{(\alpha_1+\alpha_2-\beta_1-1)} & \text{for } 0 < n < m < 1 \end{cases} \tag{4.118}$$

where $\alpha_i, \beta_i > 0$ $(i = 1, 2)$. The covariance, and marginal density functions, of n and m are given by:

Corollary 4.41.

$$Cov(n,m) = \frac{\alpha_2 \left(-(\alpha_1 (2 + \alpha_1 + \alpha_2)) + \beta_1\right) + (\alpha_1 + (\alpha_1 + \alpha_2)\beta_1)\beta_2}{(1 + \alpha_1 + \alpha_2)^2 (2 + \alpha_1 + \alpha_2)(1 + \beta_1)(1 + \beta_2)},$$

$$g_N(n) = \left(\frac{\alpha_2}{\alpha_1 + \alpha_2 - \beta_1}\right) \beta_1 n^{\beta_1 - 1}$$

$$+ \left(\frac{\alpha_1 - \beta_1}{\alpha_1 + \alpha_2 - \beta_1}\right)(\alpha_1 + \alpha_2)n^{(\alpha_1+\alpha_2)-1},$$

$$g_M(m) = \left(\frac{\alpha_1}{\alpha_1 + \alpha_2 - \beta_2}\right) \beta_2 m^{\beta_2 - 1}$$

$$+ \left(\frac{\alpha_2 - \beta_2}{\alpha_1 + \alpha_2 - \beta_2}\right)(\alpha_1 + \alpha_2)m^{(\alpha_1+\alpha_2)-1}. \tag{4.119}$$

The variables n and m are independent if and only if $\alpha_i = \beta_i$ $(i = 1, 2)$, and the marginal functions are mixture log-exponential density functions. $\square$

Directly from the definition of the Fisher metric we deduce:

Proposition 4.42. *The family of Freund bivariate mixture log-exponential density functions for random variables n, m determines a Riemannian 4-manifold which is an isometric isomorph of the Freund 4-manifold §3.1.* $\square$

4.16 Bivariate Gaussian 5-Manifold $\mathcal{N}$

The results in this section were computed in [13] and first reported in [15]. The bivariate Gaussian distribution was considered occasionally as early as the middle of the nineteenth century, and has played a predominant role in the historical development of statistical theory, so it has made its appearance in various areas of application. The differential geometrical consideration of the parameter space of the bivariate Gaussian family of density functions was considered by Sato et al. [183], who provided the density functions as a Riemannian 5-manifold. They calculated the $\alpha = 0$ geometry, i.e., the 0-connections, 0-Ricci curvature, the 0-scalar curvature etc, and they showed that the bivariate Gaussian manifold has a negative constant 0-scalar curvature and that, if the correlation coefficient vanishes, the space becomes an Einstein space.

We extend these studies by calculating the α-connections, the α-Ricci tensor, the α-scalar curvatures etc and we show that this manifold has a constant α-scalar curvature, so geometrically it constitutes part of a pseudosphere when $\alpha^2 < 1$. We derive mixture coordinates and mutually dual foliations, then we consider particular submanifolds, including the case of statistically independent random variables, and discuss their geometrical structure. We consider bivariate log-Gaussian (log-normal) density functions, with log-Gaussian marginal functions. We show that this family of density functions determines a Riemannian 5-manifold, an isometric isomorph of the bivariate Gaussian manifold, §3.1.

The probability density function with real random variables x_1, x_2 and five parameters has the form:

$$f(x, y) = \frac{1}{2\pi\sqrt{\sigma_1\sigma_2 - \sigma_{12}^2}}\, e^{-AB}, \tag{4.120}$$

$$\text{with } A = \frac{1}{2(\sigma_1\sigma_2 - \sigma_{12}^2)}$$

$$B = \left(\sigma_2(x - \mu_1)^2 - 2\sigma_{12}(x - \mu_1)(y - \mu_2) + \sigma_1(y - \mu_2)^2\right)$$

defined on $-\infty < x, y < \infty$ with parameters $(\mu_1, \mu_2, \sigma_1, \sigma_{12}, \sigma_2)$; where $-\infty < \mu_1, \mu_2 < \infty$, $0 < \sigma_1, \sigma_2 < \infty$ and σ_{12} is the covariance of X and Y.

The bivariate Gaussian distribution with central means; $\mu_1 = \mu_2 = 0$ and unit variances; $\sigma_1 = \sigma_2 = 1$, that is

$$f(x,y) = \frac{1}{2\pi\sqrt{1-\sigma_{12}{}^2}} \, e^{-\frac{1}{2(1-\sigma_{12}{}^2)}\left(x^2 - 2\,\sigma_{12}\,x\,y + y^2\right)}, \qquad (4.121)$$

is called the **standardized distribution**.

The marginal functions, §1.3, of X and Y are univariate Gaussian density functions $N_X(\mu_1, \sigma_2)$ and $N_Y(\mu_2, \sigma_2)$, §1.2.3:

$$f_X(x) = N_X(\mu_1, \sigma_1) = \frac{1}{\sqrt{2\pi}\,\sigma_1} \, e^{-\frac{(x-\mu_1)^2}{2\sigma_1}}, \qquad (4.122)$$

$$f_Y(y) = N_Y(\mu_2, \sigma_2) = \frac{1}{\sqrt{2\pi}\,\sigma_2} \, e^{-\frac{(y-\mu_2)^2}{2\sigma_2}}. \qquad (4.123)$$

The correlation coefficient is:

$$\rho(X,Y) = \frac{\sigma_{12}}{\sqrt{\sigma_1\,\sigma_2}}$$

Since $\sigma_{12}{}^2 \neq \sigma_1\,\sigma_2$ then $-1 < \rho(X,Y) < 1$; so we do not have the case when Y is a linearly increasing (or decreasing) function of X.

4.17 Bivariate Gaussian Fisher Information Metric

The information geometry of (4.120) has been studied by Sato et al. [183]; the metric tensor, §3.1, takes the following form:

$$G = [g_{ij}] = \begin{bmatrix} \frac{\sigma_2}{\triangle} & -\frac{\sigma_{12}}{\triangle} & 0 & 0 & 0 \\ -\frac{\sigma_{12}}{\triangle} & \frac{\sigma_1}{\triangle} & 0 & 0 & 0 \\ 0 & 0 & \frac{(\sigma_2)^2}{2\triangle^2} & -\frac{\sigma_{12}\,\sigma_2}{\triangle^2} & \frac{(\sigma_{12})^2}{2\triangle^2} \\ 0 & 0 & -\frac{\sigma_{12}\,\sigma_2}{\triangle^2} & \frac{\sigma_1\,\sigma_2+(\sigma_{12})^2}{\triangle^2} & -\frac{\sigma_1\,\sigma_{12}}{\triangle^2} \\ 0 & 0 & \frac{(\sigma_{12})^2}{2\triangle^2} & -\frac{\sigma_1\,\sigma_{12}}{\triangle^2} & \frac{(\sigma_1)^2}{2\triangle^2} \end{bmatrix}, \qquad (4.124)$$

where $\triangle$ is the determinant

$$\triangle = \sigma_1\,\sigma_2 - (\sigma_{12})^2$$

The inverse $[g^{ij}]$ of the metric tensor $[g_{ij}]$ defined by the relation

$$g_{ij}g^{ik} = \delta_j^k$$

is given by

$$G^{-1} = [g^{ij}] = \begin{bmatrix} \sigma_1 & \sigma_{12} & 0 & 0 & 0 \\ \sigma_{12} & \sigma_2 & 0 & 0 & 0 \\ 0 & 0 & 2\,(\sigma_1)^2 & 2\,\sigma_1\,\sigma_{12} & 2\,(\sigma_{12})^2 \\ 0 & 0 & 2\,\sigma_1\,\sigma_{12} & \sigma_1\,\sigma_2 + (\sigma_{12})^2 & 2\,\sigma_{12}\,\sigma_2 \\ 0 & 0 & 2\,(\sigma_{12})^2 & 2\,\sigma_{12}\,\sigma_2 & 2\,(\sigma_2)^2 \end{bmatrix}. \qquad (4.125)$$

4.18 Bivariate Gaussian Natural Coordinates

Proposition 4.43. *The set of all bivariate Gaussian density functions forms an exponential family, §3.2, with natural coordinate system, §3.3,* $(\theta_1, \theta_2, \theta_3, \theta_4, \theta_5) =$

$$\left(\frac{\mu_1 \sigma_2 - \mu_2 \sigma_{12}}{\triangle}, \frac{\mu_2 \sigma_1 - \mu_1 \sigma_{12}}{\triangle}, \frac{-\sigma_2}{2\triangle}, \frac{\sigma_{12}}{\triangle}, \frac{-\sigma_1}{2\triangle} \right) \qquad (4.126)$$

with corresponding potential function

$$\varphi(\theta) = \log(2\pi\sqrt{\triangle}) + \frac{\mu_2^2 \sigma_1 + \mu_1^2 \sigma_2 - 2\mu_1\mu_2\sigma_{12}}{2\triangle}$$

$$= \log(2\pi\sqrt{\triangle}) - \triangle\left(\theta_2^2\theta_3 - \theta_1\theta_2\theta_4 + \theta_1^2\theta_5\right). \qquad (4.127)$$

where

$$\triangle = \sigma_1\sigma_2 - \sigma_{12}^2 = \frac{1}{4\theta_3\theta_5 - \theta_4^2}.$$

Proof.

$$\log f(x,y) = \log\left(\frac{1}{2\pi\sqrt{\triangle}} e^{-\frac{1}{2\triangle}\left(\sigma_2(x-\mu_1)^2 - 2\sigma_{12}(x-\mu_1)(y-\mu_2) + \sigma_1(y-\mu_2)^2\right)}\right)$$

$$= \frac{\mu_1\sigma_2 - \mu_2\sigma_{12}}{\triangle} x$$

$$+ \frac{\mu_2\sigma_1 - \mu_1\sigma_{12}}{\triangle} y + \frac{-\sigma_2}{2\triangle} x^2 + \frac{\sigma_{12}}{\triangle} xy + \frac{-\sigma_1}{2\triangle} y^2 \qquad (4.128)$$

$$- \left(\log(2\pi\sqrt{\triangle}) + \frac{\mu_2^2\sigma_1 + \mu_1^2\sigma_2 - 2\mu_1\mu_2\sigma_{12}}{2\triangle} \right). \qquad (4.129)$$

Hence we have an exponential family. The line (4.128) implies that

$$\left(\frac{\mu_1\sigma_2 - \mu_2\sigma_{12}}{\triangle}, \frac{\mu_2\sigma_1 - \mu_1\sigma_{12}}{\triangle}, \frac{-\sigma_2}{2\triangle}, \frac{\sigma_{12}}{\triangle}, \frac{-\sigma_1}{2\triangle} \right)$$

is a natural coordinate system,

$$(F_1(x), F_2(x), F_3(x), F_4(x), F_5(x)) = (x, y, x^2, xy, y^2)$$

and (4.129) implies that

$$\varphi(\theta) = \log(2\pi\sqrt{\triangle}) + \frac{\mu_2^2\sigma_1 + \mu_1^2\sigma_2 - 2\mu_1\mu_2\sigma_{12}}{2\triangle}$$

is its potential function. In terms of natural coordinates by solving

$$\left\{ \theta_1 = \frac{\mu_1\sigma_2 - \mu_2\sigma_{12}}{\sigma_1\sigma_2 - \sigma_{12}^2}, \theta_2 = \frac{\mu_2\sigma_1 - \mu_1\sigma_{12}}{\sigma_1\sigma_2 - \sigma_{12}^2}, \theta_3 = \frac{-\sigma_2}{2(\sigma_1\sigma_2 - \sigma_{12}^2)} \right\},$$

$$\{\theta_4 = \frac{\sigma_{12}}{\sigma_1 \sigma_2 - \sigma_{12}{}^2}, \theta_5 = \frac{-\sigma_1}{2\left(\sigma_1 \sigma_2 - \sigma_{12}{}^2\right)}\}$$

we obtain:

$$\{\mu_1 = \frac{2\theta_1\theta_5 - \theta_2\theta_4}{\theta_4{}^2 - 4\theta_3\theta_5}, \mu_2 = \frac{2\theta_2\theta_3 - \theta_1\theta_4}{\theta_4{}^2 - 4\theta_3\theta_5}, \sigma_1 = \frac{2\theta_5}{\theta_4{}^2 - 4\theta_3\theta_5}\},$$

$$\{\sigma_{12} = \frac{\theta_4}{4\theta_3\theta_5 - \theta_4{}^2}, \sigma_2 = \frac{2\theta_3}{\theta_4{}^2 - 4\theta_3\theta_5}\}.$$

Then

$$\varphi = \log(2\pi\sqrt{\triangle}) - \triangle\left(\theta_2{}^2\theta_3 - \theta_1\theta_2\theta_4 + \theta_1{}^2\theta_5\right),$$

where $\triangle = \sigma_1\sigma_2 - \sigma_{12}{}^2 = \dfrac{1}{4\theta_3\theta_5 - \theta_4{}^2}.$

The normalization function $C(X, Y)$, required by the definition of the exponential family property §3.2, is zero here. □

4.19 Bivariate Gaussian a-Geometry

By direct calculation, §3.3 we provide the α-connections, and various α-curvature objects of $\mathcal{N}$. The analytic expressions for the α-connections and the α-curvature objects are very large if we use the natural coordinate system, so we report these components in terms of the coordinate system $(\mu_1, \mu_2, \sigma_1, \sigma_{12}, \sigma_2)$.

4.19.1 a-Connection

For each $\alpha \in \mathbb{R}$, the α (or $\nabla^{(\alpha)}$)-connection is the torsion-free affine connection with components:

$$\Gamma_{ij,k}^{(\alpha)}(\xi) = \int_{-\infty}^{\infty}\int_{-\infty}^{\infty}\left(\frac{\partial^2 \log f}{\partial \xi^i \partial \xi^j}\frac{\partial \log f}{\partial \xi^k} + \frac{1-\alpha}{2}\frac{\partial \log f}{\partial \xi^i}\frac{\partial \log f}{\partial \xi^j}\frac{\partial \log f}{\partial \xi^k}\right) f\, dx\, dy$$

Since it is difficult to derive the α-connection components with respect to the local coordinates $(\mu_1, \mu_2, \sigma_1, \sigma_{12}, \sigma_2)$ by using these integrations, we derive them with respect to the natural coordinates (θ_i), by:

$$\Gamma_{rs,t}^{(\alpha)}(\theta) = \frac{1-\alpha}{2}\,\partial_r\,\partial_s\,\partial_t\varphi(\theta),$$

where $\varphi(\theta)$ is the potential function, and $\partial_i = \frac{\partial}{\partial\theta_i}$.

Then we change the coordinates to (ξ^i):

$$\Gamma_{ij,k}^{(\alpha)}(\xi) = \left(\Gamma_{rs,t}^{(\alpha)}(\theta)\frac{\partial\theta_r}{\partial\xi^i}\frac{\partial\theta_s}{\partial\xi^j} + \frac{\partial^2\theta_t}{\partial\xi^i\partial\xi^j}\right)\frac{\partial\xi^k}{\partial\theta_t}.$$

Proposition 4.44. *The functions* $\Gamma_{ij,k}^{(\alpha)}$ *from* §3.3 *are given by:*

$$[\Gamma_{ij,1}^{(\alpha)}] = \begin{bmatrix} 0 & 0 & \frac{-(1+\alpha)\,\sigma_2{}^2}{2\,\Delta^2} & \frac{(1+\alpha)\,\sigma_2\,\sigma_{12}}{\Delta^2} & \frac{-(1+\alpha)\,\sigma_{12}{}^2}{2\,\Delta^2} \\ 0 & 0 & \frac{(1+\alpha)\,\sigma_2\,\sigma_{12}}{2\,\Delta^2} & \frac{-(1+\alpha)\,(\sigma_1\,\sigma_2+\sigma_{12}{}^2)}{2\,\Delta^2} & \frac{(1+\alpha)\,\sigma_1\,\sigma_{12}}{2\,\Delta^2} \\ \frac{-(1+\alpha)\,\sigma_2{}^2}{2\,\Delta^2} & \frac{(1+\alpha)\,\sigma_2\,\sigma_{12}}{2\,\Delta^2} & 0 & 0 & 0 \\ \frac{(1+\alpha)\,\sigma_2\,\sigma_{12}}{\Delta^2} & \frac{-(1+\alpha)\,(\sigma_1\,\sigma_2+\sigma_{12}{}^2)}{2\,\Delta^2} & 0 & 0 & 0 \\ \frac{-(1+\alpha)\,\sigma_{12}{}^2}{2\,\Delta^2} & \frac{(1+\alpha)\,\sigma_1\,\sigma_{12}}{2\,\Delta^2} & 0 & 0 & 0 \end{bmatrix}$$

$$[\Gamma_{ij,2}^{(\alpha)}] = \begin{bmatrix} 0 & 0 & \frac{(1+\alpha)\,\sigma_2\,\sigma_{12}}{2\,\Delta^2} & \frac{-(1+\alpha)\,(\sigma_1\,\sigma_2+\sigma_{12}{}^2)}{2\,\Delta^2} & \frac{(1+\alpha)\,\sigma_1\,\sigma_{12}}{2\,\Delta^2} \\ 0 & 0 & \frac{-(1+\alpha)\,\sigma_{12}{}^2}{2\,\Delta^2} & \frac{(1+\alpha)\,\sigma_1\,\sigma_{12}}{\Delta^2} & \frac{-(1+\alpha)\,\sigma_1{}^2}{2\,\Delta^2} \\ \frac{(1+\alpha)\,\sigma_2\,\sigma_{12}}{2\,\Delta^2} & \frac{-(1+\alpha)\,\sigma_{12}{}^2}{2\,\Delta^2} & 0 & 0 & 0 \\ \frac{-(1+\alpha)\,(\sigma_1\,\sigma_2+\sigma_{12}{}^2)}{2\,\Delta^2} & \frac{(1+\alpha)\,\sigma_1\,\sigma_{12}}{\Delta^2} & 0 & 0 & 0 \\ \frac{(1+\alpha)\,\sigma_1\,\sigma_{12}}{2\,\Delta^2} & \frac{-(1+\alpha)\,\sigma_1{}^2}{2\,\Delta^2} & 0 & 0 & 0 \end{bmatrix}$$

$$[\Gamma_{ij,3}^{(\alpha)}] =$$
$$\begin{bmatrix} \frac{-(\alpha-1)\,\sigma_2{}^2}{2\,\Delta^2} & \frac{(\alpha-1)\,\sigma_2\,\sigma_{12}}{2\,\Delta^2} & 0 & 0 & 0 \\ \frac{(\alpha-1)\,\sigma_2\,\sigma_{12}}{2\,\Delta^2} & \frac{-(\alpha-1)\,\sigma_{12}{}^2}{2\,\Delta^2} & 0 & 0 & 0 \\ 0 & 0 & \frac{(1+\alpha)\,\sigma_2{}^3}{-2\,\Delta^3} & \frac{(1+\alpha)\,\sigma_2{}^2\,\sigma_{12}}{\Delta^3} & \frac{(1+\alpha)\,\sigma_2\,\sigma_{12}{}^2}{-2\,\Delta^3} \\ 0 & 0 & \frac{(1+\alpha)\,\sigma_2{}^2\,\sigma_{12}}{\Delta^3} & \frac{-(1+\alpha)\,\sigma_2\,(\sigma_1\,\sigma_2+3\,\sigma_{12}{}^2)}{2\,\Delta^3} & \frac{(1+\alpha)\,\sigma_{12}\,(\sigma_1\,\sigma_2+\sigma_{12}{}^2)}{2\,\Delta^3} \\ 0 & 0 & \frac{(1+\alpha)\,\sigma_2\,\sigma_{12}{}^2}{-2\,\Delta^3} & \frac{(1+\alpha)\,\sigma_{12}\,(\sigma_1\,\sigma_2+\sigma_{12}{}^2)}{2\,\Delta^3} & \frac{(1+\alpha)\,\sigma_1\,\sigma_{12}{}^2}{-2\,\Delta^3} \end{bmatrix}$$

$$\frac{[\Gamma_{ij,4}^{(\alpha)}]}{(\alpha+1)} =$$
$$\begin{bmatrix} \frac{(\alpha-1)\sigma_2\sigma_{12}}{(\alpha+1)\Delta^2} & \frac{(\alpha-1)(\sigma_1\sigma_2+\sigma_{12}{}^2)}{-2(\alpha+1)\Delta^2} & 0 & 0 & 0 \\ \frac{(\alpha-1)(\sigma_1\sigma_2+\sigma_{12}{}^2)}{-2(\alpha+1)\Delta^2} & \frac{(\alpha-1)\sigma_1\sigma_{12}}{(\alpha+1)\Delta^2} & 0 & 0 & 0 \\ 0 & 0 & \frac{\sigma_2{}^2\sigma_{12}}{\Delta^3} & \frac{\sigma_2(\sigma_1\sigma_2+3\sigma_{12}{}^2)}{-2\,\Delta^3} & \frac{\sigma_{12}(\sigma_1\sigma_2+\sigma_{12}{}^2)}{2\,\Delta^3} \\ 0 & 0 & \frac{\sigma_2(\sigma_1\sigma_2+3\sigma_{12}{}^2)}{-2\,\Delta^3} & \frac{\sigma_{12}(3\sigma_1\sigma_2+\sigma_{12}{}^2)}{\Delta^3} & \frac{\sigma_1(\sigma_1\sigma_2+3\sigma_{12}{}^2)}{-2\,\Delta^3} \\ 0 & 0 & \frac{\sigma_{12}(\sigma_1\sigma_2+\sigma_{12}{}^2)}{2\,\Delta^3} & \frac{\sigma_1(\sigma_1\sigma_2+3\sigma_{12}{}^2)}{-2\,\Delta^3} & \frac{\sigma_1{}^2\sigma_{12}}{\Delta^3} \end{bmatrix}$$

$$[\Gamma_{ij,5}^{(\alpha)}] =$$
$$\begin{bmatrix} \frac{-(\alpha-1)\sigma_{12}{}^2}{2\Delta^2} & \frac{(\alpha-1)\sigma_1\sigma_{12}}{2\Delta^2} & 0 & 0 & 0 \\ \frac{(\alpha-1)\sigma_1\sigma_{12}}{2\Delta^2} & \frac{-(\alpha-1)\sigma_1{}^2}{2\Delta^2} & 0 & 0 & 0 \\ 0 & 0 & \frac{(1+\alpha)\sigma_2\sigma_{12}{}^2}{-2\Delta^3} & \frac{(1+\alpha)\sigma_{12}(\sigma_1\sigma_2+\sigma_{12}{}^2)}{2\Delta^3} & \frac{(1+\alpha)\sigma_1\sigma_{12}{}^2}{-2\Delta^3} \\ 0 & 0 & \frac{(1+\alpha)\sigma_{12}(\sigma_1\sigma_2+\sigma_{12}{}^2)}{2\Delta^3} & \frac{-(1+\alpha)\sigma_1(\sigma_1\sigma_2+3\sigma_{12}{}^2)}{2\Delta^3} & \frac{(1+\alpha)\sigma_1{}^2\sigma_{12}}{\Delta^3} \\ 0 & 0 & \frac{(1+\alpha)\sigma_1\sigma_{12}{}^2}{2\Delta^3} & \frac{(1+\alpha)\sigma_1{}^2\sigma_{12}}{\Delta^3} & \frac{(1+\alpha)\sigma_1{}^3}{-2\Delta^3} \end{bmatrix}.$$

$$(4.130)$$

$\square$

For each α, we have a symmetric linear connection $\nabla^{(\alpha)}$ defined by:

$$\langle \nabla^{(\alpha)}_{\partial_i}\partial_j, \partial_k \rangle = \Gamma^{(\alpha)}_{ij,k},$$

and by solving the equations

$$\Gamma^{(\alpha)}_{ij,k} = \sum_{h=1}^{3} g_{kh}\,\Gamma^{h(\alpha)}_{ij}, \, (k=1,2,3,4,5).$$

we obtain the components of $\nabla^{(\alpha)}$, §3.3.

Proposition 4.45. *The components* $\Gamma^{(\alpha)i}_{jk}$ *of the* $\nabla^{(\alpha)}$*-connections are given by:*

$$\Gamma^{(\alpha)1} = [\Gamma^{(\alpha)1}_{ij}] = \begin{bmatrix} 0 & 0 & \frac{(1+\alpha)\sigma_2}{-2\triangle} & \frac{(1+\alpha)\sigma_{12}}{2\triangle} & 0 \\ 0 & 0 & \frac{(1+\alpha)\sigma_{12}}{2\triangle} & \frac{(1+\alpha)\sigma_1}{-2\triangle} & 0 \\ \frac{(1+\alpha)\sigma_2}{-2\triangle} & \frac{(1+\alpha)\sigma_{12}}{2\triangle} & 0 & 0 & 0 \\ \frac{(1+\alpha)\sigma_{12}}{2\triangle} & \frac{(1+\alpha)\sigma_1}{-2\triangle} & 0 & 0 & 0 \\ 0 & 0 & 0 & 0 & 0 \end{bmatrix}$$

$$\Gamma^{(\alpha)2} = [\Gamma^{(\alpha)2}_{ij}] = \begin{bmatrix} 0 & 0 & 0 & \frac{(1+\alpha)\sigma_2}{-2\triangle} & \frac{(1+\alpha)\sigma_{12}}{2\triangle} \\ 0 & 0 & 0 & \frac{(1+\alpha)\sigma_{12}}{2\triangle} & \frac{(1+\alpha)\sigma_1}{-2\triangle} \\ 0 & 0 & 0 & 0 & 0 \\ \frac{(1+\alpha)\sigma_2}{-2\triangle} & \frac{(1+\alpha)\sigma_{12}}{2\triangle} & 0 & 0 & 0 \\ \frac{(1+\alpha)\sigma_{12}}{2\triangle} & \frac{(1+\alpha)\sigma_1}{-2\triangle} & 0 & 0 & 0 \end{bmatrix}$$

$$\Gamma^{(\alpha)3} = [\Gamma^{(\alpha)3}_{ij}] = \begin{bmatrix} 1-\alpha & 0 & 0 & 0 & 0 \\ 0 & 0 & 0 & 0 & 0 \\ 0 & 0 & \frac{(1+\alpha)\sigma_2}{-\triangle} & \frac{(1+\alpha)\sigma_{12}}{\triangle} & 0 \\ 0 & 0 & \frac{(1+\alpha)\sigma_{12}}{\triangle} & \frac{(1+\alpha)\sigma_1}{-\triangle} & 0 \\ 0 & 0 & 0 & 0 & 0 \end{bmatrix}$$

$$\Gamma^{(\alpha)4} = [\Gamma^{(\alpha)4}_{ij}] = \begin{bmatrix} 0 & \frac{1-\alpha}{2} & 0 & 0 & 0 \\ \frac{1-\alpha}{2} & 0 & 0 & 0 & 0 \\ 0 & 0 & 0 & \frac{(1+\alpha)\sigma_2}{-2\triangle} & \frac{(1+\alpha)\sigma_{12}}{2\triangle} \\ 0 & 0 & \frac{(1+\alpha)\sigma_2}{-2\triangle} & \frac{(1+\alpha)\sigma_{12}}{\triangle} & \frac{(1+\alpha)\sigma_1}{-2\triangle} \\ 0 & 0 & \frac{(1+\alpha)\sigma_{12}}{2\triangle} & \frac{(1+\alpha)\sigma_1}{-2\triangle} & 0 \end{bmatrix}$$

$$\Gamma^{(\alpha)5} = [\Gamma^{(\alpha)5}_{ij}] = \begin{bmatrix} 0 & 0 & 0 & 0 & 0 \\ 0 & 1-\alpha & 0 & 0 & 0 \\ 0 & 0 & 0 & 0 & 0 \\ 0 & 0 & 0 & \frac{(1+\alpha)\sigma_2}{-\triangle} & \frac{(1+\alpha)\sigma_{12}}{\triangle} \\ 0 & 0 & 0 & \frac{(1+\alpha)\sigma_{12}}{\triangle} & \frac{(1+\alpha)\sigma_1}{-\triangle} \end{bmatrix} . \quad \square \quad (4.131)$$

4.19.2 α-Curvatures

Proposition 4.46. *The components* $R_{ijkl}^{(\alpha)}$ *of the α-curvature tensor are given by:*

$$[R_{12kl}^{(\alpha)}] = (\alpha^2 - 1) \begin{bmatrix} 0 & \frac{1}{4\Delta} & 0 & 0 & 0 \\[4pt] \frac{-1}{4\Delta} & 0 & 0 & 0 & 0 \\[4pt] 0 & 0 & 0 & \frac{-\sigma_2}{4\Delta^2} & \frac{\sigma_{12}}{4\Delta^2} \\[4pt] 0 & 0 & \frac{\sigma_2}{4\Delta^2} & 0 & \frac{-\sigma_1}{4\Delta^2} \\[4pt] 0 & 0 & \frac{-\sigma_{12}}{4\Delta^2} & \frac{\sigma_1}{4\Delta^2} & 0 \end{bmatrix}$$

$$[R_{13kl}^{(\alpha)}] = (\alpha^2 - 1) \begin{bmatrix} 0 & 0 & \frac{\sigma_2^3}{-4\Delta^3} & \frac{\sigma_2^2\sigma_{12}}{2\Delta^3} & \frac{\sigma_2\sigma_{12}^2}{-4\Delta^3} \\[6pt] 0 & 0 & \frac{\sigma_2^2\sigma_{12}}{4\Delta^3} & \frac{-\sigma_2\left(\sigma_1\sigma_2+\sigma_{12}^2\right)}{4\Delta^3} & \frac{\sigma_1\sigma_2\sigma_{12}}{4\Delta^3} \\[6pt] \frac{\sigma_2^3}{4\Delta^3} & \frac{\sigma_2^2\sigma_{12}}{-4\Delta^3} & 0 & 0 & 0 \\[6pt] \frac{\sigma_2^2\sigma_{12}}{-2\Delta^3} & \frac{\sigma_2\left(\sigma_1\sigma_2+\sigma_{12}^2\right)}{4\Delta^3} & 0 & 0 & 0 \\[6pt] \frac{\sigma_2\sigma_{12}^2}{4\Delta^3} & \frac{\sigma_1\sigma_2\sigma_{12}}{-4\Delta^3} & 0 & 0 & 0 \end{bmatrix}$$

$$\frac{[R_{14kl}^{(\alpha)}]}{(\alpha^2 - 1)} = \begin{bmatrix} 0 & 0 & \frac{\sigma_2^2\sigma_{12}}{2\Delta^3} & \frac{\sigma_2\left(\sigma_1\sigma_2+3\sigma_{12}^2\right)}{-4\Delta^3} & \frac{\sigma_{12}\left(\sigma_1\sigma_2+\sigma_{12}^2\right)}{4\Delta^3} \\[6pt] 0 & 0 & \frac{\sigma_2\sigma_{12}^2}{-2\Delta^3} & \frac{\sigma_{12}\left(3\sigma_1\sigma_2+\sigma_{12}^2\right)}{4\Delta^3} & \frac{\sigma_1\left(\sigma_1\sigma_2+\sigma_{12}^2\right)}{-4\Delta^3} \\[6pt] \frac{\sigma_2^2\sigma_{12}}{-2\Delta^3} & \frac{\sigma_2\sigma_{12}^2}{2\Delta^3} & 0 & 0 & 0 \\[6pt] \frac{\sigma_2\left(\sigma_1\sigma_2+3\sigma_{12}^2\right)}{4\Delta^3} & \frac{\sigma_{12}\left(3\sigma_1\sigma_2+\sigma_{12}^2\right)}{-4\Delta^3} & 0 & 0 & 0 \\[6pt] \frac{\sigma_{12}\left(\sigma_1\sigma_2+\sigma_{12}^2\right)}{-4\Delta^3} & \frac{\sigma_1\left(\sigma_1\sigma_2+\sigma_{12}^2\right)}{4\Delta^3} & 0 & 0 & 0 \end{bmatrix}$$

$$[R_{15kl}^{(\alpha)}] = (\alpha^2 - 1) \begin{bmatrix} 0 & 0 & \frac{\sigma_2\sigma_{12}^2}{-4\Delta^3} & \frac{\sigma_{12}\left(\sigma_1\sigma_2+\sigma_{12}^2\right)}{4\Delta^3} & \frac{\sigma_1\sigma_{12}^2}{-4\Delta^3} \\[6pt] 0 & 0 & \frac{\sigma_{12}^3}{4\Delta^3} & \frac{\sigma_1\sigma_{12}^2}{-2\Delta^3} & \frac{\sigma_1^2\sigma_{12}}{4\Delta^3} \\[6pt] \frac{\sigma_2\sigma_{12}^2}{4\Delta^3} & \frac{\sigma_{12}^3}{-4\Delta^3} & 0 & 0 & 0 \\[6pt] \frac{\sigma_{12}\left(\sigma_1\sigma_2+\sigma_{12}^2\right)}{-4\Delta^3} & \frac{\sigma_1\sigma_{12}^2}{2\Delta^3} & 0 & 0 & 0 \\[6pt] \frac{\sigma_1\sigma_{12}^2}{4\Delta^3} & \frac{\sigma_1^2\sigma_{12}}{-4\Delta^3} & 0 & 0 & 0 \end{bmatrix}$$

$$[R_{23kl}^{(\alpha)}] = (\alpha^2 - 1) \begin{bmatrix} 0 & 0 & \frac{\sigma_2^2\sigma_{12}}{4\Delta^3} & \frac{\sigma_2\sigma_{12}^2}{-2\Delta^3} & \frac{\sigma_{12}^3}{4\Delta^3} \\[6pt] 0 & 0 & \frac{\sigma_2\sigma_{12}^2}{-4\Delta^3} & \frac{\sigma_{12}\left(\sigma_1\sigma_2+\sigma_{12}^2\right)}{4\Delta^3} & \frac{\sigma_1\sigma_{12}^2}{-4\Delta^3} \\[6pt] \frac{\sigma_2^2\sigma_{12}}{-4\Delta^3} & \frac{\sigma_2\sigma_{12}^2}{4\Delta^3} & 0 & 0 & 0 \\[6pt] \frac{\sigma_2\sigma_{12}^2}{2\Delta^3} & \frac{\sigma_{12}\left(\sigma_1\sigma_2+\sigma_{12}^2\right)}{-4\Delta^3} & 0 & 0 & 0 \\[6pt] \frac{\sigma_{12}^3}{-4\Delta^3} & \frac{\sigma_1\sigma_{12}^2}{4\Delta^3} & 0 & 0 & 0 \end{bmatrix}$$

$$\frac{[R^{(\alpha)}_{24kl}]}{(\alpha^2-1)} = \begin{bmatrix} 0 & 0 & \frac{\sigma_2(\sigma_1\sigma_2+\sigma_{12}^2)}{-4\triangle^3} & \frac{\sigma_{12}(3\sigma_1\sigma_2+\sigma_{12}^2)}{4\triangle^3} & \frac{\sigma_1\sigma_{12}^2}{-2\triangle^3} \\ 0 & 0 & \frac{\sigma_{12}(\sigma_1\sigma_2+\sigma_{12}^2)}{4\triangle^3} & \frac{\sigma_1(\sigma_1\sigma_2+3\sigma_{12}^2)}{-4\triangle^3} & \frac{\sigma_1^2\sigma_{12}}{2\triangle^3} \\ \frac{\sigma_2(\sigma_1\sigma_2+\sigma_{12}^2)}{4\triangle^3} & \frac{\sigma_{12}(\sigma_1\sigma_2+\sigma_{12}^2)}{-4\triangle^3} & 0 & 0 & 0 \\ \frac{\sigma_{12}(3\sigma_1\sigma_2+\sigma_{12}^2)}{-4\triangle^3} & \frac{\sigma_1(\sigma_1\sigma_2+3\sigma_{12}^2)}{4\triangle^3} & 0 & 0 & 0 \\ \frac{\sigma_1\sigma_{12}^2}{2\triangle^3} & \frac{\sigma_1^2\sigma_{12}}{-2\triangle^3} & 0 & 0 & 0 \end{bmatrix}$$

$$[R^{(\alpha)}_{25kl}] = (\alpha^2-1) \begin{bmatrix} 0 & 0 & \frac{\sigma_1\sigma_2\sigma_{12}}{4(\triangle)^3} & \frac{-\sigma_1(\sigma_1\sigma_2+\sigma_{12}^2)}{4\triangle^3} & \frac{\sigma_1^2\sigma_{12}}{4\triangle^3} \\ 0 & 0 & \frac{\sigma_1\sigma_{12}^2}{-4\triangle^3} & \frac{\sigma_1^2\sigma_{12}}{2\triangle^3} & \frac{\sigma_1^3}{-4\triangle^3} \\ \frac{\sigma_1\sigma_2\sigma_{12}}{-4\triangle^3} & \frac{\sigma_1\sigma_{12}^2}{4\triangle^3} & 0 & 0 & 0 \\ \frac{\sigma_1(\sigma_1\sigma_2+\sigma_{12}^2)}{4\triangle^3} & \frac{\sigma_1^2\sigma_{12}}{-2\triangle^3} & 0 & 0 & 0 \\ \frac{\sigma_1^2\sigma_{12}}{-4\triangle^3} & \frac{\sigma_1^3}{4\triangle^3} & 0 & 0 & 0 \end{bmatrix}$$

$$[R^{(\alpha)}_{34kl}] = (\alpha^2-1) \begin{bmatrix} 0 & \frac{-\sigma_2}{4\triangle^2} & 0 & 0 & 0 \\ \frac{\sigma_2}{4\triangle^2} & 0 & 0 & 0 & 0 \\ 0 & 0 & 0 & \frac{\sigma_2^2}{-4\triangle^3} & \frac{\sigma_2\sigma_{12}}{4\triangle^3} \\ 0 & 0 & \frac{\sigma_2^2}{4\triangle^3} & 0 & \frac{\sigma_1\sigma_2}{-4\triangle^3} \\ 0 & 0 & \frac{\sigma_2\sigma_{12}}{-4\triangle^3} & \frac{\sigma_1\sigma_2}{4\triangle^3} & 0 \end{bmatrix}$$

$$[R^{(\alpha)}_{35kl}] = (\alpha^2-1) \begin{bmatrix} 0 & \frac{\sigma_{12}}{4\triangle^2} & 0 & 0 & 0 \\ \frac{-\sigma_{12}}{4\triangle^2} & 0 & 0 & 0 & 0 \\ 0 & 0 & 0 & \frac{\sigma_2\sigma_{12}}{4\triangle^3} & \frac{\sigma_{12}^2}{-4\triangle^3} \\ 0 & 0 & \frac{\sigma_2\sigma_{12}}{-4\triangle^3} & 0 & \frac{\sigma_1\sigma_{12}}{4\triangle^3} \\ 0 & 0 & \frac{\sigma_{12}^2}{4\triangle^3} & \frac{\sigma_1\sigma_{12}}{-4\triangle^3} & 0 \end{bmatrix}$$

$$[R^{(\alpha)}_{45kl}] = (\alpha^2-1) \begin{bmatrix} 0 & \frac{-\sigma_1}{4\triangle^2} & 0 & 0 & 0 \\ \frac{\sigma_1}{4\triangle^2} & 0 & 0 & 0 & 0 \\ 0 & 0 & 0 & \frac{\sigma_1\sigma_2}{-4\triangle^3} & \frac{\sigma_1\sigma_{12}}{4\triangle^3} \\ 0 & 0 & \frac{\sigma_1\sigma_2}{4\triangle^3} & 0 & \frac{\sigma_1^2}{-4\triangle^3} \\ 0 & 0 & \frac{\sigma_1\sigma_{12}}{-4\triangle^3} & \frac{\sigma_1^2}{4\triangle^3} & 0 \end{bmatrix}. \qquad \square \quad (4.132)$$

Proposition 4.47. *The components of the α-Ricci tensor are given by the symmetric matrix $R^{(\alpha)} = [R^{(\alpha)}_{ij}]$:*

$$R^{(\alpha)} = (\alpha^2 - 1) \begin{bmatrix} \frac{\sigma_2}{2\Delta} & -\frac{\sigma_{12}}{2\Delta} & 0 & 0 & 0 \\ -\frac{\sigma_{12}}{2\Delta} & \frac{\sigma_1}{2\Delta} & 0 & 0 & 0 \\ 0 & 0 & \frac{\sigma_2{}^2}{2\Delta^2} & -\frac{\sigma_2\sigma_{12}}{\Delta^2} & \frac{3\sigma_{12}{}^2-\sigma_1\sigma_2}{4\Delta^2} \\ 0 & 0 & -\frac{\sigma_2\sigma_{12}}{\Delta^2} & \frac{3\sigma_1\sigma_2+\sigma_{12}{}^2}{2\Delta^2} & -\frac{\sigma_1\sigma_{12}}{\Delta^2} \\ 0 & 0 & \frac{3\sigma_{12}{}^2-\sigma_1\sigma_2}{4\Delta^2} & -\frac{\sigma_1\sigma_{12}}{\Delta^2} & \frac{\sigma_1{}^2}{2\Delta^2} \end{bmatrix} .$$

$$\square$$

$$(4.133)$$

Proposition 4.48. *The bivariate Gaussian manifold* $\mathcal{N}$ *has a constant* α-*scalar curvature* $R^{(\alpha)}$:

$$R^{(\alpha)} = \frac{9\left(\alpha^2 - 1\right)}{2} \tag{4.134}$$

This recovers the known result for the 0-scalar curvature $R^{(0)} = -\frac{9}{2}$. *So geometrically* $\mathcal{N}$ *constitutes part of a pseudosphere, §3.17.* $\square$

Proposition 4.49. *The* α-*sectional curvatures of* $\mathcal{N}$ *are*

$$\varrho^{(\alpha)} = [\varrho^{(\alpha)}(i,j)]$$

$$= (\alpha^2 - 1) \begin{bmatrix} 0 & -\frac{1}{4} & \frac{1}{2} & \frac{\sigma_1\sigma_2+3\sigma_{12}{}^2}{4(\sigma_1\sigma_2+\sigma_{12}{}^2)} & \frac{\sigma_{12}{}^2}{2\sigma_1\sigma_2} \\ -\frac{1}{4} & 0 & \frac{\sigma_{12}{}^2}{2\sigma_1\sigma_2} & \frac{\sigma_1\sigma_2+3\sigma_{12}{}^2}{4(\sigma_1\sigma_2+\sigma_{12}{}^2)} & \frac{1}{2} \\ \frac{1}{2} & \frac{\sigma_{12}{}^2}{2\sigma_1\sigma_2} & 0 & \frac{1}{2} & \frac{\sigma_{12}{}^2}{\sigma_1\sigma_2+\sigma_{12}{}^2} \\ \frac{\sigma_1\sigma_2+3\sigma_{12}{}^2}{4(\sigma_1\sigma_2+\sigma_{12}{}^2)} & \frac{\sigma_1\sigma_2+3\sigma_{12}{}^2}{4(\sigma_1\sigma_2+\sigma_{12}{}^2)} & \frac{1}{2} & 0 & \frac{1}{2} \\ \frac{\sigma_{12}{}^2}{2\sigma_1\sigma_2} & \frac{1}{2} & \frac{\sigma_{12}{}^2}{\sigma_1\sigma_2+\sigma_{12}{}^2} & \frac{1}{2} & 0 \end{bmatrix}$$

The α-*sectional curvatures of* $\mathcal{N}$ *can be written as a function of correlation coefficient* ρ *only, Figure 4.6:*

$$\varrho^{(\alpha)} = (\alpha^2 - 1) \begin{bmatrix} 0 & -\frac{1}{4} & \frac{1}{2} & \frac{1+3\rho^2}{4(1+\rho^2)} & \frac{\rho^2}{2} \\ -\frac{1}{4} & 0 & \frac{\rho^2}{2} & \frac{1+3\rho^2}{4(1+\rho^2)} & \frac{1}{2} \\ \frac{1}{2} & \frac{\rho^2}{2} & 0 & \frac{1}{2} & \frac{\rho^2}{1+\rho^2} \\ \frac{1+3\rho^2}{4(1+\rho^2)} & \frac{1+3\rho^2}{4(1+\rho^2)} & \frac{1}{2} & 0 & \frac{1}{2} \\ \frac{\rho^2}{2} & \frac{1}{2} & \frac{\rho^2}{1+\rho^2} & \frac{1}{2} & 0 \end{bmatrix} . \quad \square \ (4.135)$$

Proposition 4.50. *The* α-*mean curvatures* $\varrho^{(\alpha)}(\lambda)\,(\lambda = 1,2,3,4,5)$ *are given by:*

$$\varrho^{(\alpha)}(1) = \varrho^{(\alpha)}(2) = \frac{\alpha^2 - 1}{8},$$

$$\varrho^{(\alpha)}(3) = \varrho^{(\alpha)}(5) = \frac{\alpha^2 - 1}{4},$$

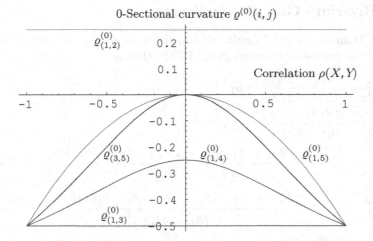

Fig. 4.6. The 0-sectional curvatures $\varrho^{(0)}(i,j)$ as a function of correlation $\rho(X,Y)$ for bivariate normal manifold $\mathcal{N}$ where: $\varrho^{(0)}(1,3) = \varrho^{(0)}(2,5) = \varrho^{(0)}(3,4) = \varrho^{(0)}(4,5) = -\frac{1}{2}$, $\varrho^{(0)}(1,2) = \frac{1}{4}$, $\varrho^{(0)}(1,4) = \varrho^{(0)}(2,4)$ and $\varrho^{(0)}(1,5) = \varrho^{(0)}(2,3)$. Note that $\varrho^{(0)}(1,4)$, $\varrho^{(0)}(1,5)$ and $\varrho^{(0)}(3,5)$ have limiting value $-\frac{1}{2}$ as $\rho \to \pm1$.

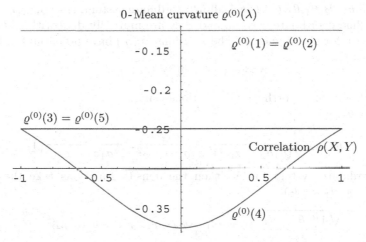

Fig. 4.7. The 0-mean curvatures $\varrho^{(\alpha)}(\lambda)$ as a function of correlation $\rho(X,Y)$ for bivariate normal manifold $\mathcal{N}$ where; $\varrho^{(0)}(1) = \varrho^{(0)}(2) = -\frac{1}{8}$, $\varrho^{(0)}(3) = \varrho^{(0)}(5) = -\frac{1}{4}$, and $\varrho^{(0)}(4)$ has limiting value $-\frac{1}{4}$ as $\rho(X,Y) \to \pm1$, and $-\frac{3}{8}$ as $\rho \to 0$.

$$\varrho^{(\alpha)}(4) = \frac{(\alpha^2 - 1)\,(3\,\sigma_1\,\sigma_2 + \sigma_{12}{}^2)}{8\,(\sigma_1\,\sigma_2 + \sigma_{12}{}^2)} = \frac{(\alpha^2 - 1)\,(3 + \rho^2)}{8\,(1 + \rho^2)}. \quad (4.136)$$

$\square$

Figure 4.7 shows a plot of the 0-mean curvatures $\varrho^{(\alpha)}$ as a function of correlation ρ for bivariate normal manifold $\mathcal{N}$.

4.20 Bivariate Gaussian Foliations

Since $\mathcal{N}$ is an exponential family, §3.2, a mutually dual coordinate system is given by the potential function $\varphi(\theta)$ (4.129), that is

$$
\eta_1 = \frac{\partial \varphi}{\partial \theta_1} = \frac{2\,\theta_1\,\theta_5 - \theta_2\,\theta_4}{\theta_4^2 - 4\,\theta_3\,\theta_5} = \mu_1 \,,
$$

$$
\eta_2 = \frac{\partial \varphi}{\partial \theta_2} = \frac{2\,\theta_2\,\theta_3 - \theta_1\,\theta_4}{\theta_4^2 - 4\,\theta_3\,\theta_5} = \mu_2 \,,
$$

$$
\eta_3 = \frac{\partial \varphi}{\partial \theta_3} = \frac{\theta_2^2\,\theta_4^2 + 2\,\theta_4\,(-2\,\theta_1\,\theta_2 + \theta_4)\,\theta_5 + 4\,\left(\theta_1^2 - 2\,\theta_3\right)\theta_5^2}{\left(\theta_4^2 - 4\,\theta_3\,\theta_5\right)^2} = \mu_1^2 + \sigma_1 \,,
$$

$$
\eta_4 = \frac{\partial \varphi}{\partial \theta_4} = -\frac{2\theta_2^2\,\theta_3\theta_4 + \theta_4^3 + 2\left(\theta_1^2 - 2\theta_3\right)\theta_4\theta_5 - \theta_1\theta_2\left(\theta_4^2 + 4\theta_3\theta_5\right)}{\left(\theta_4^2 - 4\theta_3\theta_5\right)^2}
$$

$$
= \mu_1\mu_2 + \sigma_{12} \,,
$$

$$
\eta_5 = \frac{\partial \varphi}{\partial \theta_5} = \frac{4\theta_2^2\,\theta_3^2 - 4\theta_1\,\theta_2\,\theta_3\,\theta_4 + \left(\theta_1^2 + 2\,\theta_3\right)\theta_4^2 - 8\theta_3^2\,\theta_5}{\left(\theta_4^2 - 4\theta_3\,\theta_5\right)^2} = \mu_2^2 + \sigma_2 \,.
$$

$$(4.137)$$

Then $(\theta_1, \theta_2, \theta_3, \theta_4, \theta_5)$ is a 1-affine coordinate system, $(\eta_1, \eta_2, \eta_3, \eta_4, \eta_5)$ is a (-1)-affine coordinate system, and they are mutually dual with respect to the Fisher information metric. The coordinates (η_i) have potential function λ

$$
\lambda = -\left(1 + \log(2\,\pi\,\sqrt{\triangle})\right). \tag{4.138}
$$

We obtain dually orthogonal foliations using
$(\eta_1, \eta_2, \theta_3, \theta_4, \theta_5) =$

$$
\left(\mu_1, \mu_2, \frac{-\sigma_2}{2\left(\sigma_1\sigma_2 - \sigma_{12}^2\right)}, \frac{\sigma_{12}}{\sigma_1\sigma_2 - \sigma_{12}^2}, \frac{-\sigma_1}{2\left(\sigma_1\sigma_2 - \sigma_{12}^2\right)}\right)
$$

as a coordinate system for $\mathcal{N}$, then the density functions take the form:
$f(x, y; \eta_1, \eta_2, \theta_3, \theta_4, \theta_5) =$

$$
\frac{\sqrt{4\,\theta_3\,\theta_5 - \theta_4^2}}{2\,\pi}\, e^{\theta_3\,(x-\mu_1)^2 + \theta_4\,(x-\mu_1)\,(y-\mu_2) + \theta_5\,(y-\mu_2)^2} \tag{4.139}
$$

and the Fisher metric is

$$
[g_{ij}] = \begin{bmatrix} \sigma_1 & \sigma_{12} & 0 & 0 & 0 \\ \sigma_{12} & \sigma_2 & 0 & 0 & 0 \\ 0 & 0 & \frac{\sigma_2^2}{2\triangle^2} & -\frac{\sigma_2\,\sigma_{12}}{\triangle^2} & \frac{\sigma_{12}^2}{2\triangle^2} \\ 0 & 0 & -\frac{\sigma_2\,\sigma_{12}}{\triangle^2} & \frac{\sigma_1\,\sigma_2 + \sigma_{12}^2}{\triangle^2} & -\frac{\sigma_1\,\sigma_{12}}{\triangle^2} \\ 0 & 0 & \frac{\sigma_{12}^2}{2\triangle^2} & -\frac{\sigma_1\,\sigma_{12}}{\triangle^2} & \frac{\sigma_1^2}{2\triangle^2} \end{bmatrix}. \tag{4.140}
$$

It follows that (θ_i) is a geodesic coordinate system of $\nabla^{(1)}$, and (η_i) is a geodesic coordinate system of $\nabla^{(-1)}$.

4.21 Bivariate Gaussian Submanifolds

We study three submanifolds, §2.0.5.

4.21.1 Independence Submanifold $\mathcal{N}_1$

This has definition
$$\mathcal{N}_1 \subset \mathcal{N} : \sigma_{12} = 0.$$

The density functions are of form:

$$f(x, y; \mu_1, \mu_2, \sigma_1, \sigma_2) = N_X(\mu_1, \sigma_1).N_Y(\mu_2, \sigma_2) \tag{4.141}$$

This is the case for statistical independence of X and Y, by §1.3, so the space $\mathcal{N}_1$ is the direct product of two Riemannian spaces

$$\{N_X(\mu_1, \sigma_1), \ \mu_1 \in \mathbb{R}, \ \sigma_1 \in \mathbb{R}^+\} \quad \text{and} \quad \{N_Y(\mu_2, \sigma_2), \ \mu_2 \in \mathbb{R}, \ \sigma_2 \in \mathbb{R}^+\}.$$

We report expressions for the metric, the α-connections and the α-curvature objects using the natural coordinate system

$$(\theta_1, \theta_2, \theta_3, \theta_4) = (\frac{\mu_1}{\sigma_1}, \frac{\mu_2}{\sigma_2}, -\frac{1}{2\sigma_1}, -\frac{1}{2\sigma_2})$$

and potential function $\varphi = \log(2\pi\sqrt{\triangle}) - \triangle\left(\theta_2{}^2\theta_3 + \theta_1{}^2\theta_4\right) = \log(2\pi\sqrt{\sigma_1\sigma_2})$
$+ \frac{\mu_1{}^2}{2\sigma_1} + \frac{\mu_2{}^2}{2\sigma_2}$; where $\triangle = \frac{1}{4\theta_3\theta_4}$.

Proposition 4.51. *The metric tensor is:*

$$[g_{ij}] = \begin{bmatrix} \sigma_1 & 0 & 2\,\mu_1\,\sigma_1 & 0 \\ 0 & \sigma_2 & 0 & 2\,\mu_2\,\sigma_2 \\ 2\,\mu_1\,\sigma_1 & 0 & 2\,\sigma_1\,(2\,\mu_1{}^2 + \sigma_1) & 0 \\ 0 & 2\,\mu_2\,\sigma_2 & 0 & 2\,\sigma_2\,(2\,\mu_2{}^2 + \sigma_2) \end{bmatrix}. \ \Box \ (4.142)$$

Proposition 4.52. *The nonzero independent components of the α-connection are*

$$\Gamma_{13,1}^{(\alpha)} = -(\alpha - 1)\,\sigma_1{}^2,$$
$$\Gamma_{33,1}^{(\alpha)} = -4\,(\alpha - 1)\,\mu_1\,\sigma_1{}^2,$$
$$\Gamma_{24,2}^{(\alpha)} = -(\alpha - 1)\,\sigma_2{}^2,$$
$$\Gamma_{44,2}^{(\alpha)} = -4\,(\alpha - 1)\,\mu_2\,\sigma_2{}^2,$$
$$\Gamma_{33,3}^{(\alpha)} = -4\,(\alpha - 1)\,\sigma_1{}^2\,(3\,\mu_1{}^2 + \sigma_1),$$
$$\Gamma_{11}^{(\alpha)1} = -\Gamma_{13}^{(\alpha)3} = (\alpha - 1)\,\mu_1,$$
$$\Gamma_{31}^{(\alpha)1} = (\alpha - 1)\,(2\,\mu_1{}^2 - \sigma_1),$$

$$\Gamma_{33}^{(\alpha)1} = 4\,(\alpha - 1)\,\mu_1{}^3\,,$$

$$\Gamma_{11}^{(\alpha)3} = \Gamma_{22}^{(\alpha)4} = \frac{1 - \alpha}{2}\,,$$

$$\Gamma_{33}^{(\alpha)3} = -2\,(\alpha - 1)\,\left(\mu_1{}^2 + \sigma_1\right)\,,$$

$$\Gamma_{22}^{(\alpha)2} = -\Gamma_{24}^{(\alpha)4} = (\alpha - 1)\,\mu_2\,,$$

$$\Gamma_{42}^{(\alpha)2} = (\alpha - 1)\,\left(2\,\mu_2{}^2 - \sigma_2\right)\,,$$

$$\Gamma_{44}^{(\alpha)2} = 4\,(\alpha - 1)\,\mu_2{}^3\,,$$

$$\Gamma_{44}^{(\alpha)4} = -2\,(\alpha - 1)\,\left(\mu_2{}^2 + \sigma_2\right)\,. \quad \square \qquad (4.143)$$

Proposition 4.53. *The α-curvature tensor is*

$$R_{1313}^{(\alpha)} = -\left(\alpha^2 - 1\right)\sigma_1{}^3, \quad R_{2424}^{(\alpha)} = -\left(\alpha^2 - 1\right)\sigma_2{}^3 \qquad (4.144)$$

*while the other independent components are zero. By contraction we obtain:
The α-Ricci tensor:*

$$R^{(\alpha)} = \left(\alpha^2 - 1\right) \begin{bmatrix} \frac{\sigma_1}{2} & 0 & \mu_1\,\sigma_1 & 0 \\ 0 & \frac{\sigma_2}{2} & 0 & \mu_2\,\sigma_2 \\ \mu_1\,\sigma_1 & 0 & \sigma_1\,\left(2\,\mu_1{}^2 + \sigma_1\right) & 0 \\ 0 & \mu_2\,\sigma_2 & 0 & \sigma_2\,\left(2\,\mu_2{}^2 + \sigma_2\right) \end{bmatrix}. \qquad (4.145)$$

The α-eigenvalues of the α-Ricci tensor are:

$$\left(\alpha^2 - 1\right) \begin{pmatrix} \frac{\sigma_1}{4} + \mu_1{}^2\,\sigma_1 + \frac{\sigma_1}{4}\left(\sqrt{16\,\mu_1{}^4 + (1 - 2\,\sigma_1)^2 + 8\,\mu_1{}^2\,(1 + 2\,\sigma_1)} + \frac{2}{\sigma_1}\right) \\ \frac{\sigma_1}{4} + \mu_1{}^2\,\sigma_1 - \frac{\sigma_1}{4}\left(\sqrt{16\,\mu_1{}^4 + (1 - 2\,\sigma_1)^2 + 8\,\mu_1{}^2\,(1 + 2\,\sigma_1)} - \frac{2}{\sigma_1}\right) \\ \frac{\sigma_2}{4} + \mu_2{}^2\,\sigma_2 + \frac{\sigma_2}{4}\left(\sqrt{16\,\mu_2{}^4 + (1 - 2\,\sigma_2)^2 + 8\,\mu_2{}^2\,(1 + 2\,\sigma_2)} + \frac{2}{\sigma_2}\right) \\ \frac{\sigma_2}{4} + \mu_2{}^2\,\sigma_2 - \frac{\sigma_2{}^2}{4}\left(\sqrt{16\,\mu_2{}^4 + (1 - 2\,\sigma_2)^2 + 8\,\mu_2{}^2\,(1 + 2\,\sigma_2)} - \frac{2}{\sigma_2}\right) \end{pmatrix}.$$

$$(4.146)$$

The α-scalar curvature:

$$R^{(\alpha)} = 2\,\left(\alpha^2 - 1\right) \qquad (4.147)$$

The α-sectional curvatures:

$$\varrho^{(\alpha)} = \frac{\left(\alpha^2 - 1\right)}{2} \begin{bmatrix} 0 & 0 & 1 & 0 \\ 0 & 0 & 0 & 1 \\ 1 & 0 & 0 & 0 \\ 0 & 1 & 0 & 0 \end{bmatrix} \qquad (4.148)$$

The α-mean curvatures:

$$\varrho^{(\alpha)}(1) = \varrho^{(\alpha)}(2) = \varrho^{(\alpha)}(3) = \varrho^{(\alpha)}(4) = \frac{\alpha^2 - 1}{6}. \qquad (4.149)$$

In $\mathcal{N}_1$, the α-scalar curvature, the α-sectional curvatures and the α-mean curvatures are constant; they are negative when $\alpha = 0$.   □

Proposition 4.54. *The submanifold $\mathcal{N}_1$ is an Einstein space.*

Proof. By comparison of the metric tensor (4.142) with the Ricci tensor (4.145), we deduce

$$R_{ij}^{(0)} = \frac{R^{(0)}}{k} g_{ij}$$

where k is the dimension of the space. Then the submanifold $\mathcal{N}_1$ with statistically independent random variables is an Einstein space.   □

4.21.2 Identical Marginals Submanifold $\mathcal{N}_2$

This is

$$\mathcal{N}_2 \subset \mathcal{N} : \sigma_1 = \sigma_2 = \sigma,\ \mu_1 = \mu_2 = \mu.$$

The density functions are of form $f(x, y; \mu, \sigma, \sigma_{12}) =$

$$\frac{1}{2\pi\sqrt{\sigma^2 - \sigma_{12}^2}} e^{-\frac{1}{2(\sigma^2-\sigma_{12}{}^2)}\left(\sigma(x-\mu)^2 - 2\sigma_{12}(x-\mu)(y-\mu) + \sigma(y-\mu)^2\right)} \qquad (4.150)$$

The marginal functions are $f_X = f_Y \equiv N(\mu, \sigma)$ with correlation coefficient $\rho(X, Y) = \frac{\sigma_{12}}{\sigma}$.

We report the expressions for the metric, the α-connections and the α-curvature objects using the natural coordinate system

$$(\theta_1, \theta_2, \theta_3) = (\frac{\mu}{\sigma + \sigma_{12}}, \frac{-\sigma}{2(\sigma^2 - \sigma_{12}{}^2)}, \frac{\sigma_{12}}{(\sigma^2 - \sigma_{12}{}^2)})$$

and the potential function

$$\varphi = -\frac{\theta_1{}^2}{2\theta_2 + \theta_3} + \log(2\pi) - \frac{1}{2}\log(4\theta_2{}^2 - \theta_3{}^2)$$

$$= \frac{\mu^2}{\sigma + \sigma_{12}} + \log(2\pi) + \frac{1}{2}\log(\sigma^2 - \sigma_{12}{}^2). \qquad (4.151)$$

Proposition 4.55. *The metric tensor is*
$[g_{ij}] =$

$$\begin{bmatrix} 2(\sigma + \sigma_{12}) & 4\mu(\sigma + \sigma_{12}) & 2\mu(\sigma + \sigma_{12}) \\ 4\mu(\sigma + \sigma_{12}) & 4\left(\sigma(2\mu^2 + \sigma) + 2\mu^2\sigma_{12} + \sigma_{12}{}^2\right) & 4\left(\mu^2\sigma + (\mu^2 + \sigma)\sigma_{12}\right) \\ 2\mu(\sigma + \sigma_{12}) & 4\left(\mu^2\sigma + (\mu^2 + \sigma)\sigma_{12}\right) & \sigma(2\mu^2 + \sigma) + 2\mu^2\sigma_{12} + \sigma_{12}{}^2 \end{bmatrix}.$$

$$(4.152)$$

Proposition 4.56. *The components of the α-connection are*

$$[\Gamma_{ij}^{(\alpha)1}] = (\alpha - 1) \begin{bmatrix} 2\mu & 4\mu^2 - \sigma - \sigma_{12} & \frac{4\mu^2 - \sigma - \sigma_{12}}{2} \\ 4\mu^2 - \sigma - \sigma_{12} & 8\mu^3 & 4\mu^3 \\ \frac{4\mu^2 - \sigma - \sigma_{12}}{2} & 4\mu^3 & 2\mu^3 \end{bmatrix}$$

$$[\Gamma_{ij}^{(\alpha)2}] = (\alpha - 1) \begin{bmatrix} \frac{-1}{2} & -\mu & -\frac{\mu}{2} \\ -\mu & -2\left(\mu^2 + \sigma\right) & -\left(\mu^2 + \sigma_{12}\right) \\ -\frac{\mu}{2} & -\left(\mu^2 + \sigma_{12}\right) & \frac{-\left(\mu^2 + \sigma\right)}{2} \end{bmatrix}$$

$$[\Gamma_{ij}^{(\alpha)3}] = (\alpha - 1) \begin{bmatrix} -1 & -2\mu & -\mu \\ -2\mu & -4\left(\mu^2 + \sigma_{12}\right) & -2\left(\mu^2 + \sigma\right) \\ -\mu & -2\left(\mu^2 + \sigma\right) & -\left(\mu^2 + \sigma_{12}\right) \end{bmatrix} \quad (4.153)$$

The analytic expressions for the functions $\Gamma_{ij,k}^{(\alpha)}$ are known [13] but rather long, so we do not report them here. $\qquad\square$

Proposition 4.57. *By direct calculation we have the α-curvature tensor of $\mathcal{N}_2$*

$$R_{12kl}^{(\alpha)} = (\alpha^2 - 1) \begin{bmatrix} 0 & -2\left(\sigma + \sigma_{12}\right)^3 & -\left(\sigma + \sigma_{12}\right)^3 \\ 2\left(\sigma + \sigma_{12}\right)^3 & 0 & 0 \\ \left(\sigma + \sigma_{12}\right)^3 & 0 & 0 \end{bmatrix}$$

$$R_{13kl}^{(\alpha)} = (\alpha^2 - 1) \begin{bmatrix} 0 & -\left(\sigma + \sigma_{12}\right)^3 & \frac{-\left(\sigma + \sigma_{12}\right)^3}{2} \\ \left(\sigma + \sigma_{12}\right)^3 & 0 & 0 \\ \frac{\left(\sigma + \sigma_{12}\right)^3}{2} & 0 & 0 \end{bmatrix}$$

$$(4.154)$$

while the other independent components are zero.

By contraction we obtain the
α-Ricci tensor $R^{(\alpha)} =$

$$(\alpha^2 - 1) \begin{bmatrix} \left(\sigma + \sigma_{12}\right) & 2\mu\left(\sigma + \sigma_{12}\right) & \mu\left(\sigma + \sigma_{12}\right) \\ 2\mu\left(\sigma + \sigma_{12}\right) & \left(\sigma + \sigma_{12}\right)\left(4\mu^2 + \sigma + \sigma_{12}\right) & \frac{\left(\sigma + \sigma_{12}\right)\left(4\mu^2 + \sigma + \sigma_{12}\right)}{2} \\ \mu\left(\sigma + \sigma_{12}\right) & \frac{\left(\sigma + \sigma_{12}\right)\left(4\mu^2 + \sigma + \sigma_{12}\right)}{2} & \frac{\left(\sigma + \sigma_{12}\right)\left(4\mu^2 + \sigma + \sigma_{12}\right)}{4} \end{bmatrix}$$

$$(4.155)$$

The α-eigenvalues of the α-Ricci tensor are

$$\left(\begin{array}{c} 0 \\ \dfrac{10(\alpha^2 - 1)(\sigma + \sigma_{12})^2}{4(1 + 5\mu^2) + 5 - \sqrt{400\mu^4 + (4 - 5\sigma)^2 + 40\mu^2(4 + 5\sigma) + 5\sigma_{12}\left(-8 + 40\mu^2 + 10\sigma + 5\sigma_{12}\right)}} \\ \dfrac{10(\alpha^2 - 1)(\sigma + \sigma_{12})^2}{4(1 + 5\mu^2) + 5 + \sqrt{400\mu^4 + (4 - 5\sigma)^2 + 40\mu^2(4 + 5\sigma) + 5\sigma_{12}(-8 + 40\mu^2 + 10\sigma + 5\sigma_{12})}} \end{array} \right)$$

$$(4.156)$$

The α-scalar curvature:

$$R^{(\alpha)} = (\alpha^2 - 1) \qquad (4.157)$$

The α-sectional curvatures:

$$\varrho^{(\alpha)} = (\alpha^2 - 1) \begin{bmatrix} 0 & \frac{(\sigma+\sigma_{12})^2}{4\,(\sigma^2+\sigma_{12}{}^2)} & \frac{(\sigma+\sigma_{12})^2}{4\,(\sigma^2+\sigma_{12}{}^2)} \\ \frac{(\sigma+\sigma_{12})^2}{4\,(\sigma^2+\sigma_{12}{}^2)} & 0 & 0 \\ \frac{(\sigma+\sigma_{12})^2}{4\,(\sigma^2+\sigma_{12}{}^2)} & 0 & 0 \end{bmatrix} \qquad (4.158)$$

The α-mean curvatures:

$$\varrho^{(\alpha)}(1) = \frac{1}{4}\left(\alpha^2 - 1\right),$$

$$\varrho^{(\alpha)}(2) = \varrho^{(\alpha)}(3) = \frac{(\alpha^2 - 1)\,(\sigma + \sigma_{12})\,(4\,\mu^2 + \sigma + \sigma_{12})}{8\,(\sigma\,(2\,\mu^2 + \sigma) + 2\,\mu^2\,\sigma_{12} + \sigma_{12}{}^2)}. \quad \square \quad (4.159)$$

4.21.3 Central Mean Submanifold $\mathcal{N}_3$

This is defined as

$$\mathcal{N}_3 \subset \mathcal{N} : \mu_1 = \mu_2 = 0.$$

The density functions are
$$f(x, y; \sigma_1, \sigma_2, \sigma_{12}) =$$

$$\frac{1}{2\,\pi\,\sqrt{\sigma_1\,\sigma_2 - \sigma_{12}{}^2}}\, e^{-\frac{1}{2\,(\sigma_1\,\sigma_2 - \sigma_{12}{}^2)}\left(\sigma_2 x^2 - 2\,\sigma_{12}\,x\,y + \sigma_1 y^2\right)} \qquad (4.160)$$

with marginal functions

$$f_X(x) = N_X(0, \sigma_1), \text{ and } f_Y(y) = N_Y(0, \sigma_2)$$

and correlation coefficient

$$\rho(X, Y) = \frac{\sigma_{12}}{\sqrt{\sigma_1\,\sigma_2}}.$$

We report the expressions for the metric, the α-connections and the α-curvature objects using the natural coordinate system

$$(\theta_1, \theta_2, \theta_3) = \left(-\frac{\sigma_2}{2\,(\sigma_1\,\sigma_2 - \sigma_{12}{}^2)}, \frac{\sigma_{12}}{\sigma_1\,\sigma_2 - \sigma_{12}{}^2}, -\frac{\sigma_1}{2\,(\sigma_1\,\sigma_2 - \sigma_{12}{}^2)}\right)$$

and the potential function

$$\varphi = \log(2\,\pi) - \frac{1}{2}\,\log(\sqrt{4\,\theta_1\,\theta_3 - \theta_4{}^2}) = \log(2\,\pi) + \frac{1}{2}\,\log(\sigma_1\,\sigma_2 - \sigma_{12}{}^2).$$

Proposition 4.58. *The metric tensor is as follows:*

$$[g_{ij}] = \begin{bmatrix} 2\sigma_1{}^2 & 2\sigma_1\sigma_{12} & 2\sigma_{12}{}^2 \\ 2\sigma_1\sigma_{12} & \sigma_1\sigma_2+\sigma_{12}{}^2 & 2\sigma_2\sigma_{12} \\ 2\sigma_{12}{}^2 & 2\sigma_2\sigma_{12} & 2\sigma_2{}^2 \end{bmatrix}. \quad \square \tag{4.161}$$

Proposition 4.59. *The components of the α-connection are*

$$[\Gamma^{(\alpha)}_{ij,1}] = (\alpha-1) \begin{bmatrix} -4\sigma_1{}^3 & -4\sigma_1{}^2\sigma_{12} & -4\sigma_1\sigma_{12}{}^2 \\ -4\sigma_1{}^2\sigma_{12} & -\sigma_1\left(\sigma_1\sigma_2+3\sigma_{12}{}^2\right) & -2\left(\sigma_1\sigma_2\sigma_{12}+\sigma_{12}{}^3\right) \\ -4\sigma_1\sigma_{12}{}^2 & -2\left(\sigma_1\sigma_2\sigma_{12}+\sigma_{12}{}^3\right) & -4\sigma_2\sigma_{12}{}^2 \end{bmatrix}$$

$$\frac{[\Gamma^{(\alpha)}_{ij,2}]}{(\alpha-1)} = \begin{bmatrix} -4\sigma_1{}^2\sigma_{12} & -\sigma_1\left(\sigma_1\sigma_2+3\sigma_{12}{}^2\right) & -2\left(\sigma_1\sigma_2\sigma_{12}+\sigma_{12}{}^3\right) \\ -\sigma_1\left(\sigma_1\sigma_2+3\sigma_{12}{}^2\right) & -3\sigma_1\sigma_2\sigma_{12}-\sigma_{12}{}^3 & -\sigma_2\left(\sigma_1\sigma_2+3\sigma_{12}{}^2\right) \\ -2\left(\sigma_1\sigma_2\sigma_{12}+\sigma_{12}{}^3\right) & -\sigma_2\left(\sigma_1\sigma_2+3\sigma_{12}{}^2\right) & -4\sigma_2{}^2\sigma_{12} \end{bmatrix}$$

$$[\Gamma^{(\alpha)}_{ij,3}] = (\alpha-1) \begin{bmatrix} -4\sigma_1\sigma_{12}{}^2 & -2\left(\sigma_1\sigma_2\sigma_{12}+\sigma_{12}{}^3\right) & -4\sigma_2\sigma_{12}{}^2 \\ -2\left(\sigma_1\sigma_2\sigma_{12}+\sigma_{12}{}^3\right) & -\sigma_2\left(\sigma_1\sigma_2-3\sigma_{12}{}^2\right) & -4\sigma_2{}^2\sigma_{12} \\ -4\sigma_2\sigma_{12}{}^2 & -4\sigma_2{}^2\sigma_{12} & -4\sigma_2{}^3 \end{bmatrix}$$

$$[\Gamma^{(\alpha)1}_{ij}] = (\alpha-1) \begin{bmatrix} -2\sigma_1 & -\sigma_{12} & 0 \\ -\sigma_{12} & \frac{-\sigma_2}{2} & 0 \\ 0 & 0 & 0 \end{bmatrix}$$

$$[\Gamma^{(\alpha)2}_{ij}] = (\alpha-1) \begin{bmatrix} 0 & -\sigma_1 & -2\sigma_{12} \\ -\sigma_1 & -\sigma_{12} & -\sigma_2 \\ -2\sigma_{12} & -\sigma_2 & 0 \end{bmatrix}$$

$$[\Gamma^{(\alpha)3}_{ij}] = (\alpha-1) \begin{bmatrix} 0 & 0 & 0 \\ 0 & \frac{-\sigma_1}{2} & -\sigma_{12} \\ 0 & -\sigma_{12} & -2\sigma_2 \end{bmatrix}. \quad \square \tag{4.162}$$

Proposition 4.60. *By direct calculation we have the nonzero independent components of the α-curvature tensor of $\mathcal{N}_3$*

$$R^{(\alpha)}_{12kl} = \left(\alpha^2-1\right) \begin{bmatrix} 0 & -\sigma_1{}^2\triangle & -2\sigma_1\sigma_{12}\triangle \\ \sigma_1{}^2\triangle & 0 & -\sigma_1\sigma_2\triangle \\ 2\sigma_1\sigma_{12}\triangle & \sigma_1\sigma_2\triangle & 0 \end{bmatrix}$$

$$R^{(\alpha)}_{13kl} = \left(\alpha^2-1\right) \begin{bmatrix} 0 & -2\sigma_1\sigma_{12}\triangle & -4\sigma_{12}{}^2\triangle \\ 2\sigma_1\sigma_{12}\triangle & 0 & -2\sigma_2\sigma_{12}\triangle \\ 4\sigma_{12}{}^2\triangle & 2\sigma_2\sigma_{12}\triangle & 0 \end{bmatrix}$$

$$R^{(\alpha)}_{23kl} = (\alpha^2 - 1) \begin{bmatrix} 0 & -\sigma_1 \sigma_2 \triangle & -2\sigma_2 \sigma_{12} \triangle \\ \sigma_1 \sigma_2 \triangle & 0 & -\sigma_2{}^2 \triangle \\ 2\sigma_2 \sigma_{12} \triangle & \sigma_2{}^2 \triangle & 0 \end{bmatrix}. \qquad (4.163)$$

By contraction:
The α- Ricci tensor:

$$R^{(\alpha)} = (\alpha^2 - 1) \begin{bmatrix} \sigma_1{}^2 & \sigma_1 \sigma_{12} & 2\sigma_{12}{}^2 - \sigma_1 \sigma_2 \\ \sigma_1 \sigma_{12} & \sigma_1 \sigma_2 & \sigma_2 \sigma_{12} \\ 2\sigma_{12}{}^2 - \sigma_1 \sigma_2 & \sigma_2 \sigma_{12} & \sigma_2{}^2 \end{bmatrix}. \qquad (4.164)$$

The α-eigenvalues of the α-Ricci tensor are given by:

$$\frac{(\alpha^2 - 1)}{2} \begin{pmatrix} 0 \\ \sigma_1{}^2 + \sigma_1 \sigma_2 + \sigma_2{}^2 - S \\ \sigma_1{}^2 + \sigma_1 \sigma_2 + \sigma_2{}^2 + S \end{pmatrix} \qquad (4.165)$$

where

$$S = \sqrt{(\sigma_1{}^2 - \sigma_1 \sigma_2 + \sigma_2{}^2)^2 + 4(\sigma_1{}^2 - 4\sigma_1 \sigma_2 + \sigma_2{}^2)\sigma_{12}{}^2 + 16\sigma_{12}{}^4}.$$

The α-scalar curvature:

$$R^{(\alpha)} = 2(\alpha^2 - 1) \qquad (4.166)$$

The α-sectional curvatures:

$$\varrho^{(\alpha)} = (\alpha^2 - 1) \begin{bmatrix} 0 & \frac{1}{2} & \frac{\rho^2}{1+\rho^2} \\ \frac{1}{2} & 0 & \frac{1}{2} \\ \frac{\rho^2}{1+\rho^2} & \frac{1}{2} & 0 \end{bmatrix} \qquad (4.167)$$

The α-mean curvatures:

$$\varrho^{(\alpha)}(1) = \varrho^{(\alpha)}(3) = \frac{1}{4}(\alpha^2 - 1),$$

$$\varrho^{(\alpha)}(2) = \frac{(\alpha^2 - 1)\sigma_1 \sigma_2}{2(\sigma_1 \sigma_2 + \sigma_{12}{}^2)} = \frac{(\alpha^2 - 1)}{2(1 + \rho^2)}. \qquad (4.168)$$

For $\mathcal{N}_3$ the α-mean curvatures have limiting value $\frac{(\alpha^2-1)}{4}$ as $\rho^2 \to 1$. $\square$

4.21.4 Affine Immersion

Proposition 4.61. *Let $\mathcal{N}$ be the bivariate Gaussian manifold with the Fisher metric g and the exponential connection $\nabla^{(1)}$. Denote by (θ_i) the natural coordinate system (4.128). Then $\mathcal{N}$ can be realized in $\mathbb{R}^6$ by the graph of a potential function, §3.4, via the affine immersion $\{f, \xi\}$:*

$$f : \mathcal{N} \to \mathbb{R}^6 : [\theta_i] \mapsto \begin{bmatrix} \theta_i \\ \varphi(\theta) \end{bmatrix}, \quad \xi = \begin{bmatrix} 0 \\ 0 \\ 0 \\ 0 \\ 0 \\ 1 \end{bmatrix}, \tag{4.169}$$

where $\varphi(\theta)$ is the potential function

$$\varphi(\theta) = \log(2\,\pi\,\sqrt{\triangle}) - \triangle\,(\theta_2{}^2\,\theta_3 - \theta_1\,\theta_2\,\theta_4 + \theta_1{}^2\,\theta_5)\,. \quad \square$$

The submanifold consisting of the independent case with zero means and identical standard deviations is represented by the curve:

$$(-\infty, 0) \to \mathbb{R}^3 : (\theta_1) \mapsto (\theta_1, 0, \log(-4\,\pi\,\triangle\,\theta_1)), \quad \xi = (0, 0, 1)$$

$$: (-\frac{1}{2\,\sigma}) \mapsto (-\frac{1}{2\,\sigma}, 0, \log(2\,\pi\,\sigma)), \quad \xi = (0, 0, 1).$$

4.22 Bivariate Log-Gaussian Manifold

The bivariate log-Gaussian distribution has log-Gaussian marginal functions. These arise from the bivariate Gaussian density functions (4.120) for the non-negative random variables $x = log\frac{1}{n}$ and $y = log\frac{1}{m}$, or equivalently, $n = e^{-x}$ and $m = e^{-y}$.

Their probability density functions are given for $n, m > 0$ by:

$$g(n, m) = \frac{1}{2\pi nm\sqrt{\sigma_1\sigma_2 - \sigma_{12}{}^2}} e^{-VW} \tag{4.170}$$

$$V = \frac{1}{2\,(\sigma_1\sigma_2 - \sigma_{12}{}^2)}$$

$$W = \left(\sigma_2(\log(n) + \mu_1)^2 - 2\sigma_{12}\,(\log(n) + \mu_1)\,(\log(m) + \mu_2)\right.$$

$$\left. + \sigma_1(\log(m) + \mu_2)^2\right).$$

Corollary 4.62. *The covariance, correlation coefficient, and marginal density functions are:*

$$Cov_g(n, m) = (e^{\sigma_{12}} - 1)\,e^{-(\mu_1 + \mu_2) + \frac{1}{2}\,(\sigma_1 + \sigma_2)}\,,$$

$$\rho_g(n, m) = \frac{(e^{\sigma_{12}} - 1)}{\sqrt{\sigma_1\,\sigma_2}}\,e^{-(\mu_1 + \mu_2) + \frac{1}{2}\,(\sigma_1 + \sigma_2)}\,,$$

$$g_n(n) = \frac{1}{n\,\sqrt{2\,\pi\,\sigma_1}}\,e^{-\frac{(\log(n) + \mu_1)^2}{2\,\sigma_1}}\,,$$

$$g_m(m) = \frac{1}{m\,\sqrt{2\,\pi\,\sigma_2}}\,e^{-\frac{(\log(m) + \mu_2)^2}{2\,\sigma_2}}\,. \tag{4.171}$$

Note that the variables n and m are independent if and only if $\sigma_{12} = 0$, and the marginal functions are log-Gaussian density functions. $\quad \square$

Directly from the definition of the Fisher metric we deduce:

Proposition 4.63. *The family of bivariate log-Gaussian density functions for random variables* n, m *determines a Riemannian 5-manifold which is an isometric isomorph of the bivariate Gaussian 5-manifold, §3.1.* ☐

5

Neighbourhoods of Poisson Randomness, Independence, and Uniformity

As we have mentioned before, colloquially in applications, it is very common to encounter the usage of 'random' to mean the specific case of a Poisson process §1.1.3 whereas formally in statistics, the term random has a more general meaning: probabilistic, that is dependent on random variables. When we speak of neighbourhoods of randomness we shall mean neighbourhoods of a Poisson process and then the neighbourhoods contain perturbations of the Poisson process. Similarly, we consider processes that are perturbations of a process controlled by a uniform distribution on a finite interval, yielding neighbourhoods of uniformity. The third situation of interest is when we have a bivariate process controlled by independent exponential, gamma or Gaussian distributions; then perturbations are contained in neighbourhoods of independence. These neighbourhoods all have well-defined metric structures determined by information theoretic maximum likelihood methods. This allows trajectories in the space of processes, commonly arising in practice by altering input conditions, to be studied unambiguously with geometric tools and to present a background on which to describe the output features of interest of processes and products during changes.

The results here augment our information geometric measures for distances in smooth spaces of probability density functions, §1.2, by providing explicit geometric representations with distance measures of neighbourhoods for each of these important states of statistical processes:

- (Poisson) randomness, §1.1.3
- independence, §1.3
- uniformity, §1.2.1.

Such results are significant theoretically because they are very general, and practically because they are topological and so therefore stable under perturbations.

K. Arwini, C.T.J. Dodson, *Information Geometry.*
Lecture Notes in Mathematics 1953,
© Springer-Verlag Berlin Heidelberg 2008

5.1 Gamma Manifold $\mathcal{G}$ and Neighbourhoods of Randomness

The univariate gamma density function, §1.4.1, is widely used to model processes involving a continuous positive random variable. It has important uniqueness properties [106]. Its information geometry is known and has been applied recently to represent and metrize departures from Poisson of, for example, the processes that allocate gaps between occurrences of each amino acid along a protein chain within the *Saccharomyces cerevisiae* genome, see Cai et al [34], clustering of galaxies and communications, and cryptographic attacks Dodson [63, 64, 61, 77]. We have made precise the statement that around every Poisson random process there is a neighbourhood of processes subordinate to the gamma distribution, so gamma density functions can approximate any small enough departure from Poisson randomness.

Theorem 5.1. *Every neighbourhood of a Poisson random process contains a neighbourhood of processes subordinate to gamma density functions.*

Proof. Dodson and Matsuzoe [68] have provided an affine immersion in Euclidean $\mathbb{R}^3$ for $\mathcal{G}$, the manifold of gamma density functions, §3.5 with Fisher information metric, §3.1. The coordinates (ν, κ) form a natural coordinate system (cf Amari and Nagaoka [11]) for the gamma manifold $\mathcal{G}$ of density functions (3.15):

$$p(x; \nu, \kappa) = \nu^\kappa \frac{x^{\kappa-1} e^{-x\nu}}{\Gamma(\kappa)}.$$

Then $\mathcal{G}$ can be realized in Euclidean $\mathbb{R}^3$ as the graph of the affine immersion, §3.4, §3.5.5, $\{h, \xi\}$ where ξ is a transversal vector field along h [11, 68]:

$$h : \mathcal{G} \to \mathbb{R}^3 : \begin{pmatrix} \nu \\ \kappa \end{pmatrix} \mapsto \begin{pmatrix} \nu \\ \kappa \\ \log \Gamma(\kappa) - \kappa \log \nu \end{pmatrix}, \quad \xi = \begin{pmatrix} 0 \\ 0 \\ 1 \end{pmatrix}.$$

The submanifold, §2.0.5, of exponential density functions, §1.2.2 is represented by the curve

$$(0, \infty) \to \mathbb{R}^3 : \nu \mapsto \{\nu, 1, \log \frac{1}{\nu}\}$$

and for this curve, a tubular neighbourhood in $\mathbb{R}^3$ such as that bounded by the surface

$$\left\{ \{\nu - \frac{0.6 \cos \theta}{\sqrt{1 + \nu^2}}, 1 - 0.6 \sin \theta, \frac{-0.6 \nu \cos \theta}{\sqrt{1 + \nu^2}} - \log \nu \} \; \theta \in [0, 2\pi) \right\} \quad (5.1)$$

will contain all immersions for small enough perturbations of exponential density functions. In Figure 5.1 this is depicted in natural coordinates ν, κ. The tubular neighbourhood (5.1) intersects with the gamma manifold immersion to yield the required neighbourhood in the manifold of gamma density functions, which completes our proof. $\qquad \square$

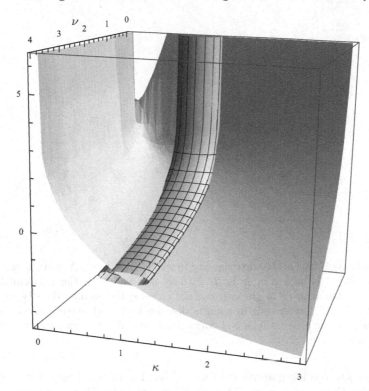

Fig. 5.1. Tubular neighbourhood of univariate Poisson random processes. Affine immersion in natural coordinates ν, κ as a surface in $\mathbb{R}^3$ for the gamma manifold $\mathcal{G}$; the tubular neighbourhood surrounds all exponential density functions—these lie on the curve $\kappa = 1$ in the surface. Since the log-gamma manifold $\mathcal{L}$ is an isometric isomorph of $\mathcal{G}$, this figure represents also a tubular neighbourhood in $\mathbb{R}^3$ of the uniform density function from the log-gamma manifold.

5.2 Log-Gamma Manifold $\mathcal{L}$ and Neighbourhoods of Uniformity

The family of log-gamma density functions discussed in §3.6 has probability density functions for random variable $N \in (0, 1]$ given by equation (3.30):

$$q(N, \nu, \tau) = \frac{\frac{1}{N}^{1-\frac{\tau}{\nu}} \left(\frac{\tau}{\nu}\right)^\tau \left(\log \frac{1}{N}\right)^{\tau-1}}{\Gamma(\tau)} \quad \text{for } \nu > 0 \text{ and } \tau > 0 .$$

This family has the uniform density function, §1.2.1, as a limit

$$\lim_{\nu, \kappa \to 1} q(N, \nu, \kappa) = q(N, 1, 1) = 1 .$$

Figure 5.2 shows some log-gamma density functions around the uniform density function and Figure 3.3 shows part of the continuous family from which these are drawn.

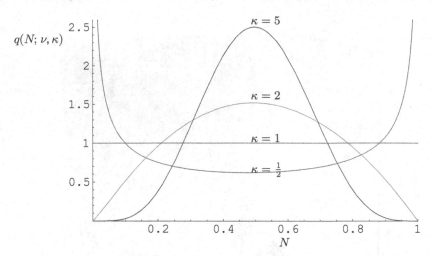

Fig. 5.2. Log-gamma probability density functions $q(N; \nu, \kappa)$, $N \in (0, 1]$, with central mean $\bar{N} = \frac{1}{2}$, and $\kappa = \frac{1}{2}, 1, 2, 5$. The case $\kappa = 1$ is the uniform density function, complementary to the exponential case for the gamma density function. The regime $\kappa < 1$ corresponds in gamma distributions to clustering in an underlying spatial process; conversely, $\kappa > 1$ corresponds to dispersion and greater evenness than random.

From §3.6, the log-gamma manifold $\mathcal{L}$ has information metric (3.36), isometric with the gamma manifold, $\mathcal{G}$ by Proposition 3.9. Hence, from the result of Dodson and Matsuzoe [68] the immersion of $\mathcal{G}$ in $\mathbb{R}^3$, Figure 5.1, represents also the log-gamma manifold $\mathcal{L}$. Then, since the isometry, §2.0.5 sends the exponential density function to the uniform density function on $[0, 1]$, we obtain a general deduction

Theorem 5.2. *Every neighbourhood of the uniform density function contains a neighbourhood of log-gamma density functions.*   □

Equivalently,

Theorem 5.3. *Every neighbourhood of a uniform process contains a neighbourhood of processes subordinate to log-gamma density functions.*   □

5.3 Freund Manifold $\mathcal{F}$ and Neighbourhoods of Independence

Let $\mathcal{F}$ be the manifold of Freund bivariate mixture exponential density functions (4.63), §4.9, so with positive parameters $\alpha_i, \beta_i,$

$$\mathcal{F} \equiv \left\{ f \mid f(x, y; \alpha_1, \beta_1, \alpha_2, \beta_2) = \begin{cases} \alpha_1 \beta_2 e^{-\beta_2 y - (\alpha_1 + \alpha_2 - \beta_2) x} & \text{for } 0 \le x < y \\ \alpha_2 \beta_1 e^{-\beta_1 x - (\alpha_1 + \alpha_2 - \beta_1) y} & \text{for } 0 \le y \le x \end{cases} \right\}.$$

$$(5.2)$$

5.3.1 Freund Submanifold F_2

This is
$$F_2 \subset \mathcal{F} : \alpha_1 = \alpha_2, \ \beta_1 = \beta_2.$$

The densities are of form:

$$f(x, y; \alpha_1, \beta_1) = \begin{cases} \alpha_1 \beta_1 \, e^{-\beta_1 y - (2\alpha_1 - \beta_1)x} \text{ for } 0 < x < y \\ \alpha_1 \beta_1 \, e^{-\beta_1 x - (2\alpha_1 - \beta_1)y} \text{ for } 0 < y < x \end{cases} \tag{5.3}$$

with parameters $\alpha_1, \beta_1 > 0$. The covariance, correlation coefficient and marginal density functions, of X and Y are given by:

$$Cov(X, Y) = \frac{1}{4} \left(\frac{1}{\alpha_1{}^2} - \frac{1}{\beta_1{}^2} \right), \tag{5.4}$$

$$\rho(X, Y) = 1 - \frac{4\alpha_1{}^2}{3\alpha_1{}^2 + \beta_1{}^2}, \tag{5.5}$$

$$f_X(x) = \left(\frac{\alpha_1}{2\alpha_1 - \beta_1} \right) \beta_1 \, e^{-\beta_1 x}$$
$$+ \left(\frac{\alpha_1 - \beta_1}{2\alpha_1 - \beta_1} \right) (2\alpha_1) \, e^{-2\alpha_1 x}, \ x \geq 0, \tag{5.6}$$

$$f_Y(y) = \left(\frac{\alpha_1}{2\alpha_1 - \beta_1} \right) \beta_1 \, e^{-\beta_1 y}$$
$$+ \left(\frac{\alpha_1 - \beta_1}{2\alpha_1 - \beta_1} \right) (2\alpha_1) \, e^{-2\alpha_1 y}, \ y \geq 0. \tag{5.7}$$

Proposition 5.4. F_2 *forms an exponential family, §3.2, with parameters* (α_1, β_1) *and potential function*

$$\psi = -\log(\alpha_1 \beta_1) \tag{5.8}$$

Proposition 5.5. $Cov(X, Y) = \rho(X, Y) = 0$ *if and only if* $\alpha_1 = \beta_1$ *and in this case the density functions are of form*

$$f(x, y; \alpha_1, \alpha_1) = \alpha_1^2 \, e^{\alpha_1 |y - x|} = f_X(x) f_Y(y) \tag{5.9}$$

so that here we do have independence of these exponentials if and only if the covariance is zero.

Neighbourhoods of Independence in F_2

An important practical application of the Freund submanifold F_2 is the representation of a bivariate proces for which the marginals are identical exponentials. The next result is important because it provides topological neighbourhoods of that subspace W in F_2 consisting of the bivariate processes that have zero covariance: we obtain neighbourhoods of independence for Poisson random (ie exponentially distributed §1.2.2) processes.

Theorem 5.6. *Every neighbourhood of an independent pair of identical random processes contains a neighbourhood of bivariate processes subordinate to Freund density functions.*

Proof. Let $\{F_2, g, \nabla^{(1)}, \nabla^{(-1)}\}$ be the manifold F_2 with Fisher metric g and exponential connection $\nabla^{(1)}$. Then F_2 can be realized in Euclidean $\mathbb{R}^3$ by the graph of a potential function, via the affine immersion

$$h : \mathcal{G} \to \mathbb{R}^3 : \begin{pmatrix} \alpha_1 \\ \beta_1 \end{pmatrix} \mapsto \begin{pmatrix} \alpha_1 \\ \beta_1 \\ -\log(\alpha_1\,\beta_1) \end{pmatrix}.$$

In F_2, the submanifold W consisting of the independent case $(\alpha_1 = \beta_1)$ is represented by the curve

$$W : (0, \infty) \to \mathbb{R}^3 : (\alpha_1) \mapsto (\alpha_1, \alpha_1, -2\log\alpha_1). \tag{5.10}$$

which has tubular neighbourhoods of form

$$\left\{ t - \frac{r\cos(\theta)}{\sqrt{1+\frac{1}{t^2}}\sqrt{\frac{1}{t^4}}t^3}, t - \frac{r\sin(\theta)}{\sqrt{\frac{1}{t^4}}t^2}, -\frac{r\cos(\theta)}{\sqrt{1+\frac{1}{t^2}}\sqrt{\frac{1}{t^4}}t^2} - 2\log(t) \right\}$$

This is illustrated in Figure 5.3 which shows an affine embedding of F_2 as a surface in $\mathbb{R}^3$, and an $\mathbb{R}^3$-tubular neighbourhood of W, the curve $\alpha_1 = \beta_1$ in the surface. This curve W represents all bivariate density functions having identical exponential marginals and zero covariance; its tubular neighourhood represents all small enough departures from independence. $\qquad\square$

5.4 Neighbourhoods of Independence for Gaussians

The bivariate Gaussian density function, §4.16 has the form:

$$f(x,y) = \frac{1}{2\pi\sqrt{\sigma_1\sigma_2 - \sigma_{12}{}^2}}\, e^{-\frac{1}{2(\sigma_1\sigma_2 - \sigma_{12}{}^2)}\left(\sigma_2(x-\mu_1)^2 - 2\sigma_{12}(x-\mu_1)(y-\mu_2) + \sigma_1(y-\mu_2)^2\right)},$$

$$\tag{5.11}$$

defined on $-\infty < x, y < \infty$ with parameters $(\mu_1, \mu_2, \sigma_1, \sigma_{12}, \sigma_2)$; where $-\infty < \mu_1, \mu_2 < \infty$, $0 < \sigma_1, \sigma_2 < \infty$ and σ_{12} is the covariance of X and Y.

The marginal functions, of X and Y are univariate Gaussian density functions, §1.2.3:

$$f_X(x, \mu_1, \sigma_1) = \frac{1}{\sqrt{2\pi}\,\sigma_1}\, e^{-\frac{(x-\mu_1)^2}{2\sigma_1}}, \tag{5.12}$$

$$f_Y(y, \mu_2, \sigma_2) = \frac{1}{\sqrt{2\pi}\,\sigma_2}\, e^{-\frac{(y-\mu_2)^2}{2\sigma_2}}. \tag{5.13}$$

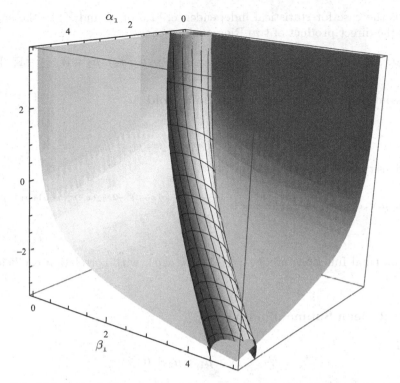

Fig. 5.3. Tubular neighbourhood of independent Poisson random processes. An affine immersion in natural coordinates (α_1, β_1) as a surface in $\mathbb{R}^3$ for the Freund submanifold F_2; the tubular neighbourhood surrounds the curve ($\alpha_1 = \beta_1$ in the surface) consisting of all bivariate density functions having identical exponential marginals and zero covariance.

The correlation coefficient is, §1.3:

$$\rho(X, Y) = \frac{\sigma_{12}}{\sqrt{\sigma_1 \sigma_2}}$$

Since $\sigma_{12}^2 < \sigma_1 \sigma_2$ then $-1 < \rho(X, Y) < 1$; so we do not have the case when Y is a linearly increasing (or decreasing) function of X. The space of bivariate Gaussians becomes a Riemannian 5-manifold $\mathcal{N}$, §4.17, with Fisher information metric, §3.1.

Gaussian Independence Submanifold $\mathcal{N}_1$

This is $\mathcal{N}_1 \subset \mathcal{N} : \sigma_{12} = 0$. The density functions are of form:

$$f(x, y; \mu_1, \mu_2, \sigma_1, \sigma_2) = f_X(x, \mu_1, \sigma_1) . f_Y(y, \mu_2, \sigma_2) \tag{5.14}$$

This is the case for statistical independence, §1.3 of X and Y, so the space $\mathcal{N}_1$ is the direct product of two Riemannian spaces

$$\{f_X(x,\mu_1,\sigma_1),\ \mu_1 \in \mathbb{R},\ \sigma_1 \in \mathbb{R}^+\} \quad \text{and} \quad \{f_Y(y,\mu_2,\sigma_2),\ \mu_2 \in \mathbb{R},\ \sigma_2 \in \mathbb{R}^+\}.$$

Gaussian Identical Marginals Submanifold $\mathcal{N}_2$

This is

$$\mathcal{N}_2 \subset \mathcal{N} : \sigma_1 = \sigma_2 = \sigma,\ \mu_1 = \mu_2 = \mu.$$

The density functions are of form:

$$f(x,y;\mu,\sigma,\sigma_{12}) = \frac{1}{2\pi\sqrt{\sigma^2 - \sigma_{12}{}^2}}\, e^{-\frac{1}{2(\sigma^2-\sigma_{12}{}^2)}\left(\sigma(x-\mu)^2 - 2\sigma_{12}(x-\mu)(y-\mu) + \sigma(y-\mu)^2\right)}.$$

$$(5.15)$$

The marginal functions are $f_X = f_Y \equiv N(\mu,\sigma)$, with correlation coefficient $\rho(X,Y) = \frac{\sigma_{12}}{\sigma}$.

Central Mean Submanifold $\mathcal{N}_3$

This is

$$\mathcal{N}_3 \subset \mathcal{N} : \mu_1 = \mu_2 = 0.$$

The density functions are of form:

$$f(x,y;\sigma_1,\sigma_2,\sigma_{12}) = \frac{1}{2\pi\sqrt{\sigma_1\sigma_2 - \sigma_{12}{}^2}}\, e^{-\frac{1}{2(\sigma_1\sigma_2-\sigma_{12}{}^2)}\left(\sigma_2 x^2 - 2\,\sigma_{12}\,x\,y + \sigma_1 y^2\right)}.$$

$$(5.16)$$

The marginal functions are $f_X(x,0,\sigma_1)$ and $f_Y(y,0,\sigma_2)$, with correlation coefficient $\rho(X,Y) = \frac{\sigma_{12}}{\sqrt{\sigma_1\sigma_2}}$.

By similar methods to that used for Freund density functions, the following results are obtained [15] for the case of Gaussian marginal density functions [15].

Theorem 5.7. *The bivariate Gaussian 5-manifold admits a 2-dimensional submanifold through which can be provided a neighbourhood of independence for bivariate Gaussian processes.*   □

Corollary 5.8. *Via the Central Limit Theorem, by continuity the tubular neighbourhoods of the curve of zero covariance will contain all immersions of limiting bivariate processes sufficiently close to the independence case for all processes with marginals that converge in density function to Gaussians.*   □

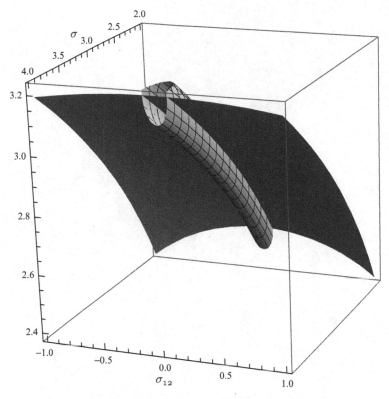

Fig. 5.4. Continuous image, as a surface in $\mathbb{R}^3$ using standard coordinates, of an affine immersion for the bivariate Gaussian density functions with zero means and identical standard deviation σ. The tubular neighbourhood surrounds the curve of independence cases ($\sigma_{12} = 0$) in the surface.

Figure 5.4 shows explicitly a tubular neighbourhood

$$\left\{ t - \frac{r\cos(\theta)}{\sqrt{1 + \frac{1}{t^2}}\sqrt{\frac{1}{t^4}}t^3}, -\frac{r\sin(\theta)}{\sqrt{\frac{1}{t^4}}t^2}, \log\left(\frac{\pi}{\sqrt{t^2}}\right) - \frac{r\cos(\theta)}{\sqrt{1 + \frac{1}{t^2}}\sqrt{\frac{1}{t^4}}t^2} \right\}$$

for the curve of zero covariance processes ($\sigma_{12} = 0$,) in the submanifold of bivariate Gaussian density functions with zero means and identical standard deviation σ.

6

Cosmological Voids and Galactic Clustering

For a general account of large-scale structures in the universe, see, for example, Peebles [162] and Fairall [82], the latter providing a comprehensive atlas. See also Cappi et al [39], Coles [42], Labini et al. [128, 129], Vogeley et al. [208] and van der Weygaert [202] for further recent discussion of large structures. The Las Campanas Redshift Survey was a deep survey, providing some 26,000 data points in a slice out to $500\,h^{-1}Mpc$. Doroshkevich et al. [79] (cf. also Fairall [82] §5.4 and his Figure 5.5) revealed a rich texture of filaments, clusters and voids and suggested that it resembled a composite of three Poisson processes, §1.1.3, consisting of sheets and filaments:

- **Superlarge-scale sheets:**
  60 percent of galaxies, characteristic separation about $77\,h^{-1}Mpc$
- **Rich filaments:**
  20 percent of galaxies, characteristic separation about $30\,h^{-1}Mpc$
- **Sparse filaments:**
  20 percent of galaxies, characteristic separation about $13\,h^{-1}Mpc$.

Most recently, the data from the 2-degree field Galaxy Redshift Survey (2dF-GRS), cf Croton et al. [49, 50] can provide improved statistics of counts in cells and void volumes.

In this chapter we outline some methods whereby such statistical properties may be viewed in an information geometric way. First we look at Poisson processes of extended objects then at coupled processes that relate void and density statistics, somewhat heuristically but intended to reveal the way the information geometry can be used to represent such near-Poisson spatial processes. The applications to cosmology here are based on the publications [63, 62, 64, 65].

K. Arwini, C.T.J. Dodson, *Information Geometry.*
Lecture Notes in Mathematics 1953,
© Springer-Verlag Berlin Heidelberg 2008

6.1 Spatial Stochastic Processes

There is a body of theory that provides the means to calculate the variance of density in planar Poisson processes of arbitrary rectangular elements, using arbitrary finite cells of inspection [58]. We provide details of this method. In principle, it may be used to interpret the survey data by finding a best fit for filament and sheet sizes—and perhaps their size distributions—and for detecting departures from Poisson processes. For analyses using 'counts in cells' in other surveys, see Efstathiou [81] and Szapudi et al. [197]. A hierachy of N-point correlation functions needed to represent clustering of galaxies in a complete sense was derived by White [210] and he provided explicit formulae, including their continuous limit.

The basic random model, §1.1.3, for spatial stochastic processes representing the distribution of galaxies in space is that arising from a Poisson process of mean density $\overline{N}$ galaxies per unit volume in a large box—the region covered by the catalogue data to be studied. Then, the probability of finding exactly m galaxies in a given sample region of volume V, is

$$P_m = \frac{(\overline{N}V)^m}{m!} e^{-\overline{N}V} \quad \text{for } m = 0, 1, 2, \dots \tag{6.1}$$

The Poisson probability distribution (6.1) has mean equal to its variance, $\overline{m} = Var(m) = \overline{N}V$, and this is used as a reference case for comparison with observational data. Complete sampling of the available space using cells of volume V will reveal clustering if the variance of local density over the cells exceeds $\overline{N}$. Moreover, the covariance, §1.3, of density between cells encodes correlation information about the spatial process being observed. The correlation function, §1.3 cf. Peebles [162], is the ratio of the covariance of density of galaxies in cells separated by distance r, divided by the variance of density for the chosen cells

$$\xi(r) = \frac{Cov(r)}{Cov(0)} = \frac{< m(r_0)m(r_0 + r) >}{\overline{m}^2} - 1 \tag{6.2}$$

In the absence of correlation, we expect $\xi(r)$ to decay rapidly to zero with the separation distance r. In practice, we find that, for r not too large, $\xi(r)$ resembles an exponential decay $e^{-r/d}$ with d of the order of the smallest dimension of the characteristic structural feature.

Another way to detect clustering is to use an increasing sequence $V_0 < V_1 < V_2 < \dots$ of sampling cell volumes; in the absence of correlation we expect that the variance of numbers found using these cells will be the average numbers of galaxies in them, $\overline{N}V_0 < \overline{N}V_1 < \overline{N}V_2 < \dots$, respectively. Suppose that a sampling cell of volume V_1 contains exactly k sampling cells of volume V_0, then Var_1, the variance of density of galaxies found using a cell of volume V_1, is expressible as

$$Var_1 = \frac{1}{k} Var_0 + \frac{k-1}{k} Cov_{0,1} \tag{6.3}$$

where Var_0 is the variance found using the smaller cells and $Cov_{0,1}$ is the average covariance among the smaller cells in the larger cells. As $k \to \infty$, so $\frac{1}{k} Var_0 \to 0$ and Var_1 tends to the mean covariance among points inside the V_1 cells. Now, the mean covariance among points inside V_1 cells is the expectation of the covariance between pairs of points separated by distance r, taken over all possible values for r inside a V_1 cell. Explicitly

$$Var_1 = \int_0^D Cov(r)\, b(r)\, dr \qquad (6.4)$$

where b is the probability density function, §1.2, for the distance r between pairs of points chosen independently and at random in a V_1 cell and D is the diameter or maximum dimension of such a cell.

Ghosh [91] gave examples of different functions b and some analytic results are known for covariance functions arising from spatial point processes—by representing the clusters as 'smoothed out' lumps of matter—see [58] for the case of arbitrary rectangles in planar situations. It is convenient to normalize equation (6.4) by division through by $Cov(0) = Var(0)$, which is known for a Poisson process; this gives the 'between cell' variance for complete sampling using v_1 cells. Then we obtain

$$Var_1 - Var(0) \int_0^D a(r)\, b(r)\, dr \qquad (6.5)$$

where a is the point autocorrelation function for the particular type of lumps of matter being used to represent a cluster of galaxies; typically, $a(r) \approx e^{-r/d}$, for 'small' r and d is of the order of the smallest dimension of a cluster. Since it involves finite cells, Var_1 is in principle measurable so (6.5) can be compared with observational data, once the type of sampling cell and representative extended matter object are chosen. We return to this in the sequel and provide examples for a two dimensional model. From Labini et al. [129], we note that experimentally for clusters of galaxies

$$\xi(r) \approx \left(\frac{25}{r}\right)^{1.7}, \quad \text{with } r \text{ in } h^{-1}Mpc \qquad (6.6)$$

which for $2 < r < 10$ resembles $e^{-r/d}$ for suitable d near 1.8.

6.2 Galactic Cluster Spatial Processes

From the atlases shown in Fairall [82] and surveys discussed by Labini et al. [129], one may estimate in a planar slice a representative galactic 'wall' filament thickness of about $5\,h^{-1}Mpc$ and a wall 'thickness-to-length' aspect ratio A in the range $10 < A < 50$. Then, in order to represent galactic clustering as a Poisson process of wall filaments of length λ and width ω, we need the point autocorrelation function a for such filaments. In two dimensions it was shown in [58] that the function a is given in three parts for rectangles of length λ and width ω by the following.

For $0 < r \leq \omega$

$$a_1(r) = 1 - \frac{2}{\pi}\left(\frac{r}{\lambda} + \frac{r}{\omega} - \frac{r^2}{2\omega\lambda}\right).$$ (6.7)

For $\omega < r \leq \lambda$

$$a_2(r) = \frac{2}{\pi}\left(\arcsin\frac{\omega}{r} - \frac{\omega}{2\lambda} - \frac{r}{\omega} + \sqrt{(\frac{r^2}{\omega^2} - 1)}\right).$$ (6.8)

For $\lambda < r \leq \sqrt{(\lambda^2 + \omega^2)}$

$$a_3(r) = \frac{2}{\pi}\left(\arcsin\frac{\omega}{r} - \arccos\frac{\lambda}{r} - \frac{\omega}{2\lambda} - \frac{\lambda}{2\omega} - \frac{r^2}{2\lambda\omega} + \sqrt{(\frac{r^2}{\lambda^2} - 1)} + \sqrt{(\frac{r^2}{\omega^2} - 1)}\right)$$ (6.9)

For small r, as expected even in three dimensions, $a(r) \approx e^{-2r/\pi\omega}$.

Note that for Poisson random *squares* of side length s, $\omega = \lambda = s$ and we have only two cases:

For $0 < r \leq s$

$$a_1(r) = 1 - \frac{2}{\pi}\left(\frac{2r}{s} - \frac{r^2}{2s^2}\right).$$ (6.10)

For $s < r \leq \sqrt{(2s^2)}$

$$a_3(r) = \frac{2}{\pi}\left(\arcsin\frac{s}{r} - \arccos\frac{s}{r} - 1 - \frac{r^2}{2s^2} + 2\sqrt{(\frac{r^2}{s^2} - 1)}\right)$$ (6.11)

This case may be used to represent in two dimensions clusters of galaxies as a Poisson process of smoothed out squares of matter—the sheet-like elements of Doroshkevich et al. [79].

Next we need b, the probability density function for the distance r between pairs of points chosen independently and at random in a suitable inspection cell. From [91], for square inspection cells of side length x,

for $0 \leq r \leq x$

$$b(r, x) = \frac{4r}{x^4}\left(\frac{\pi x^2}{2} - 2rx + \frac{r^2}{2}\right).$$ (6.12)

For $x \leq r \leq D = \sqrt{2}x$

$$b(r, x) = \frac{4r}{x^4}\left(x^2\left(\arcsin\left(\frac{x}{r}\right) - \arccos\left(\frac{x}{r}\right)\right) + 2x\sqrt{(r^2 - x^2)} - \frac{1}{2}(r^2 + 2x^2)\right).$$ (6.13)

A plot of this function is given in Figure 6.1.

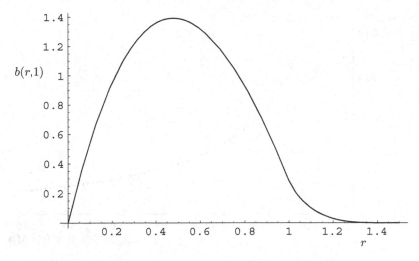

Fig. 6.1. Probability density function $b(r, 1)$ for the distance r between two points chosen independently and at random in a unit square.

Ghosh [91] gave also the form of b for other types of cells; for arbitrary rectangular cells those expressions can be found in [58]. It is of interest to note that for small values of r, so $r \ll D$, the formulae for plane convex cells of area A and perimeter P all reduce to

$$b(r, A, P) = \frac{2\pi r}{A} - \frac{2Pr^2}{A^2}$$

which would be appropriate to use when the filaments are short compared with the dimensions of the cell. The filaments are supposed to be placed independently by a Poisson process in the plane and hence their variance contributions can be summed in a cell to give the variance for zonal averages—that is the between cell variance for complete sampling schemes. So, the variance between cells is the expectation of the covariance function, taken over all possible pairs of points in the cell, as given in (6.5). We re-write this for square cells of side length x as

$$Var(x) = Var(0) \int_0^{\sqrt{2}x} a(r) \, b(r, x) \, dr \qquad (6.14)$$

Using this equation, in Figure 6.2 we plot $Var(x)/Var(0)$ against inspection cell size $x \, h^{-1} Mpc$ for the case of filaments with width $\omega = 5 \, h^{-1} Mpc$ and length $\lambda = 100 \, h^{-1} Mpc$. Note that $Var(x)$ is expressible also as an integral of the (point) power spectrum over wavelength interval $[x, \infty)$ and that Landy et al. [133] detected evidence of a strong peak at $100 \, h^{-1} Mpc$ in the power spectrum of the Las Campanas Redshift Survey, cf. also Lin et al. [136].

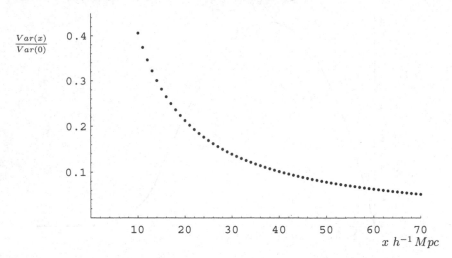

Fig. 6.2. Relative between cell variance for a planar Poisson process of filaments with width $\omega = 5 \, h^{-1} Mpc$ and length $\lambda = 100 \, h^{-1} Mpc$ for complete sampling using square cells of side length $x \, h^{-1} Mpc$ from equation (6.14).

These spatial statistical models may be used in two distinct ways. If observational data is available for $Var(x)$ for a range of x values, for example by digitizing catalogue data on 2-dimensional slices, then attempts may be made to find the best fit for λ and ω. That would give statistical estimates of filament sizes on the presumption that the underlying process of filaments is Poisson. On the other hand, given observed $Var^{obs}(x)$ for a range of x, the variance ratio of this to (6.14)

$$VR(x) = \frac{Var^{obs}(x)}{Var(x)} \qquad (6.15)$$

will be an increasing function of x if there is a tendency of the filaments to cluster.

According to the Las Campanas Redshift Survey, some 40 percent of galaxies out to $500 \, h^{-1} Mpc$ are contained in filaments and the remainder in 'sheets', which we may interpret perhaps as rectangles and squares, respectively, both apparently following a Poisson process. Such a composite spatial structure may be represented easily with our model, if the individual Poisson processes are independent; then the net variance for any choice of inspection cells is the weighted sum of the variances for the individual processes. So the between cell variance (6.14) becomes a weighted sum of integrals, using the appropriate a functions for the constituent representative lumps of matter—perhaps squares for sheets and two kinds of rectangles for filaments, dense and light.

6.3 Cosmological Voids

A number of recent studies have estimated the inter-galactic void probability function and investigated its departure from various statistical models. We study a family of parametric statistical models based on gamma distributions, which do give realistic descriptions for other stochastic porous media. Gamma distributions, §1.4.1, contain as a special case the exponential distributions, §1.2.2, which correspond to the 'random' void size probability arising from Poisson processes, §1.1.3. The space of parameters is a surface with a natural Riemannian metric structure, §2.0.5, §3.1. This surface contains the Poisson processes as an isometric embedding, §2.0.5, and our recent theorem [14] cf. §5.1, shows that it contains neighbourhoods of all departures from Poisson randomness. The method provides thereby a geometric setting for quantifying such departures and on which may be formulated cosmological evolutionary dynamics for galactic clustering and for the concomitant development of the void size distribution.

Several years ago, the second author presented an information geometric approach to modelling a space of perturbations of the Poisson random state for galactic clustering and cosmological void statistics [62]. Here we update somewhat and draw attention to this approach as a possible contribution to the interpretation of the data from the 2-degree field Galaxy Redshift Survey (2dFGRS), cf Croton et al. [49, 50]. The new 2dFGRS data offers the possibility of more detailed investigation of this approach than was possible when it was originally suggested [62, 63, 64] and some parameter estimations are given.

The classical random model is that arising from a Poisson process (cf. §1.1.3) of mean density $\overline{N}$ galaxies per unit volume in a large box. Then, in a region of volume V, the probability of finding exactly m galaxies is given by equation (6.1). So the probability that the given region is devoid of galaxies is $P_0 = e^{-\overline{N}V}$. It follows that the probability density function for the continuous random variable V in the Poisson case is

$$p_{random}(V) = \overline{N}\, e^{-\overline{N}V} \qquad (6.16)$$

In practice of course, measurements will depend on algorithms that specify threshold values for density ranges of galaxies in cells and the lowest range will represent the 'underdense' regions which include the voids; Benson et al. [22] discuss this.

A hierarchy of N-point correlation functions needed to represent clustering of galaxies in a complete sense was devised by White [210] and he provided explicit formulae, including their continuous limit. In particular, he made a detailed study of the probability that a sphere of radius R is empty and showed that formally it is symmetrically dependent on the whole hierarchy of correlation functions. However, White concentrated his applications on the case when the underlying galaxy distribution was a Poisson process, the starting point for the present approach which is concerned with geometrizing the parameter space of *departures* from a Poisson process. Croton et al. [50] found that the negative binomial model for galaxy clustering gave a very good approximation to the 2dFGRS, pointing out that this model is a discrete version of the gamma distribution, §1.4.1.

6.4 Modelling Statistics of Cosmological Void Sizes

For a general account of large-scale structures in the universe, see Fairall [82]. Kauffmann and Fairall [114] developed a catalogue search algorithm for larger nearly spherical regions devoid of bright galaxies and obtained a spectrum for radii of significant voids. This indicated a peak radius near $4\ h^{-1}Mpc$, a long tail stretching at least to $32\ h^{-1}Mpc$, and is compatible with the recent extrapolation models of Baccigalupi et al [18] which yield an upper bound on void radii of about $50\ h^{-1}Mpc$. This data has of course omitted the expected very large numerical contribution of smaller voids. More recent work, notably of Croton et al. [50] provide much larger samples with improved estimates of void size statistics and Benson et al. [22] gave a theoretical analysis in anticipation of the 2dFGRS survey data, including the evaluation of the void and underdense probability functions. Hoyle and Vogeley [104] provided detailed results for the statistics of voids larger than $10\ h^{-1}Mpc$ in the 2dFGRS survey data; they concluded that such voids constitute some 40% of the universe and have a mean radius of about $15\ h^{-1}Mpc$.

The count density $N(V)$ of galaxies observed in zones using a range of sampling schemes each with a fixed zone volume V results in a decreasing variance $Var(N(V))$ of count density with increasing zone size, roughly of the form

$$Var(N(V)) \approx V(0)\, e^{-V/V_k} \ \text{ as } V \to 0 \qquad (6.17)$$

where V_k is some characteristic scaling parameter. This monotonic decay of variance with zone size is a natural consequence of the monotonic decay of the covariance function, roughly isotropically and of the form

$$Cov(r) \approx e^{-r/r_k} \ \text{ as } r \to 0 \qquad (6.18)$$

where r_k is some characteristic scaling parameter of the order of magnitude of the diameter of filament structures; this was discussed in [63]. Then

$$Var(N(V)) \approx \int_0^\infty Cov(r)\, b(r)\, dr \qquad (6.19)$$

where $b(r)$ is the probability density of finding two points separated by distance r independently and at random in a zone of volume V. The power spectrum using, say, cubical cells of side lengths R is given by the family of integrals

$$Pow(N(R)) \approx \int_R^\infty Cov(r)\, b(r)\, dr. \tag{6.20}$$

Fairall [82] (page 124) reported a value $\sigma^2 = 0.25$ for the ratio of variance $Var(N(1))$ to mean squared $\overline{N}^2$ for counts of galaxies in cubical cells of unit side length. In other words, the coefficient of variation, §1.2, for sampling with cells of unit volume is

$$cv(N(1)) = \frac{\sqrt{Var(N(1))}}{\overline{N}} = 0.5 \tag{6.21}$$

and this is dimensionless.

We choose a family of parametric statistical models for void volumes that includes the Poisson random model (6.16) as a special case. There are of course many such families, but we take one that our recent theorem [14], cf. §5.1, has shown contains neighbourhoods of all departures from Poisson randomness and it has been successful in modelling void size distributions in terrestrial stochastic porous media with similar departures from randomness [72, 73]. Also, the complementary logarithmic version has been used in the representation of clustering of galaxies [63, 64].

The family of gamma distributions has event space $\Omega = \mathbb{R}^+$, parameters $\mu, \alpha \in \mathbb{R}^+$ and probability density functions given by, Figure 6.3,

$$f(V; \mu, \alpha) = \left(\frac{\alpha}{\mu}\right)^\alpha \frac{V^{\alpha-1}}{\Gamma(\alpha)} e^{-V\alpha/\mu} \tag{6.22}$$

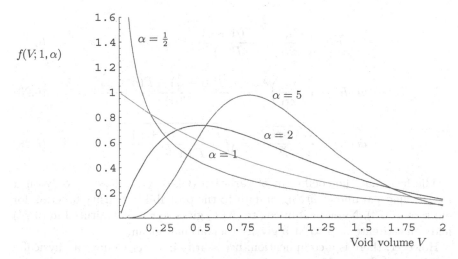

Fig. 6.3. Gamma probability density functions, $f(V; \mu, \alpha)$, from (6.22) representing the inter-galactic void volumes V with unit mean $\mu = 1$, and $\alpha = \frac{1}{2}$, 1, 2, 5. The case $\alpha = 1$ corresponds to the 'random' case from an underlying Poisson process of galaxies; $\alpha < 1$ corresponds to clustering and $\alpha > 1$ corresponds to dispersion.

Then $\overline{V} = \mu$ and $Var(V) = \mu^2/\alpha$ and we see that μ controls the mean of the distribution while the spread and shape is controlled by $1/\alpha$, the square of the coefficient of variation, §1.2.

The special case $\alpha = 1$ corresponds to the situation when V represents the Poisson process in (6.16). The family of gamma distributions can model a range of statistical processes corresponding to non-independent 'clumped' events, for $\alpha < 1$, and dispersed events, for $\alpha > 1$, as well as the Poisson random case $\alpha = 1$ (cf. [14, 72, 73]). Thus, if we think of this range of processes as corresponding to the possible distributions of centroids of extended objects such as galaxies that are initially distributed according to a Poisson process with $\alpha = 1$, then the three possibilities are:

Chaotic—Poisson random structure: no interactions among constituents, $\alpha = 1$;

Clustered structure: mutually attractive type interactions, $\alpha < 1$;

Dispersed structure: mutually repulsive type interactions, $\alpha > 1$.

For our gamma-based void model we consider the radius R of a spherical void with volume $V = \frac{4}{3}\pi R^3$ having distribution (6.22). Then the probability density function for R is given by

$$p(R; \mu, \alpha) = \frac{4\pi R^2}{\Gamma(\alpha)} \left(\frac{\alpha}{\mu}\right)^\alpha \left(\frac{4\pi R^3}{3}\right)^{\alpha-1} e^{\frac{-4\pi R^3 \alpha}{3\mu}} \qquad (6.23)$$

The mean $\overline{R}$, variance $Var(R)$ and coefficient of variation $cv(R)$ of R are given, respectively, by

$$\overline{R} = \left(\frac{3\mu}{4\pi\alpha}\right)^{\frac{1}{3}} \frac{\Gamma(\alpha + \frac{1}{3})}{\Gamma(\alpha)} \qquad (6.24)$$

$$Var(R) = \left(\frac{3\mu}{4\pi\alpha}\right)^{\frac{2}{3}} \frac{\Gamma(\alpha)\,\Gamma(\alpha + \frac{2}{3}) - \Gamma(\alpha + \frac{1}{3})^2}{\Gamma(\alpha)^2} \qquad (6.25)$$

$$cv(R) = \frac{\sqrt{Var(R)}}{\overline{R}} = \sqrt{\frac{\Gamma(\alpha)\,\Gamma\left(\alpha + \frac{2}{3}\right)}{\Gamma\left(\alpha + \frac{1}{3}\right)} - 1} \qquad (6.26)$$

The fact that the coefficient of variation (6.26), §1.2, depends only on α gives a rapid parameter fitting of data to the probability density function for void radii (6.23). Numerical fitting to (6.26) gives α; this substituted in (6.24) yields an estimate of μ to fit a given observational mean.

However, there is a complication: necessarily in order to have a physically meaningful definition for voids, observational measurements introduce a minimum threshold size for voids. For example, Hoyle and Vogeley [104] used algorithms to obtain statistics on 2dFGRS voids with radius $R > R_{min} = 10\,h^{-1}Mpc$; for voids above this threshold they found their mean size is about

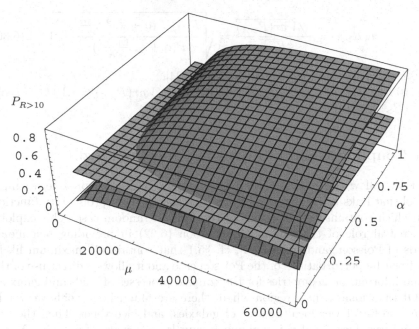

Fig. 6.4. Probability that a void will have radius $R > 10\,h^{-1}Mpc$ as a function of parameters μ, α from equation (6.27). The range $\alpha < 1$ corresponds to clustering regimes. The plane at level $P_{R>10} = 0.4$ corresponds to the fraction 40% of the universe filled by voids, as reported by Hoyle and Vogeley [104].

$15\,h^{-1}Mpc$ with a variance of about 8.1. This of course is not directly comparable with the above distribution for R in equation (6.23) since the latter has domain $R > 0$. Now, from (6.23), the probability that a void has radius $R > A$ is, Figure 6.4,

$$P_A = \frac{\Gamma(\alpha, \frac{4A^3\pi\alpha}{3\mu})}{\Gamma(\alpha)} \tag{6.27}$$

where

$$\Gamma(\alpha, A) = \int_A^\infty t^{\alpha-1}\, e^{-t}\, dt$$

is the incomplete gamma function, with $\Gamma(\alpha) = \Gamma(\alpha, 0)$.

Hence the mean, variance and coefficient of variation, §1.2, for the void distribution with $R > A$ become:

$$\overline{R_{>A}} = \left(\frac{3\mu}{4\pi\alpha}\right)^{\frac{1}{3}} \frac{\Gamma(\alpha+\frac{1}{3}, \frac{4A^3\pi\alpha}{3\mu})}{\Gamma(\alpha, \frac{4A^3\pi\alpha}{3\mu})} \tag{6.28}$$

$$Var(R_{>A}) = \left(\frac{3\mu}{4\pi\alpha}\right)^{\frac{2}{3}} \frac{\Gamma(\alpha)\,\Gamma(\alpha+\frac{2}{3}, \frac{4A^3\pi\alpha}{3\mu}) - \Gamma(\alpha+\frac{1}{3}, \frac{4A^3\pi\alpha}{3\mu})^2}{\Gamma(\alpha, \frac{4A^3\pi\alpha}{3\mu})\,\Gamma(\alpha)} \tag{6.29}$$

$$cv(R_{>A}) = \frac{\sqrt{Var(R_{>A})}}{R_{>A}} = \sqrt{\frac{\Gamma(\alpha)\,\Gamma(\alpha + \frac{2}{3}, \frac{4A^3\pi\alpha}{3\mu})}{\Gamma(\alpha + \frac{1}{3}, \frac{4A^3\pi\alpha}{3\mu})^2} - 1} \quad (6.30)$$

Summarizing from Hoyle and Vogeley [104]:
$A = 10\ h^{-1}Mpc$, $P_A \approx 0.4$, $\overline{R_{>A}} \approx 15\ h^{-1}Mpc$, $Var(R_{>A}) \approx 8.1\ (h^{-1}Mpc)^2$
so $cv(R_{>A}) \approx 0.19$.

6.5 Coupling Galaxy Clustering and Void Sizes

Next we follow the methodology introduced in [63, 64] to provide a simple model that links the number counts in cells and the void probability function and which contains perturbations of the Poisson random case. This exploits the central role of the gamma distribution (6.22) in providing neighbour-hoods of Poisson randomness [14], cf. §5.1 that contain all maximum likeli-hood nearby perturbations of the Poisson case and it allows a direct use of the linked information geometries for the coupled processes of voids and galaxies.

Observationally, in a region where there are found large voids we would expect to find lower local density of galaxies, and vice versa. Then the two random variables, local void volume V and local density of galaxies N, are presumably inversely related. Many choices are possible; we take a simple func-tional form using an exponential and normalize the local density of galaxies to be bounded above by 1. Denoting the random variable representing this normalized local density by N, we put:

$$N(V) = e^{-V} \quad (6.31)$$

This model was explored in [63, 64] and it is easy to show that the probability density function for N is given by the log gamma distribution from equation (3.30) with $\nu = \frac{\alpha}{\mu}$ and $\tau = \alpha$:

$$g(N; \mu, \alpha) = \left(\frac{\alpha}{\mu}\right)^\alpha \frac{N^{\alpha/\mu-1}}{\Gamma(\alpha)} |\log N|^{1-\alpha} \quad (6.32)$$

This distribution (cf. Figure 3.3) for local galactic number density has mean $\overline{N}$, variance $Var(N)$ and coefficient of variation, §1.2, $cv(N) = \sqrt{Var(N)}/\overline{N}$ given by

$$\overline{N} = \left(\frac{\alpha}{\alpha + \mu}\right)^\alpha \quad (6.33)$$

$$Var(N) = \left(\frac{\alpha}{\alpha + 2\mu}\right)^\alpha - \left(\frac{\alpha}{\alpha + \mu}\right)^{2\alpha} \quad (6.34)$$

$$cv(N) = \frac{\sqrt{Var(N)}}{\overline{N}} = \sqrt{\left(\frac{\alpha}{\alpha + \mu}\right)^{-2\alpha} \left(\frac{\alpha}{\alpha + 2\mu}\right)^\alpha - 1}. \quad (6.35)$$

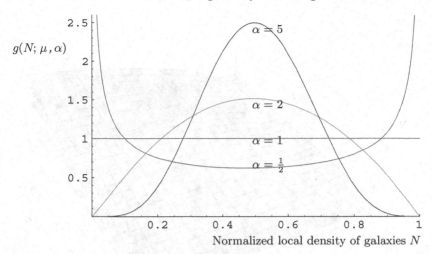

Fig. 6.5. Log-gamma probability density functions $g(N; \mu, \alpha)$, from (6.32) representing the normalized local density of galaxies, $N \in (0, 1]$, with central mean $\overline{N} = \frac{1}{2}$, and $\alpha = \frac{1}{2}, 1, 2, 5$. The case $\alpha = 1$ is the uniform distribution. The cases $\alpha < 1$ correspond to clustering in the underlying spatial process of galaxies, so there are probability density peaks of high and low density; conversely, $\alpha > 1$ corresponds to dispersion.

Figure 6.5 shows the distribution (6.32) for mean normalized density $\overline{N} = \frac{1}{2}$ and $\alpha = \frac{1}{2}, 1, 2, 5$. Note that as $\alpha \to 1$ so the distribution (6.32) tends to the uniform distribution. For $\alpha < 1$ we have clustering in the underlying process, with the result that the population has high and low density peaks. Other choices of functional relationship between local void volume and local density of galaxies would lead to different distributions; for example, $N(V) = e^{-V^k}$ for $k = 2, 3, \ldots$, would serve. However, the persisting qualitative observational feature that would discriminate among the parameters is the prominence of a central modal value—indicating a smoothed or dispersed structure, or the prominence of high and low density peaks—indicating clustering.

Using the reported value $cv(N(1)) = 0.5$ from Fairall [82] (page 124) for cubical volumes with side length $R = 1\ h^{-1}Mpc$, the curve so defined in the parameter space for the log gamma distributions (6.32), has maximum clustering for $\alpha \approx 0.6, \mu \approx 0.72$.

From the 2-degree field Galaxy Redshift Survey (2dFGRS), Croton et al. [49] in their Figure 2 reported the decay of normalised variance, $\overline{\xi_2} = cv(N(R))^2$ with scale radius R and the associated departure from Poisson randomness. in the form

$$\chi = -\frac{\log_{10} P_0(R)}{\overline{N}},$$

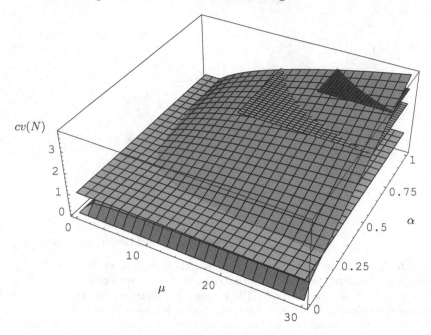

Fig. 6.6. Coefficient of variation of counts in cells N for log gamma distribution equation (6.32). The range $\alpha < 1$ corresponds to clustering regimes. The three planes show the levels $cv(N) = 1, \sqrt{6}, \sqrt{10}$ as reported by Fairall [82] and Croton et al. [49, 50].

where $P_0(R)$ is the probability of finding zero galaxies in a spherical region of radius R when the mean number is $\overline{N}$. From Figure 2 in [49] we see that, for the data of the Volume Limited Catalogue with magnitude range -20 to -21 and $\overline{N}(1) = 1.46 : cv(N(1))^2 \approx 6$ and $\chi \approx 0.9$ at $R \approx 1$ also $cv(N(7))^2 \approx 1$ and $\chi \approx 0.4$ at $R \approx 7$.

Croton et al. [50] in Table 1 reported $\overline{N}$ values for cubical volumes with side length $R = 1 \ h^{-1}Mpc$ in the range $0.11 \leq \overline{N} \leq 11$. From Figure 3 in that paper we see that, at the scale $R = 1 \ h^{-1}Mpc$, $\log_{10} \overline{\xi}_2 \approx 1$ which gives a coefficient of variation $cv(N(1)) \approx \sqrt{10}$.

The above-mentioned observations $cv(N) = 1, \sqrt{6}, \sqrt{10}$ are shown as horizontal planes in Figure 6.6, a plot of the coefficient of variation, §1.2, for the number counts in cells from the log gamma family of distributions equation (6.32). The range $\alpha < 1$ corresponds to clustering regimes.

6.6 Representation of Cosmic Evolution

Theoretical models for the evolution of galactic clustering and consequential void statistics can be represented crudely as curves on the gamma manifold of parameters with the information metric (3.19). So we have a means of interpreting the parameter changes with time through an evolving process subordinate to the log-gamma distribution (6.32). The coupling with the void probability function controlled by the gamma distribution (6.22) allows the corresponding void evolution to be represented. It is of course very unlikely that this simple model is realistic in all respects but it may provide a convenient model with qualitatively stable properties that are realistic. Moreover, given a different family of distributions the necessary information geometry can be computed for the representation of evolutionary processes.

The entropy for the gamma probability density function (6.22) was given in equation (1.57) and it was shown in Figure 1.4. At unit mean, the maximum entropy (or maximum uncertainty) occurs at $\alpha = 1$, which is the Poisson random case, and then $S_f(\mu, 1) = 1 + \log \mu$.

The 'maximum likelihood' estimates $\hat{\mu}, \hat{\alpha}$ of μ, α can be expressed in terms of the mean and mean logarithm of a set of independent observations $X = \{X_1, X_2, \ldots, X_n\}$. These estimates are obtained in terms of the properties of X by maximizing the 'log-likelihood' function

$$l_X(\mu, \alpha) = \log lik_X(\mu, \alpha) = \log \left(\prod_{i=1}^{n} p(X_i; \mu, \alpha) \right)$$

with the following result

$$\hat{\mu} = \bar{X} = \frac{1}{n} \sum_{i=1}^{n} X_i \qquad (6.36)$$

$$\log \hat{\alpha} - \psi(\hat{\alpha}) = \overline{\log X} - \log \bar{X} \qquad (6.37)$$

where $\overline{\log X} = \frac{1}{n} \sum_{i=1}^{n} \log X_i$ and $\psi(\alpha) = \frac{\Gamma'(\alpha)}{\Gamma(\alpha)}$ is the digamma function, the logarithmic derivative of the gamma function

The Riemannian information metric on the 2-dimensional parameter space

$$\mathcal{G} = \{(\mu, \alpha) \in \mathbb{R}^+ \times \mathbb{R}^+\}$$

has arc length function given by

$$ds^2 = \frac{\alpha}{\mu^2} d\mu^2 + \left(\psi'(\alpha) - \frac{1}{\alpha} \right) d\alpha^2 \quad \text{for } \mu, \alpha \in \mathbb{R}^+. \qquad (6.38)$$

Moreover, as we have seen above in Proposition 3.9, the manifold of log-gamma density functions has the same information metric as the gamma manifold.

The 1-dimensional subspace parametrized by $\alpha = 1$ corresponds to the available Poisson processes. A path through the parameter space $\mathcal{G}$ of gamma models determines a curve

$$c : [a, b] \rightarrow \mathcal{G} : t \mapsto (c_1(t), c_2(t)) \qquad (6.39)$$

with tangent vector $\dot{c}(t) = (\dot{c}_1(t), \dot{c}_2(t))$ and norm $||\dot{c}||$ given via (6.38) by

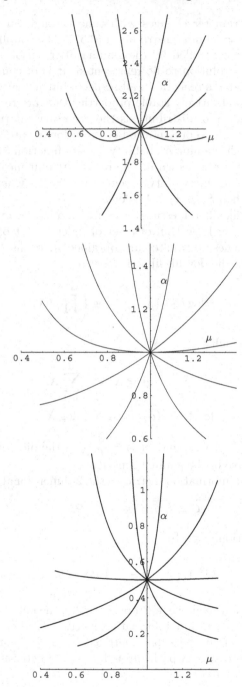

Fig. 6.7. Geodesic sprays radiating from the points with unit mean $\mu = 1$, and $\alpha = 0.5,\ 1,\ 2$. The case $\alpha = 1$ corresponds to an exponential distribution from an underlying Poisson process of galaxies; $\alpha < 1$ corresponds to clustering and α increasing above 1 corresponds to dispersion and greater uniformity.

$$||\dot{c}(t)||^2 = \frac{c_2(t)}{c_1(t)^2}\,\dot{c}_1(t)^2 + \left(\psi'(c_2(t)) - \frac{1}{c_2(t)}\right)\dot{c}_2(t)^2. \qquad (6.40)$$

The information length of the curve is

$$L_c(a,b) = \int_a^b ||\dot{c}(t)||\,dt \qquad (6.41)$$

and the curve corresponding to an underlying Poisson process has $c(t) = (t,1)$, so $t = \mu$ and $\alpha = 1 = constant$, and the information length is $\log\frac{b}{a}$.

Presumably, at early times the universe was dense, so $\overline{N} \to 1$ hence μ was small, and essentially chaotic and Poisson so $\alpha \to 1$; in the current epoch we observe clustering of galaxies so $\alpha < 1$. What is missing is some cosmological physics to prescribe the dynamics of progress from early times to the present. In the absence of such physical insight, there is a distinguished candidate trajectory from an information geometric viewpoint: that of a geodesic, satisfying the condition (2.7). Some examples of sprays of geodesics are drawn in Figure 6.7 from numerical solutions to the equation $\nabla_{\dot{c}}\dot{c} = 0$.

In Figure 6.8 we show two sets of three geodesics, passing through $(\mu,\alpha) = (1,1)$ and through $(\mu,\alpha) = (0.5,1)$ respectively; for each set we have maximally extended the geodesics in the parameter space $\mathbb{R}^+ \times \mathbb{R}^+$. In Figure 6.9 the horizontal geodesic through $(\mu,\alpha) = (0.1,1)$ begins in a Poisson random state **A** at high mean density, $\overline{N} \approx 1$ from equation (6.33), and evolves with decreasing α to lower mean density **B** through increasingly

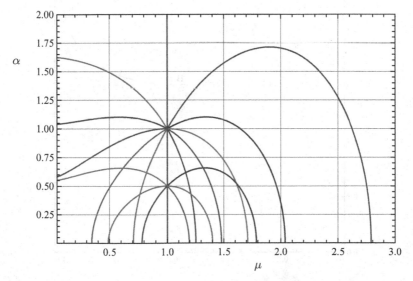

Fig. 6.8. Examples of geodesics passing through the points $(\mu,\alpha) = (1,1)$ and $(\mu,\alpha) = (0.5,1)$ then maximally extended to the edge of the available space.

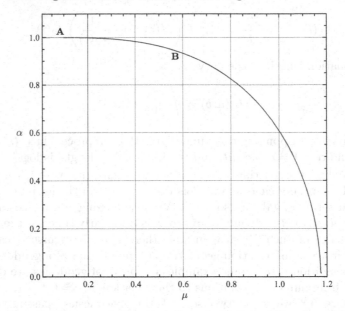

Fig. 6.9. A maximally extended horizontal geodesic through $(\mu, \alpha) = (0.1, 1)$ begins at **A** with high mean density $\overline{N} \approx 1$ in a Poisson state and evolves to lower mean density through increasingly clustered states until **B** where $\alpha \approx 0.6$, after which $\overline{N}$ increases again along the geodesic, as shown in Figure 6.10.

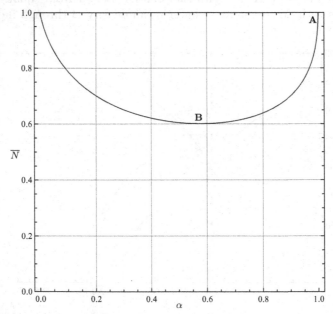

Fig. 6.10. Plot of the mean density $\overline{N}$ along the horizontal geodesic through $(\mu, \alpha) = (0.1, 1)$ in Figure 6.9. It begins at high mean density in a Poisson state **A** and evolves to lower mean density through increasingly clustered states until **B** where $\alpha \approx 0.6$, after which $\overline{N}$ increases again along the geodesic.

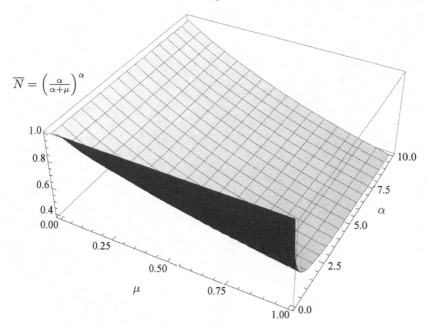

Fig. 6.11. Mean density $\overline{N}$ of counts in cells from equation (6.33) for log gamma distribution (6.32).

clustered states. Being horizontal as it passes through $(\mu, \alpha) = (0.1, 1)$ means that it is directed towards future Poisson random states; but we see that in fact it curves down towards clustered states as a result of the information geometric curvature.

Figure 6.10 shows an approximate numerical plot of the mean density $\overline{N}$ along the horizontal geodesic through $(\mu, \alpha) = (0.1, 1)$ in Figure 6.9. The Figure 6.11 shows a surface plot of $\overline{N}$ as a function of (μ, α).

7

Amino Acid Clustering

With A.J. Doig

In molecular biology a fundamental problem is that of relating functional effects to structural features of the arrangement of amino acids in protein chains. Clearly, there are some features that have localized deterministic origin from the geometrical organization of the helices; other features seem to be of a more stochastic character with a degree of stability persisting over long sequences that approximates to stationarity. These latter features were the subject of our recent study [34], which we outline in this chapter. We make use of gamma distributions to model the spacings between occurrences of each amino acid; this is an approximation because the molecular process is of course discrete. However, the long protein chains and the large amount of data lead us to believe that the approximation is justified, particularly in light of the clear qualitative features of our results.

7.1 Spacings of Amino Acids

We analysed for each of the 20 amino acids X the statistics of spacings between consecutive occurrences of X within the *Saccharomyces cerevisiae* genome, which has been well characterised elsewhere [95]. These occurrences of amino acids may exhibit near Poisson random, clustered or smoothed out behaviour, like 1-dimensional spatial statistical processes along the protein chain. If amino acids are distributed independently and with uniform probability within a sequence then they follow a Poisson process and a histogram of the number of observations of each gap size would asymptotically follow a negative exponential distribution. The question that arises then is how 20 different approximately Poisson processes constrained in finite intervals be arranged along a protein. We used differential geometric methods to quantify information on sequencing structures of amino acids and groups of amino acids, via the sequences of intervals between their occurrences. The differential geometry arises from the information-theoretic distance function on the 2-dimensional

K. Arwini, C.T.J. Dodson, *Information Geometry.*
Lecture Notes in Mathematics 1953,
© Springer-Verlag Berlin Heidelberg 2008

Table 7.1. Experimental data from over 3 million amino acid occurrences in sequences of length up to $n = 6294$ for protein chains of the Saccharomyces cerevisiae genome. Also shown are relative abundance, mean spacing, variance, and maximum likelihood gamma parameter for each amino acid, fitting the interval distribution data to equation (7.3). The grand mean relative abundance was $\bar{p}_i \approx 0.05$ and the grand mean interval separation was $\bar{\mu}_i \approx 18$.

Amino Acid i	Occurrences	Abundance p_i	Mean separation μ_i	Variance σ_i^2	α_i
A Alanine	163376	0.055	17	374	0.81
C Cysteine	38955	0.013	55	5103	0.59
D Aspartate	172519	0.058	16	346	0.77
E Glutamate	192841	0.065	15	292	0.73
F Phenylalanine	133737	0.045	21	554	0.78
G Glycine	147416	0.049	19	487	0.74
H Histidine	64993	0.022	39	1948	0.79
I Isoleucine	195690	0.066	15	222	0.95
K Lysine	217315	0.073	14	240	0.74
L Leucine	284652	0.095	10	122	0.85
M Methionine	62144	0.021	46	2461	0.85
N Asparagine	182314	0.061	16	277	0.87
P Proline	130844	0.044	21	587	0.77
Q Glutamine	116976	0.039	24	691	0.81
R Arginine	132789	0.045	21	565	0.78
S Serine	269987	0.091	11	136	0.85
T Threonine	176558	0.059	16	307	0.85
V Valine	166092	0.056	17	315	0.93
W Tryptophan	31058	0.010	62	6594	0.58
Y Tyrosine	100748	0.034	27	897	0.79

space processes subordinate to gamma distributions—which latter include the Poisson random process as a special case.

Table 7.1 summarizes some 3 million experimentally observed occurrences of the 20 different amino acids within the *Saccharomyces cerevisiae* genome from the analysis of 6294 protein chains with sequence lengths up to $n = 4092$. Listed also for each amino acid are the relative abundances p_i and mean separation μ_i; the grand mean relative abundance was $p \approx 0.05$ and the grand mean interval separation was $\bar{\mu}_i \approx 17$. We found that maximum-likelihood estimates of parametric statistics showed that all 20 amino acids tend to cluster, some substantially. In other words, the frequencies of short gap lengths tends to be higher and the variance of the gap lengths is greater than expected by chance. Our information geometric approach allowed quantitative comparisons to be made. The results contribute to the characterisation of whole amino acid sequences by extracting and quantifying stable statistical features; further information may be found in [34] and references therein.

7.2 Poisson Spaced Sequences

First we consider a simple reference model by computing the statistics of the separation between consecutive occurrences of each amino acid X along a protein chain for a Poisson random disposition of the amino acids.

For example, in the sequence fragment AKLMAATWPFDA, for amino acid A (denoting Ala or Alanine) there are gaps of 1, 4 and 6 since the successive Ala residues are spaced i,i+4, i,i+1 and i,i+6, respectively. In the random case, of haphazard allocation of events along a line, the result is an exponential distribution of inter-event gaps when the line is infinite. For finite length processes it is more involved and we analyse this first in order to provide our reference structure.

Consider a protein chain simply as a sequence of amino acids among which we distinguish one, represented by the letter X, while all others are represented by ?. The relative abundance of X is given by the probability p that an arbitrarily chosen location has an occurrence of X. Then $(1-p)$ is the probability that the location contains a different amino acid from X. All locations are occupied by some amino acid. If the X locations are chosen with uniform probability subject to the constraint that the net density of X in the chain is p, then either X happens or it does not; so we have a binomial process.

Then in a sequence of n amino acids, the mean or expected number of occurrences of X is np and its variance is $np(1-p)$, but the distribution of lengths of spaces between consecutive occurrences of X is less clear. The distribution of such lengths r, measured in units of one location length also is controlled by the underlying binomial distribution.

We seek the probability of finding in a sequence of n amino acids a subsequence of form

$$\underbrace{\cdots ? X \overbrace{? \cdots ?}\, X ? \cdots}_{},$$

where the overbrace $\frown$ encompasses precisely r amino acids that are not X and the underbrace $\smile$ encompasses precisely n amino acids, the whole sequence.

In a sequence of n locations filled by amino acids we seek the probability of finding a subsequence containing two X's separated by exactly r non-X ?'s, that is the occurrence of an inter-X space length r.

The probability distribution function $P(r, p, n)$ for inter-X space length r reduces to the first expression below (7.1), which is a geometric distribution and simplifies to (7.2)

$$P(r, p, n) = \frac{\left(p^2(1-p)^r(n-r-2)\right)}{\sum_{r=0}^{n-2}\left(p^2(1-p)^r(n-r-2)\right)}, \tag{7.1}$$

$$= \frac{(1-p)^{1+r}\, p^2\,(n-r-2)}{-1+(1-p)^n + p\,(n+p-np)}, \tag{7.2}$$

$$\text{for } r = 0, 1, \ldots, (n-2).$$

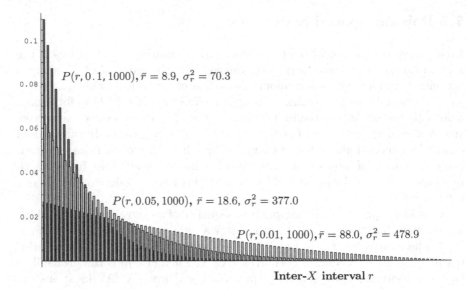

$P(r, 0.1, 1000), \bar{r} = 8.9, \sigma_r^2 = 70.3$

$P(r, 0.05, 1000), \bar{r} = 18.6, \sigma_r^2 = 377.0$

$P(r, 0.01, 1000), \bar{r} = 88.0, \sigma_r^2 = 478.9$

Inter-X interval r

Fig. 7.1. Sample probability distributions $P(r, p, n)$ from (7.2) of interval length r between occurrences of amino acid X shown for $0 \le r \le 100$ in Poisson random sequences of length $n = 1000$.

Three sample distributions are shown in Figure 7.1, for a sequence of $n = 1000$ amino acids in which X has mean abundance probability values $p = 0.01, 0.05, 0.10$. The mean $\bar{r}$ and standard deviation σ_r of the distribution (7.2) are known analytically for $r = 0, 1, \ldots, (n-2)$ and their expressions may be found in [34]. As $n \to \infty$, $\bar{r} \to \frac{1}{p} - 1$ and $\alpha = \frac{\bar{r}^2}{\sigma_r^2} \to (1 - p)$.

The main variables of interest are: the number n of amino acids in the sequence, and the relative abundance probability p of occurrence of X for each amino acid X. Their effects on the statistics of the distribution of intervals between consecutive occurrences of X are illustrated in Figure 7.2 and Figure 7.3, respectively. As might be expected for a Poisson random process, the standard deviation is approximately equal to the mean. For we know that the Poisson distribution is a good approximation to the binomial distribution when n is large and p is small. In the case of a Poisson process along a line the distribution of interval lengths is exponential with standard deviation equal to the mean.

7.3 Non-Poisson Sequences as Gamma Processes

Here we model the spacings between occurrences of amino acids (viewed as a renewal process of occurrences [150]) by supposing that the spacing distribution is of gamma type; this model includes the Poisson case and perturbations of that case.

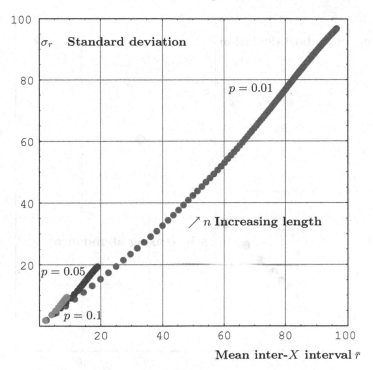

Fig. 7.2. Finite Poisson random sequences. Effect of sequence length $10 \leq n \leq 4000$ in steps of 10 on standard deviation σ_r versus mean $\bar{r}$ for inter-X interval distributions (7.2) with abundance probabilities $p = 0.01$, 0.05 and 0.1, corresponding to the cases in Figure 7.1. The standard deviation is roughly equal to the mean; mean and standard deviation increase monotonically with increasing n.

The family of gamma distributions, §1.4.1, with event space $\mathbb{R}^+$, parameters $\mu, \alpha \in \mathbb{R}^+$ has probability density functions given by

$$f(t; \mu, \alpha) = \left(\frac{\alpha}{\mu}\right)^{\alpha} \frac{t^{\alpha-1}}{\Gamma(\alpha)} e^{-t\alpha/\mu}. \tag{7.3}$$

Then μ is the mean and $Var(t) = \mu^2/\alpha$ is the variance, so the coefficient of variation, §1.2, $\sqrt{Var(t)}/\mu = 1/\sqrt{\alpha}$ is independent of the mean. As mentioned before, this latter property characterizes gamma distributions as shown recently by Hwang and Hu [106] (cf. their concluding remark). For independent positive random variables $x_1, x_2, \ldots, x_n$ with a common continuous probability density function h, that having independence of the sample mean $\bar{x}$ and sample coefficient of variation $cv = S/\bar{x}$ is equivalent to h being a gamma distribution.

The special case $\alpha = 1$ corresponds to the situation of the random or Poisson process with mean inter-event interval μ. In fact, for *integer* $\alpha = 1, 2, \ldots$, equation (7.3) models a process that is Poisson but with $(\alpha - 1)$

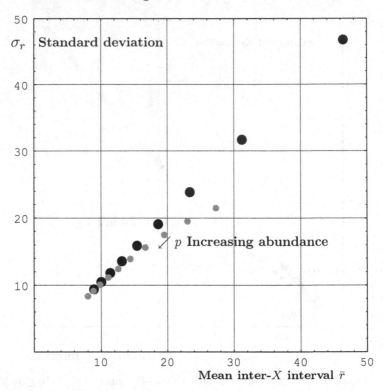

Fig. 7.3. Finite Poisson random sequences. Effect of relative abundance probability $0.01 \le p \le 0.1$ in steps of 0.01 on standard deviation σ_r versus mean $\bar{r}$ for inter-X interval in Poisson random sequences of length $n = 100$ (lower points) and length $n = 1000$ (upper points). The standard deviation is roughly equal to the mean; both decrease monotonically with increasing p.

intermediate events removed to leave only every α^{th}. Gamma distributions can model a range of statistical processes corresponding to non-independent clustered events, for $\alpha < 1$, and dispersed or smoothed events, for $\alpha > 1$, as well as the Poisson random case $\alpha = 1$. Figure 7.4 shows sample gamma distributions, all of unit mean, representing clustering, Poisson and dispersed spacing distributions, respectively, with $\alpha = 0.4, 0.6, 0.8, 1, 1.2, 2$.

From §3.4, the Riemannian information metric on the parameter space $\mathcal{G} = \{(\mu, \alpha) \in \mathbb{R}^+ \times \mathbb{R}^+\}$ for the gamma distributions (7.3) is given by the arc length function from equation (3.19)

$$ds_{\mathcal{G}}^2 = \frac{\alpha}{\mu^2} \, d\mu^2 + \left(\psi'(\alpha) - \frac{1}{\alpha} \right) d\alpha^2 \quad \text{for } \mu, \alpha \in \mathbb{R}^+, \qquad (7.4)$$

where $\psi(\alpha) = \frac{\Gamma'(\alpha)}{\Gamma(\alpha)}$ is the logarithmic derivative of the gamma function. The 1-dimensional subspace parametrized by $\alpha = 1$ corresponds to all possible 'random' (Poisson) processes, or equivalently, exponential distributions.

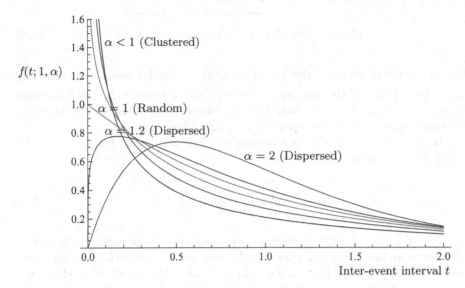

Fig. 7.4. Gamma probability density functions, $f(t; \mu, \alpha)$, (7.3) for inter-event intervals t with unit mean $\mu = 1$, and descending from the top left $\alpha = 0.4, 0.6, 0.8, 1, 1.2, 2$. The case $\alpha = 1$ corresponds to randomness via an exponential distribution from an underlying Poisson process; $\alpha \neq 1$ represents some non-Poisson clustering or dispersion.

7.3.1 Local Geodesic Distance Approximations

A path through the parameter space $\mathcal{G}$ of gamma models determines a curve, parametrized by t in some interval $a \leq t \leq b$, given by

$$c : [a, b] \rightarrow \mathcal{G} : t \mapsto (c_1(t), c_2(t)) \qquad (7.5)$$

and its tangent vector $\dot{c}(t) = (\dot{c}_1(t), \dot{c}_2(t))$ has norm $||\dot{c}||$ given via (7.4) by

$$||\dot{c}(t)||^2 = \frac{c_2(t)}{c_1(t)^2} \dot{c}_1(t)^2 + \left(\psi'(c_2(t)) - \frac{1}{c_2(t)} \right) \dot{c}_2(t)^2 \qquad (7.6)$$

and the information length of the curve is

$$L_c(a, b) = \int_a^b ||\dot{c}(t)|| \, dt \quad \text{for } a \leq b. \qquad (7.7)$$

For example, the curve $c(t) = (t, 1)$, which passes through processes with $t = \mu$ and $\alpha = 1 = constant$, has information length $\log \frac{b}{a}$. Locally, minimal paths in $\mathcal{G}$ are given by the geodesics [70] from equation (2.7) using the Levi-Civita connection ∇ (2.5) induced by the Riemannian metric (7.4).

In a neighbourhood of a given point we can obtain a locally bilinear approximation to distances in the space of gamma models. From (7.4) for small variations $\Delta \mu, \Delta \alpha$, near $(\mu_0, \alpha_0) \in \mathcal{G}$; it is approximated by

$$\Delta s_{\mathcal{G}} \approx \sqrt{\frac{\alpha_0}{\mu_0^2}\,\Delta\mu^2 + \left(\psi'(\alpha_0) - \frac{1}{\alpha_0}\right)\Delta\alpha^2}\,. \qquad (7.8)$$

As α_0 increases from 1, the factor $(\psi'(\alpha_0) - \frac{1}{\alpha_0})$ decreases monotonically from $\frac{\pi^2}{6} - 1$. So, in the information metric, the difference $\Delta\mu$ has increasing prominence over $\Delta\alpha$ as the standard deviation reduces with increasing α_0—corresponding to increased spatial smoothing of occurrences.

In particular, near the exponential distribution, where $(\mu_0, \alpha_0) = (1, 1)$, (7.8) is approximated by

$$\Delta s_{\mathcal{G}} \approx \sqrt{\Delta\mu^2 + \left(\frac{\pi^2}{6} - 1\right)\Delta\alpha^2}\,. \qquad (7.9)$$

For a practical implementation we need to obtain rapid estimates of distances in larger regions than can be represented by quadratics in incremental coordinates. This can be achieved using the result of Dodson and Matsuzoe [68] that established geodesic foliations for the gamma manifold. Now, a geodesic curve is locally minimal and so a network of two non-parallel sets of geodesics provides a mesh of upper bounds on distances by using the triangle inequality about any point. Distance bounds using such a geodesic mesh are shown in Figure 7.8 using the geodesic curves $\mu = \alpha$ and $\alpha = constant$, which foliate $\mathcal{G}$ [68].

Explicitly, the arc length along the geodesic curves $\mu = \alpha$ from (μ_0, α_0) to $(\mu = \alpha, \alpha)$ is

$$\left| \frac{d^2 \log \Gamma}{d\alpha^2}(\alpha) - \frac{d^2 \log \Gamma}{d\alpha^2}(\alpha_0) \right|$$

and the distance along curves of constant $\alpha = \alpha_0$ from (μ_0, α_0) to (μ, α_0) is

$$\left| \alpha_0 \log \frac{\mu_0}{\mu} \right|\,.$$

In Figure 7.8 we use the base point $(\mu_0, \alpha_0) = (18, 1) \in \mathcal{G}$ and combine the above two arc lengths of the geodesics to obtain an upper bound on distances from (μ_0, α_0) as

$$Distance[(\mu_0, \alpha_0), (\mu, \alpha)] \leq \left| \frac{d^2 \log \Gamma}{d\alpha^2}(\alpha) - \frac{d^2 \log \Gamma}{d\alpha^2}(\alpha_0) \right| + \left| \alpha_0 \log \frac{\mu_0}{\mu} \right|\,.$$
$$(7.10)$$

The gamma distribution fitted the experimental data quite well and Figure 7.7 shows the histogram for the first 30 data points for all 20 amino acids, and their residuals. Figure 7.5 shows that the expected values for the gamma parameter α would exceed 0.97 in Poisson random sequences whereas all 20 amino acids had maximum likelihood fits of the gamma parameter in the range $0.59 \leq \alpha \leq 0.95$.

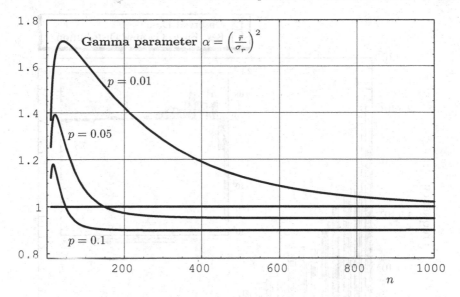

Fig. 7.5. Finite Poisson random sequences. Effect of sequence length n on gamma parameter $\alpha = \left(\frac{\bar{r}}{\sigma_r}\right)^2$ for relative abundances $p = 0.01, 0.05, 0.1$. The 20 maximum likelihood fits of the gamma parameter had $0.59 \le \alpha \le 0.95$.

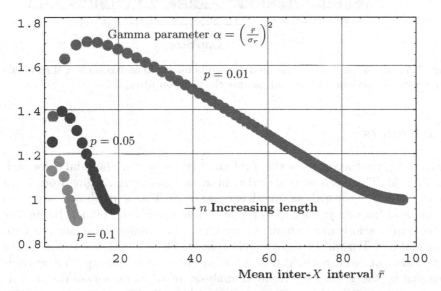

Fig. 7.6. Finite Poisson random sequences. Effect of sequence length $10 \le n \le 4000$ in steps of 10 on gamma parameter α from (7.3) versus mean $\bar{r}$ for inter-X interval distributions (7.2). The mean probabilities for the occurrences of X are $p = 0.1$ (left), $p = 0.05$ (centre) and $p = 0.01$ (right), corresponding to the cases in Figures 7.1 and 7.2.

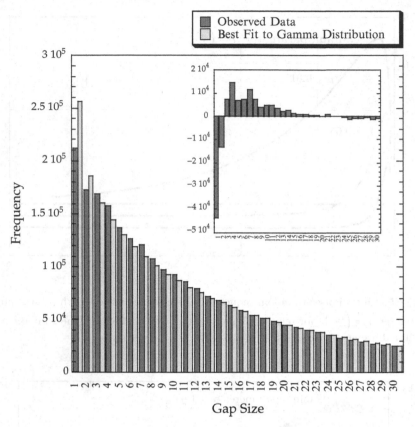

Fig. 7.7. Histograms of the first 30 data points for all 20 amino acids and residuals compared to maximum likelihood gamma distribution fitting.

7.4 Results

Table 7.1 gives the values of the total number, mean, variance and α for each amino acid. There is a large variation in mean gap size μ, ranging from 10.75 (Ser) to 61.82 (Trp) with an overall average of 25. This is attributed largely to amino acid frequency; the gap to the next amino acids will tend to be smaller if the amino acid is more abundant. Similarly, rare amino acids, such as Cys, His, Met and Trp will be more widely spaced. There is a therefore a negative correlation between mean gap size and amino acid frequency. Clustering is revealed by the gamma distribution analysis. In all cases, we see that $\alpha < 1$; hence every amino acid tends to cluster with itself. There is some variation, with Cys and Trp most clustered and Ile and Val very close to a Poisson random process in their spatial statistics.

Figure 7.8 shows a plot of all 20 amino acids as points on a surface over the space of (μ, α) values; the height of the surface represents the information-

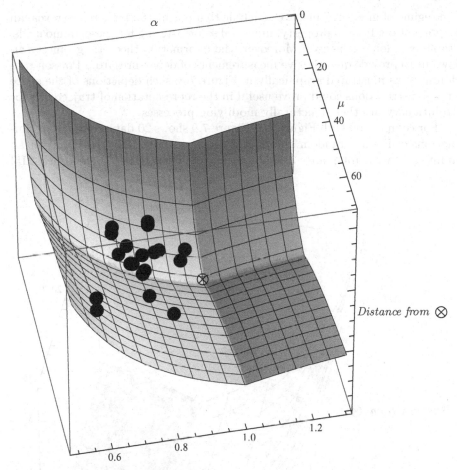

Fig. 7.8. Distances in the space of gamma models, using a geodesic mesh. The surface height represents upper bounds on distances from the grand mean point $(\mu, \alpha) = (18, 1)$, the Poisson random case with mean $\mu = 18$, marked with $\otimes$. Depicted also are the 20 data points for the amino acid sequences from Table 7.1. All amino acids show clustering to differing degrees by lying to the left of the Poisson random line $\alpha = 1$, some substantially so.

theoretic distance using (7.10) from the point marked as $\otimes$, this point is the case of Poisson randomly distributed amino acids with mean $\mu = 18$ and $\alpha = 1$.

If an exponential distribution gave the maximum likelihood fit then it would yield $\alpha \approx 1$. This is arguably within experimental tolerance for I,N and V, but unlikely in the other cases which have maximum likelihood gamma parameter $\alpha \leq 0.85$. However, we find no case of $\alpha > 0.97$ and the analytic results for the case of finite Poisson random sequences did not yield $\alpha > 0.95$ in the regime of interest. Thus, we conclude that our methods therefore reveal an important qualitative property: universal self-clustering for these amino acids, stable over long sequences. Moreover, the information-theoretic geometry allows us to provide quantitative measurements of departures from Poisson randomness, as illustrated graphically in Figure 7.8; such depictions of the space of gap distributions could prove useful in the representation of trajectories for evolutionary or other structurally modifying processes.

For comparison with Figure 7.8, Figure 7.9 shows 20 data points from simulations of Poisson random amino acid sequences of length $n = 10000$ for an amino acid with abundance probability $p = 0.05$, using the *Mathematica* [215]

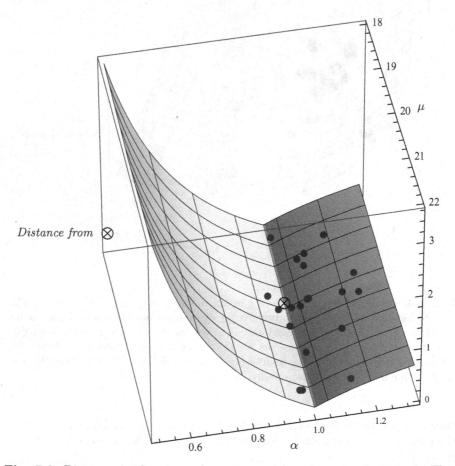

Fig. 7.9. Distances in the space of gamma models, using a geodesic mesh. The surface height represents upper bounds on distances from the nominal target point, $(\mu, \alpha) = (20, 1)$, for 20 data points from simulations of Poisson random amino acid sequences of length $n = 10,000$ for an amino acid with abundance probability $p = 0.05$. Whereas the observations of real sequences, Figure 7.8, showed variation $0.59 \leq \alpha \leq 0.95$, for these Poisson random simulations we find $0.95 \leq \alpha \leq 1.23$.

pseudorandom number generator. Whereas the observations of real sequences, Figure 7.8 showed variation $0.59 \leq \alpha \leq 0.95$, we find for Poisson random simulations that $0.95 \leq \alpha \leq 1.23$.

7.5 Why Would Amino Acids Cluster?

Clustering could arise from secondary structure preferences. α-Helices are typically 4-15 and β-strands 2-8 amino acids in length [163]. For any secondary structural element to form, most amino acids within its sequence must have a high propensity for that structure. Identical amino acids will therefore cluster over these length ranges as this will favour a sequence with a high preference for forming one particular secondary structure. For example, Ala has a high preference for the α-helix. Hence evolution will select sequences where Alanines are clustered in order to favour α-helix formation. If amino acids were Poisson distributed, the probability that a stretch of amino acids would contain a high preference for a secondary structural element would be decreased. A second possibility is that amino acids of similar hydrophobicity cluster in order to produce a hydrophobic membrane spanning the sequence or water exposed polar loop.

Some deterministic effects arise from the preferred spatial configurations, as visible in Figure 7.7. Gap sizes of 1 or 2 are disfavoured, 1 strongly so. The only exceptions to this are Gln and Ser, which strongly favour short gaps of 1, 2 or 3. Poly(Gln) sequences give a high frequency of gaps of 1 and are a well known feature of a number of proteins, implicated in several diseases, including Huntington's disease [115]. Gaps of 3-12 are generally favoured, perhaps because this is the usual length of secondary structure. There are also local preferences for gaps of 4 and 7 that can be attributed to α-helices. Side chains spaced i,i+4 and i,i+7 are on the same side of an α-helix so can bond to one another. Sequences are favoured that have identical side chains close in space in the α-helix. In particular, a favoured gap of 7 for Leu can be attributed to coiled-coils that are characterised by pairs of α-helices held together by hydrophobic faces with Leu spaced i,i+7 [132], [137], [160].

Clearly, the maximum likelihood gamma distributions fit only sttistical features and in that respect view the data as exhibiting transient behaviour at small gap sizes—we recall from Table 7.1 that the overall mean interval is about 18—other methods are available for interpretation of deterministic features. We concentrate here on the whole sequences by extracting and quantifying stable statistical features and we find that all 20 amino acids tend to self-cluster in protein sequences.

8

Cryptographic Attacks and Signal Clustering

Typical public-key encryption methods involve variations on the RSA procedure devised by Rivest, Shamir and Adleman [174]. This employs modular arithmetic with a very large modulus in the following manner. We compute

$$R \equiv y^e \ (mod \, m) \quad \text{or} \quad R \equiv y^d \ (mod \, m) \tag{8.1}$$

depending respectively on whether we are encoding or decoding a message y. The (very large) modulus m and the encryption key e are made public; the decryption key d is kept private. The modulus m is chosen to be the product of two large prime numbers p, q which are also kept secret and we choose d, e such that

$$ed \equiv 1 \ (mod \, (p-1)(q-1)). \tag{8.2}$$

8.1 Cryptographic Attacks

It is evident that both encoding and decoding will involve repeated exponentiation procedures. Then, some knowledge of the design of an implementation and information on the timing or power consumption during the various stages could yield clues to the decryption key d. Canvel and Dodson [38, 37] have shown how timing analyses of the modular exponentiation algorithm quickly reveal the private key, regardless of its length. In principle, an incorporation of obscuring procedures could mask the timing information but that may not be straightforward for some devices. Nevertheless, it is important to be able to assess departures from Poisson randomness of underlying or overlying procedures that are inherent in devices used for encryption or decryption and here we outline some information geometric methods to add to the standard tests [179].

In a review, Kocher et al. [119] showed the effectiveness of Differential Power Analysis (DPA) in breaking encryption procedures using correlations between power consumption and data bit values during processing, claiming

K. Arwini, C.T.J. Dodson, *Information Geometry.*
Lecture Notes in Mathematics 1953,
© Springer-Verlag Berlin Heidelberg 2008

that most smartcards reveal their keys using fewer than 15 power traces. Power consumption information can be extracted from even noisy recordings using inductive probes external to the device.

Chari et al. [41] provided a probabilistic encoding (secret sharing) scheme for effectively secure computation. They obtained lower bounds on the number of power traces needed to distinguish distributions statistically, under certain assumptions about Gaussian noise functions. DPA attacks depend on the assumption that power consumption in a given clock cycle will have a distribution depending on the initial state; the attacker needs to distinguish between different 'nearby' distributions in the presence of noise. Zero-Knowledge proofs allow verification of secret-based actions without revealing the secrets. Goldreich et al. [94] discussed the class of promise problems in which interaction may give additional information in the context of Statistical Zero-Knowlege (SZK). They invoked two types of difference between distributions: the 'statistical difference' and the 'entropy difference' of two random variables. In this context, typically, one of the distributions is the uniform distribution.

Thus, in the contexts of DPA and SZK tests, it is necessary to compare two nearby distributions on bounded domains. This involves discrimination between noisy samples drawn from pairs of closely similar distributions. In some cases the distributions resemble truncated Gaussians; sometimes one distribution is uniform. Dodson and Thompson [77] have shown that information geometry can help in evaluating devices by providing a metric on a suitable space of distributions.

8.2 Information Geometry of the Log-gamma Manifold

The log-gamma family of probability density functions §3.6 provides a 2-dimensional metric space of distributions with compact support on $[0, 1]$, ranging from the uniform distribution to symmetric unimodular distributions of arbitrarily small variance, as may be seen in Figure 3.3 and Figure 3.4.

Information geometry provided the metric for a discrimination procedure reported by Dodson and Thompson [77] exploiting the geometry of the manifold of log-gamma distributions, which we have seen above has these useful properties:

• it contains the uniform distribution
• it contains approximations to truncated Gaussian distributions
• as a Riemannian 2-manifold it is an isometric isomorph of the manifold of gamma distributions.

The log-gamma probability density functions discussed in § 3.6 for random variable $N \in (0, 1]$ were given in equation (3.38), Figure 8.1,

$$g(N; \gamma, \tau) = \frac{1}{\Gamma(\tau)} \left(\frac{\tau}{\gamma}\right)^{\tau} N^{\frac{\tau}{\gamma}-1} \left(\log \frac{1}{N}\right)^{\tau-1} \quad \text{for } \gamma > 0 \text{ and } \tau > 0 . \quad (8.3)$$

These coordinates (γ, τ) are actually orthogonal for the Fisher information metric on the parameter space $\mathcal{L} = \{(\gamma, \tau) \in (0, \infty) \times (0, \infty)\}$. Its arc length

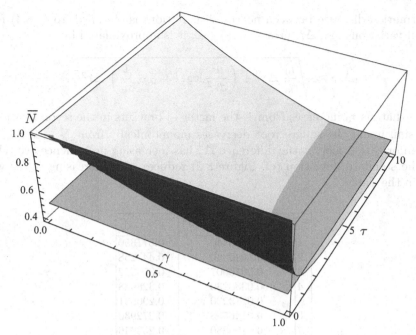

Fig. 8.1. *Mean value* $\overline{N} = \left(\frac{\tau}{\tau+\gamma}\right)^{\tau}$ *as a surface with a horizontal section at the central value* $\overline{N} = \frac{1}{2}$, *which intersects the* $\overline{N}$ *surface in the curve* $\gamma = \tau(2^{1/\tau} - 1)$.

function is given from equation (3.39) by

$$ds^2 = \sum_{ij} g_{ij} \, dx^i dx^j = \frac{\tau}{\gamma^2} \, d\gamma^2 + \left(\frac{d^2}{d\tau^2} \log(\Gamma) - \frac{1}{\tau}\right) d\tau^2. \qquad (8.4)$$

In fact, (8.3) arises from the gamma family, §1.4.1,

$$f(x, \gamma, \tau) = \frac{x^{\tau-1} \left(\frac{\tau}{\gamma}\right)^{\tau}}{\Gamma(\tau)} \, e^{-\frac{x\tau}{\gamma}} \qquad (8.5)$$

for the non-negative random variable $x = \log \frac{1}{N}$ with mean $\bar{x} = \gamma$. It is known that the gamma family (8.5) has also the information metric (8.4) so the identity map on the space of coordinates (γ, τ) is not only a diffeomorphism but also an isometry of Riemannian manifolds.

8.3 Distinguishing Nearby Unimodular Distributions

Log-gamma examples of unimodular distributions resembling truncated Gaussians are shown on the right of Figure 8.3. Such kinds of distributions can arise in practical situations for bounded random variables. A measure of information distance between nearby distributions is obtained from (8.4) for small variations $\Delta\gamma, \Delta\tau$, near $(\gamma_0, \tau_0) \in \mathcal{L}$; it is approximated by

$$\Delta s_{\mathcal{L}} \approx \sqrt{\frac{\tau_0}{\gamma_0^2}\,\Delta\gamma^2 + \left(\frac{d^2}{d\tau^2}\log(\Gamma)_{|\tau 0} - \frac{1}{\tau_0}\right)\Delta\tau^2} \; . \tag{8.6}$$

Note that, as τ_0 increases from 1, the factor in brackets in the second part of the sum under the square root decreases monotonically from $\frac{\pi^2}{6} - 1$. So, in the information metric, the difference $\Delta\gamma$ has increasing prominence over $\Delta\tau$ as the standard deviation (cf. Figure 8.2) reduces with increasing τ_0, as we see in the Table.

τ_0	$\left(\frac{d^2}{d\tau^2}\log(\Gamma)_{|\tau 0} - \frac{1}{\tau_0}\right)$	$cv_N(\tau_0)^\dagger$
1	0.6449340	0.577350
2	0.1449340	0.443258
3	0.0616007	0.373322
4	0.0338230	0.328638
5	0.0213230	0.296931
6	0.0146563	0.272930
7	0.0106880	0.253946
8	0.0081370	0.238442
9	0.0064009	0.225472
10	0.0051663	0.214411
		†At $\overline{N} = \frac{1}{2}$

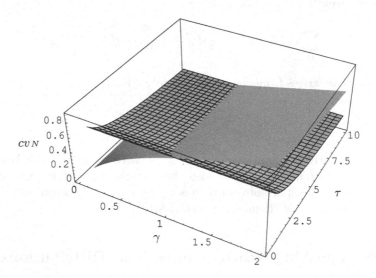

Fig. 8.2. Coefficient of variation $cv_N = \frac{\sigma_N}{N}$ for the log-gamma distribution as a smooth surface with a hatched surface at the central mean case $\overline{N} = \frac{1}{2}$.

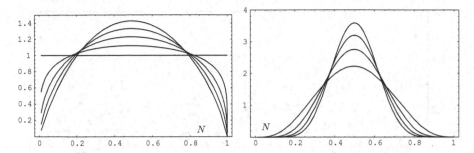

Fig. 8.3. Examples from the log-gamma family of probability densities with central mean $\overline{N} = \frac{1}{2}$. Left: $\tau = 1, 1.2, 1.4, 1.6, 1.8$. Right: $\tau = 4, 6, 8, 10$.

For example,some data on power measurements from a smartcard leaking information during processing of a '0' and a '1', at a specific point in process time, yielded two data sets C, D. These had maximum likelihood parameters $(\gamma_C = 0.7246, \tau_C = 1.816)$ and $(\gamma_D = 0.3881, \tau_D = 1.757)$. We see that here the dominant parameter in the information metric is γ. In terms of the underlying gamma distribution, from which the log-gamma is obtained, γ is the mean.

8.4 Difference From a Uniform Distribution

The situation near to the uniform distribution $\tau = 1$ is shown on the left in Figure 8.3. In this case we have $(\gamma_0, \tau_0) = (1, 1)$ and for nearby distributions, (8.6) is approximated by

$$\Delta s_{\mathcal{L}} \approx \sqrt{\Delta\gamma^2 + \left(\frac{\pi^2}{6} - 1\right) \Delta\tau^2} \ . \tag{8.7}$$

We see from (8.7) that, in the information metric, $\Delta\tau$ is given about 80% of the weight of $\Delta\gamma$, near the uniform distribution.

The information-theoretic metric and these approximations may be an improvement on the areal-difference comparator used in some recent SZK studies [57, 94] and as an alternative in testing security of devices like smartcards.

8.5 Gamma Distribution Neighbourhoods of Randomness

In a variety of contexts in cryptology for encoding, decoding or for obscuring procedures, sequences of pseudorandom numbers are generated. Tests for randomness of such sequences have been studied extensively and the NIST

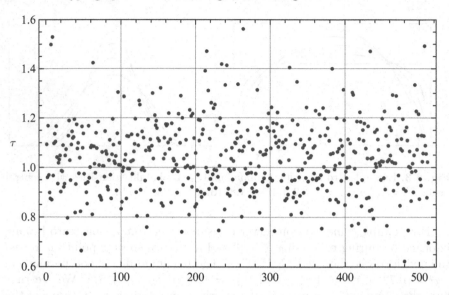

Fig. 8.4. Maximum likelihood gamma parameter τ fitted to separation statistics for simulations of Poisson random sequences of length 100000 for an element with expected parameters $(\gamma, \tau) = (511, 1)$. These simulations used the pseudorandom number generator in Mathematica [215].

Suite of tests [179] for cryptological purposes is widely employed. Information theoretic methods also are used, for example see Grzegorzewski and Wieczorkowski [101] also Ryabko and Monarev [180] and references therein for recent work. Here we can show how pseudorandom sequences may be tested using information geometry by using distances in the gamma manifold to compare maximum likelihood parameters for separation statistics of sequence elements.

Mathematica [215] simulations were made of Poisson random sequences with length $n = 100000$ and spacing statistics were computed for an element with abundance probability $p = 0.00195$ in the sequence. Figure 8.4 shows maximum likelihood gamma parameter τ data points from such simulations. In the data from 500 simulations the ranges of maximum likelihood gamma distribution parameters were $419 \leq \gamma \leq 643$ and $0.62 \leq \tau \leq 1.56$.

The surface height in Figure 8.5 represents upper bounds on information geometric distances from $(\gamma, \tau) = (511, 1)$ in the gamma manifold. This employs the geodesic mesh function we developed in the previous Chapter (7.10)

$$Distance[(511,1),(\gamma,\tau)] \leq \left| \frac{d^2 \log \Gamma}{d\tau^2}(\tau) - \frac{d^2 \log \Gamma}{d\tau^2}(1) \right| + \left| \log \frac{511}{\gamma} \right|. \quad (8.8)$$

Also shown in Figure 8.5 are data points from the *Mathematica* simulations of Poisson random sequences of length 100000 for an element with expected separation $\gamma = 511$.

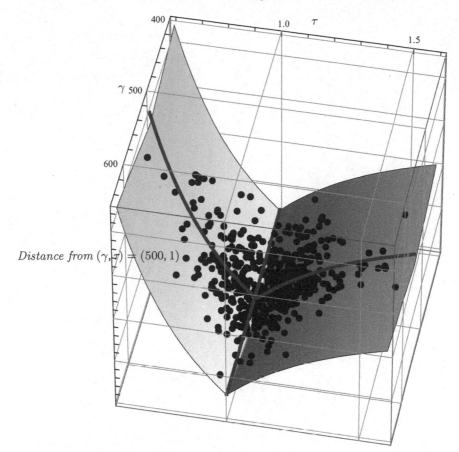

Fig. 8.5. Distances in the space of gamma models, using a geodesic mesh. The surface height represents upper bounds on distances from $(\gamma, \tau) = (511, 1)$ from Equation (8.8). Also shown are data points from simulations of Poisson random sequences of length 100000 for an element with expected separation $\gamma = 511$. In the limit as the sequence length tends to infinity and the element abundance tends to zero we expect the gamma parameter τ to tend to 1.

In the limit, as the sequence length tends to infinity and the abundance of the element tends to zero, we expect the gamma parameter τ to tend to 1. However, finite sequences must be used in real applications and then provision of a metric structure allows us, for example, to compare real sequence generating procedures against an ideal Poisson random model.

9

Stochastic Fibre Networks

With W.W. Sampson

There is considerable interest in the materials science community in the structure of stochastic fibrous materials and the influence of structure on their
mechanical, optical and transport properties. We have common experience
of such materials in the form of paper, filters, insulating layers and support-
ing matrices for composites. The reference model for such stochastic fibre
networks is the 2-dimensional array of line segments with centres following a
Poisson process in the plane and axis orientations following a uniform process;
that structure is commonly called a *random fibre network* and we study this
before considering departures from it.

9.1 Random Fibre Networks

Micrographs of four stochastic fibrous materials are shown in Figure 9.1. The
carbon fibre network on the top left of Figure 9.1 is used in fuel cell applications and provides the backbone of the electromagnetic shielding used in
stealth aerospace technologies; the glass fibre network on the top right is of
the type used in laboratory and industrial filtration applications; the network
on the bottom right is a sample of paper formed from softwood fibres; on the
left of the bottom row is an electrospun nylon nanofibrous network. Examples of the latter type are the focus of worldwide research activity since such
materials have great potential for application as cell culture scaffolds in tissue
engineering, see *e.g.* [172, 165, 32]. Although the micrographs in Figure 9.1
are manifestly different from each other, it is equally evident that they exhibit
strikingly similar structural characteristics.

A classical reference structure for modelling is an isotropic planar network
of infinite random lines. So the angles of lines relative to a given fixed direction
are uniformly distributed and on each line the locations of the intersections
with other lines in the network form a Poisson point process. A graphical representation of part of an infinite line network is shown on the left of Figure 9.2;
the graphic on the right of this figure shows a network having the same total

K. Arwini, C.T.J. Dodson, *Information Geometry.*   161
Lecture Notes in Mathematics 1953,
© Springer-Verlag Berlin Heidelberg 2008

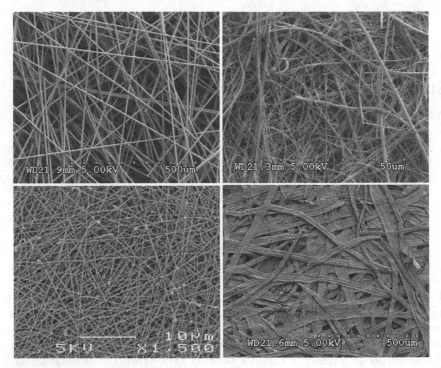

Fig. 9.1. Micrographs of four stochastic fibrous materials. Top left: Nonwoven carbon fibre mat; Top right: glass fibre filter; Bottom left: electrospun nylon nanofibrous network (Courtesy S.J. Eichhorn and D.J. Scurr); Bottom right: paper.

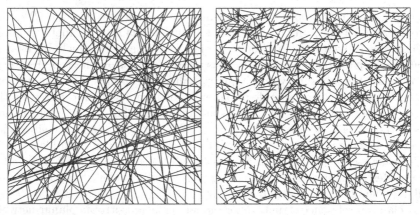

Fig. 9.2. Graphical representations of planar random networks of lines with infinite and finite length; both have the same total length of lines.

line length per unit area but made from lines of finite length. We discuss networks of these types in detail in the sequel, but for now we observe qualitative similarity among the networks in the above Figures 9.1,9.2, particularly in the sizes and shapes of the polygons enclosed by the intersections of local groups of lines.

The polygons generated by the intersections of lines have been studied by many workers and several analytic results are known. There are results of Miles [147, 148] and Tanner [198] (cf. also Stoyan et al. [196]) for random lines in a plane, for example:

- Expected number of sides per polygon

$$\bar{n} = 4.$$

- Variance of the number of sides per polygon

$$\sigma^2(n) = \frac{\pi^2 + 24}{2}.$$

- Perimeter P of polygons with n sides has a χ^2 distribution with $2(n-2)$ degrees of freedom and probability density function q given by

$$q(P, n) = \frac{P^{n-3} e^{-P/2}}{2^{n-2} \Gamma(n-2)}, \quad n = 3, 4, \ldots \tag{9.1}$$

where P is given as a multiple of the mean polygon side length and the case of $n = 3$ for perimeter of triangles coincides with an exponential distribution.

- Probability of triangles

$$p_3 = (2 - \frac{\pi^2}{6}) \approx 0.355$$

- Probability of quadrilaterals

$$p_4 = \frac{1}{3} - \frac{7\pi^2}{36} + 4 \int_0^{\pi/2} x^2 \cot x \, dx \approx 0.381.$$

Stoyan et al. [196] p325 collect further results from Monte Carlo methods:

$$p_5 \approx 0.192, \ p_6 \approx 0.059, \ p_7 \approx 0.013, \ p_8 \approx 0.002 \tag{9.2}$$

and mention the empirical approximation obtained by Crain and Miles for the distribution of the number of sides per random polygon

$$p_n \approx \frac{e^{-1}}{(n-3)!}. \tag{9.3}$$

In Dodson and Sampson [75] we develop this a little more by providing new analytic approximations to the distributions of the areas and local line densities for random polygons and we compute various limiting properties of random polygons.

9.2 Random Networks of Rectangular Fibres

Most real fibrous materials consist of fibres of finite width and length with distributed morphologies. However, we note also the important result of Miles [147] that the distribution of the diameters of inscribed circles is exponential and unaltered by changing the infinite random lines to infinite random rectangles for arbitrary distributions of width. This characteristic arises from the fact that as the lines change to rectangles with increasing width, so the area of polygons decreases and some small polygons disappear; accordingly, we expect the same independence of width for the distances between adjacent fibre crossings on a given line and in the polygon area distribution.

Although the modelling of fibres as lines of infinite length is convenient when considering the statistics of the void structure and porous properties of stochastic fibrous materials, there is an influence of fibre length on network uniformity and other properties that must be considered. For such properties, our reference structure for modelling is a two dimensional random fibre network where fibres are considered to be rectangles of given aspect ratio with their centroids distributed according to a planar Poisson point process and the orientation of their major axes to any fixed direction are uniformly distributed; such a network is represented graphically on the right of Figure 9.2. As an example for a familiar material, the typical range of aspect ratios for natural cellulose fibres in paper is from about 20 to perhaps 100, as may be seen by tearing the edge of a a sheet of writing paper. Figure 9.3 shows areal density radiographs for three wood fibre networks made with the same mean areal density but with different spatial distributions of fibres, from Oba [156].

In the case of fibres of finite width, we call the number of fibres covering a point in the plane the *coverage*, c. The coverage is distributed according to a Poisson distribution so has probability function

$$P(c) = \frac{\bar{c}^c\, e^{-\bar{c}}}{c!} \qquad \text{where } c = 0, 1, 2, \ldots \qquad (9.4)$$

and $\bar{c}$ is the expected coverage.

Referring to Figure 9.2, we observe that if we partitioned the network into square zones of side length d, then the total fibre area in each zone and hence the local average coverage $\tilde{c}$ there would vary from zone to zone. For a random network of fibres of uniform length λ and uniform width ω, the variance of the local average coverage, $\sigma_d^2(\tilde{c})$ for such random fibre networks was derived by Dodson [58]:

$$\sigma_d^2(\tilde{c}) = \overline{(\tilde{c} - \bar{c})^2} = \bar{c} \int_0^{\sqrt{2}\,d} a(r, \omega, \lambda)\, b(r, d)\, dr. \qquad (9.5)$$

Here $a(r, \omega, \lambda)$ is the point autocorrelation function for coverage at points separated by a distance r and $b(r, d)$ is the probability density function for the distance r between two points chosen independently and at random within

Fig. 9.3. Density maps of coverage for three wood fibre networks with constant mean coverage, $\bar{c} \approx 20$ fibres, but different distributions of fibres. Each image represents a square region of side length 5 cm; darker regions correspond to higher coverage. Top: $cv_{1mm} = 0.08$; centre: $cv_{1mm} = 0.11$; bottom: $cv_{1mm} = 0.15$. The top image is similar to that expected for a Poisson process of the same fibres.

the a zone. The point autocorrelation function was derived for arbitrary rectangular zones [58] and for square zones of side length d, it is given by,

$$a(r,\omega,\lambda) = \begin{cases} 1 - \frac{2}{\pi}\left(\frac{r}{\lambda} + \frac{r}{\omega} - \frac{r^2}{2\omega\lambda}\right) & \text{for } 0 < r \le \omega \\ \frac{2}{\pi}\left(\arcsin\left(\frac{\omega}{r}\right) - \frac{\omega}{2\lambda} - \frac{r}{\omega} + \sqrt{\frac{r^2}{\omega^2}-1}\right) & \text{for } \omega < r \le \lambda \\ \frac{2}{\pi}\left(\arcsin\left(\frac{\omega}{r}\right) - \arccos\left(\frac{\lambda}{r}\right) - \frac{\omega}{2\lambda} - \frac{\lambda}{2\omega}\right. \\ \qquad \left. -\frac{r^2}{2\lambda\omega} + \sqrt{\frac{r^2}{\lambda^2}-1} + \sqrt{\frac{r^2}{\omega^2}-1}\right) & \text{for } \lambda < r \le \sqrt{\lambda^2+\omega^2} \\ 0 & \text{for } r > \sqrt{\lambda^2+\omega^2} \end{cases}$$

$$(9.6)$$

Also, Ghosh [91] had provided

$$b(r,d) = \begin{cases} \frac{4r}{d^4}\left(\frac{\pi d^2}{2} - 2rd + \frac{r^2}{2}\right) & \text{for } 0 \le r \le d \\ \frac{4r}{d^4}\left(d^2\left(\arcsin\left(\frac{d}{r}\right) - \arccos\left(\frac{d}{r}\right)\right)\right. \\ \qquad \left. +2d\sqrt{r^2-d^2} - \frac{1}{2}\left(r^2+2d^2\right)\right) & \text{for } d \le r \le \sqrt{2}d \\ 0 & \text{for } r > \sqrt{2}d. \end{cases}$$

$$(9.7)$$

The integral term in equation (9.5) is the fractional between zones variance and is plotted in Figure 9.4. We observe that it increases with increasing fibre length and width, and decreases with increasing zone size. The actual distribution of local zonal averages of coverage for a random network of rectangles would by the Central Limit Theorem be a (truncated) Gaussian, being the result of a large number of independent Poisson events.

Knowing the fractional between zones variance allows us to compare the measured distribution of mass of a real fibre network with that of a random fibre network formed from the same constituent fibres. There is a large archive

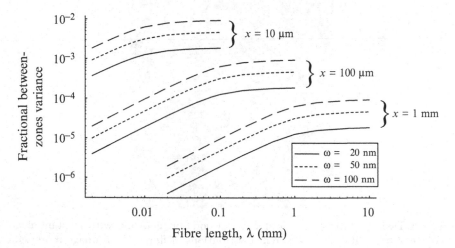

Fig. 9.4. Dependence of the fractional between zones variance on fibre length λ, fibre width ω and side length x of square inspection zones, equation (9.5).

of such analyses from radiographic imaging, particularly for paper—arguably the most common and familiar stochastic fibrous material used in society and industry. Such data reveals that industrially formed networks invariably exhibit a higher variance of local coverage, at all scales above a few fibre widths, than the corresponding random structure [56, 182, 155].

In characterizing fibre networks, the random case is taken as a well-defined reference structure and then we proceed to consider how the effects of fibre orientation and fibre clumping combine to yield structures with non-random features—clustering of fibre centres and preferential orientation of fibre axes. These departures from randomness represent in practical applications the component of variability within the structure which has the potential to be influenced through intervention and control in manufacturing processes. In later sections we model the effects of non-randomness in fibre networks on the distribution of in-plane pore dimensions, which is important for fluid transfer properties. We deal first with the case of clustering of fibres in an isotropic network, then proceed to the case of clustering and anisotropy.

Typically, the coefficient of variation, §1.2, of local areal density at the one millimetre scale for papers varies in the range $0.02 < cv(\tilde{c})_{1mm} < 0.20$, the lower end corresponding to very fine papers and the upper end corresponding to papers with very clumpy, contrasty appearance when viewed in transmitted light. Distributions of such types are shown in Figure 9.5 with $\mathbf{A}$ corresponding to $cv(\tilde{c})_{1mm} \approx 0.02$ and $\mathbf{B}$ corresponding to $cv(\tilde{c})_{1mm} \approx 0.20$.

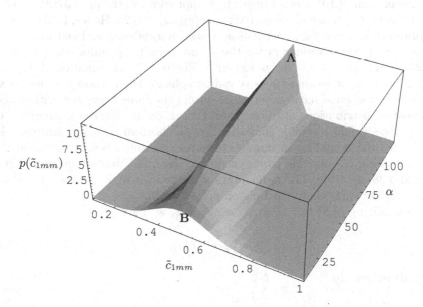

Fig. 9.5. The log-gamma family of probability densities (9.10) for $\tilde{c}_{1mm} \in (0, 1]$, representing the local areal density at the 1mm scale observed in paper, normalized with central mean $\bar{c} = \frac{1}{2}$. Here, $cv(\tilde{c})_{1mm}$ ranges from about 0.02 at $\mathbf{A}$ to about 0.20 at $\mathbf{B}$; about half way between these two points corresponds to a structure with areal density map like the central image in Figure 9.3.

A larger variance than for the random case for the same fibres corresponds to a higher degree of fibre clumping or clustering (also called 'flocculation' by paper scientists) in real networks than in random networks. This arises from non-independence in the deposition of fibres through interactions during manufacturing. In continuous manufacturing processes of papermaking type, including non-woven textiles, this is largely because fibres are flexible and can entangle during mixing in aqueous suspensions from which the network is made by forced filtration. These interactions cause departure from the Poisson assumption that the locations of fibre centres are independent of each other.

9.3 Log-Gamma Information Geometry for Fibre Clustering

Using the log-gamma densities as approximations to the truncated Gaussians from Proposition 3.9, we have approximate probability density functions for $\tilde{c}$ given by

$$p(\tilde{c}; \mu, \alpha) \approx \frac{1}{\Gamma(\alpha)} \left(\frac{1}{\tilde{c}}\right)^{1-\frac{\alpha}{\mu}} \left(\frac{\alpha}{\mu}\right)^{\alpha} \log^{\alpha-1}\left(\frac{1}{\tilde{c}}\right). \qquad (9.8)$$

Figure 9.5 shows the log-gamma probability density functions(9.8) with central mean from (9.10) over a range which approximates the probability density functions for the range of values $0.02 < cv(\tilde{c})_{1mm} < 0.20$. Hence, in the regime of practical interest for paper-like networks, normalising at fixed mean coverage $\bar{c} = \frac{1}{2}$, the parameters for the appropriate log-gamma densities have $\mu \approx 0.7$ and $50 < \alpha < 120$, as shown in Figure 9.6—cf. equation (9.11) below. Figure 9.3 shows areal density radiographs for three wood fibre networks made with the same mean coverage and from the same fibres but with different spatial distributions of fibres, from Oba [156]; an electron micrograph of the surfaces would resemble that shown at the bottom right in Figure 9.1. In Figure 9.3 the actual mean number of fibres covering a point was about 20, corresponding to a writing paper grade and the fibres were on average 2 mm long and 38 μm wide. The radiograph at the top resembles that expected for a random network, that is, a Poisson process of fibres.

Normalizing (9.8) to a central mean $\bar{c} = \frac{1}{2}$, we have

$$\mu = \left(2^{\frac{1}{\alpha}} - 1\right)\alpha, \qquad (9.9)$$

so (9.8) reduces to

$$p(\tilde{c}; \left(2^{\frac{1}{\alpha}} - 1\right)\alpha, \alpha) \approx \frac{1}{\Gamma(\alpha)} \left(\frac{1}{-1 + 2^{\frac{1}{\alpha}}}\right)^{\alpha} \left(\frac{1}{\tilde{c}}\right)^{1+\frac{1}{1-2^{\frac{1}{\alpha}}}} \log^{\alpha-1}\left(\frac{1}{\tilde{c}}\right). \qquad (9.10)$$

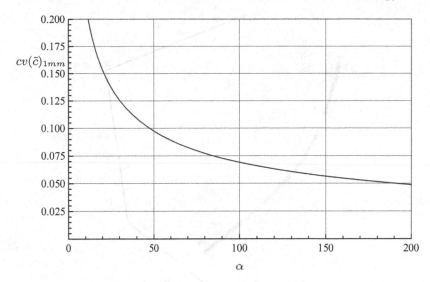

Fig. 9.6. Coefficient of variation $cv(\tilde{c})_{1mm}$ when $\bar{c} = \frac{1}{2}$ for the log-gamma density functions (9.10) as shown in Figure 9.5. A random structure would have $\mu \approx 0.7$ and $cv(\tilde{c})_{1mm} \approx 0.07$, with areal density map rather like that at the top of Figure 9.3.

We observe that from (9.9)

$$\lim_{\alpha \to \infty} \left(2^{\frac{1}{\alpha}} - 1 \right) \alpha = \log 2 \approx 0.7. \tag{9.11}$$

Figure 9.7 gives a surface plot of $cv(\tilde{c})_{1mm}$ for the regime of practical interest, with the curve (9.9) passing through the points having $\bar{c} = \frac{1}{2}$. Essentially, this curve represents the range of fibre network structures that can typically be manufactured by a forced filtration process. A strong brown bag paper could correspond to a point towards **B** with areal density map rather like that at the bottom of Figure 9.3. A very fine glass fibre filter could correspond to a point towards **A** and a random structure would be between these having $(\mu, \alpha) \approx (0.7, 100)$ and $cv(\tilde{c})_{1mm} \approx 0.07$, with areal density map rather like that at the top of Figure 9.3.

Some geodesics in the log-gamma manifold are shown in Figure 9.8 passing through $(\mu, \alpha) = (0.7, 50)$, and in Figure 9.9 passing through the point $(\mu, \alpha) = (0.7, 100)$. Both sets of geodesics have initial directions around the $\alpha = constant$ direction.

9.4 Bivariate Gamma Distributions for Anisotropy

Intuitively, the degree of clumping in the network will influence also the distribution of inter-crossing distances in the network and hence the polygon size distribution. Most processes for the manufacture of stochastic fibrous

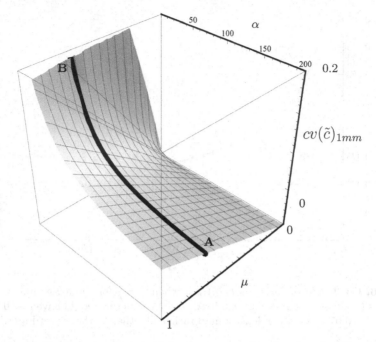

Fig. 9.7. Coefficient of variation of the log-gamma family of probability densities (9.8) approximating the range of distributions of local areal density at the 1mm scale observed for paper. The curve in the surface is given by (9.9) and passes through points having central mean $\bar{c} = \frac{1}{2}$ and for $cv(\tilde{c})_{1mm}$ ranging from about 0.02 at **A** to about 0.20 at **B**. A random structure would be between these with $(\mu, \alpha) \approx (0.7, 100)$ and $cv(\tilde{c})_{1mm} \approx 0.07$.

materials are continuous and yield a web suitable for reeling. Accordingly, the processes tend to impart some directionality to the structure since fibres exhibit a preferential orientation along the direction of manufacture [185, 45]. Several probability densities have been used to model the fibre orientation distribution including von Mises [139] and wrapped-Cauchy distributions [185]. However, in practice a simple one-parameter cosine distribution is sufficient to represent the orientation distribution for most industrially formed networks; this has probability density

$$f(\theta) = \frac{1}{\pi} - \nu \cos(2\,\theta) \tag{9.12}$$

where $0 \leq \nu \leq \frac{1}{\pi}$ is a free parameter controlling the extent of orientation, such that when $\nu = 0$, θ has a uniform distribution. Equation (9.12) is plotted in Figure 9.10 for most of the applicable range of ν; for most machine made papers $0.1 \leq \nu \leq 0.2$.

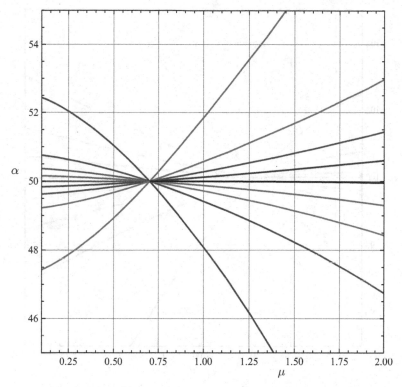

Fig. 9.8. Examples of geodesics in the log-gamma manifold passing through the point $(\mu, \alpha) = (0.7, 50)$ with initial directions around the $\alpha = constant$ direction.

9.5 Independent Polygon Sides

For a two-dimensional random network of lines the distribution of inter-crossing distances, g can be considered as the distribution of intervals between Poisson events on a line and so has an exponential distribution with probability density,

$$f(g) = \frac{1}{\overline{g}}\, e^{-g/\overline{g}}, \tag{9.13}$$

with coefficient of variation, §1.2, $cv(g) = \sigma_g/\overline{g} = 1$.

We expect that the effect of clumping over and above that observed for a random process will be to increase the variance of these inter-crossing distances without significantly affecting the mean. Conversely, we might expect preferential orientation of lines to reduce the number of crossings between lines and hence increase the mean inter-crossing distance and to increase or decrease the variance. A convenient distribution to characterise the inter-crossing distances in such near-random networks is the gamma distribution as suggested by Deng and Dodson [56].

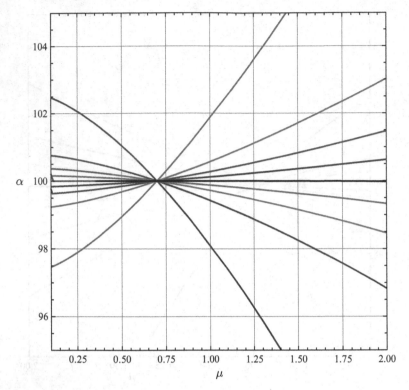

Fig. 9.9. Examples of geodesics in the log-gamma manifold passing through $(\mu, \alpha) = (0.7, 100)$ with initial directions around the $\alpha = constant$ direction.

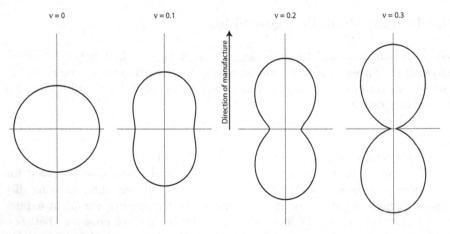

Fig. 9.10. Probability densities for fibre orientation according to the 1-parameter cosine distribution as given by equation (9.12).

The gamma distribution, §1.4.1, has probability density

$$f(g) = \left(\frac{\alpha}{\mu}\right)^\alpha \frac{g^{\alpha-1}}{\Gamma(\alpha)} e^{-\alpha g/\mu} \tag{9.14}$$

with mean $\bar{g} = \mu$ and coefficient of variation, §1.2, $cv(g) = 1/\sqrt{\alpha}$. When $\alpha = 1$ we recover the probability density function for the exponential distribution as given by equation (9.13).

The inter-crossing distances in the network generate the perimeters of the polygonal voids and, from the work of Miles [147], the expected number of sides per polygon is four. This led Corte and Lloyd [48] to model the polygon area distribution for a random line network as a system of rectangular pores with sides given by independent and identical exponential distributions. Interestingly, that model for the void structure of a random fibre network predated by three years the solution in 1968 [58] of the corresponding problem for the matter distribution; so, for paper-like materials, there was an analytic model for the statistics of where matter was not, before there was one for where matter was. Some thirty years later, we extended the treatment given by Corte and Lloyd and derived the probability density of pore area for rectangular pores with sides given by independent and identical gamma distributions [71, 72], where the free parameters of the gamma distribution are assumed to represent the influence of fibre clumping and fibre orientation.

Once again, the uniqueness property of the gamma distribution that was proved by Hwang and Hu [106] (cf. their concluding remark) given as Theorem 1.1 above is relevant here. Now, it is commonly found in experimental measurements of pore size distributions for papers and non-woven textiles made under different conditions that the standard deviation is proportional to the mean. This encourages the view that the gamma distribution is an appropriate generalization of the exponential distribution for adjacent polygon side lengths in near-random fibre networks. Miles [147] had proved that for planar random lines the distribution of the diameters of inscribed circles in polygons is exponential and unaltered by changing the infinite random lines to infinite random rectangles for arbitrary distributions of width. This suggests that gamma distributions may well be appropriate to model the inscribed circles in non-random cases.

We seek the probability density for the areas, a of rectangles with sides g_x and g_y such that $a = g_x g_y$ and the probability densities of g_x and g_y are given by equation (9.14). The probability density of a is given by

$$g(a) = \int_0^\infty f(g_x) f(g_y) dg_x \tag{9.15}$$

$$= \int_0^\infty \frac{1}{g_x} f(g_x) f(a/g_x) dg_x \tag{9.16}$$

$$= \frac{2}{\Gamma(\alpha)^2} a^{\alpha-1} \alpha^{2\alpha} \mu^{-2\alpha} K_0(\zeta) \quad \text{where } \zeta = \frac{2\alpha}{\mu} \sqrt{a} \tag{9.17}$$

and $K_0(\zeta)$ is the zeroth order modified Bessel function of the second kind.

The mean, variance and coefficient of variation, §1.2, of rectangle area are

$$\bar{a} = \mu^2 \tag{9.18}$$

$$\sigma^2(a) = \frac{1 + 2\alpha}{\alpha^2} \mu^4 \tag{9.19}$$

$$cv(a) = \frac{\sqrt{1 + 2\alpha}}{\alpha} \tag{9.20}$$

respectively. Then it follows that we can solve for α in terms of $cv(a)$:

$$\alpha = \frac{1 \pm \sqrt{1 + cv(a)^2}}{cv(a)^2} \tag{9.21}$$

so we take the positive root here and note that we recover the random case, $\alpha = 1$, precisely if $cv(a) = \sqrt{3}$.

Experimental analyses of porous materials using mercury porosimetry or fluid displacement porometry typically infer pore sizes and their distributions in terms of an equivalent pore diameter. Such measures are convenient as they provide an accessible measure of, for example, the sizes of particles that might be captured by a fibrous filter. Following Corte and Lloyd [48], we define the equivalent radius r_p of a rectangular pore as the radius of a circle with the same area, such that

$$r_p = \sqrt{\frac{a}{\pi}}. \tag{9.22}$$

From equation (9.17) we have

$$p(r) = 2\pi r_p \, g(\pi r_p^2) \tag{9.23}$$

$$= \frac{4\pi^\alpha \alpha^{2\alpha} \mu^{-2\alpha} r_p^{2\alpha-1} K_0(\zeta)}{\Gamma(\alpha)^2} \quad \text{where } \zeta = 2\sqrt{\pi} r_p \alpha/\mu \tag{9.24}$$

and the mean, variance and coefficient of variation, §1.2, of pore radius are

$$\bar{r}_p = \frac{\mu}{\sqrt{\pi}} \frac{\Gamma(\alpha + 1/2)^2}{\alpha \Gamma(\alpha)^2} \tag{9.25}$$

$$\sigma^2(r_p) = \frac{\mu^2}{\pi} \left(1 - \frac{\Gamma(\alpha + 1/2)^4}{\alpha^2 \Gamma(\alpha)^4} \right) \tag{9.26}$$

$$cv(r_p) = \frac{\alpha \Gamma(\alpha)^2}{\Gamma(\alpha + 1/2)^2} \sqrt{1 - \frac{\Gamma(\alpha + 1/2)^4}{\alpha^2 \Gamma(\alpha)^4}} \tag{9.27}$$

We see that here the coefficient of variation is independent of the mean, a property that characterises the gamma distribution, and it is easily shown that the probability density function (9.24) for pore radii is well-approximated by a gamma distribution of the same mean and variance [73].

Recall that the mean of the underlying gamma distributions representing the polygon side lengths is μ and the coefficient of variation is $1/\sqrt{\alpha}$,

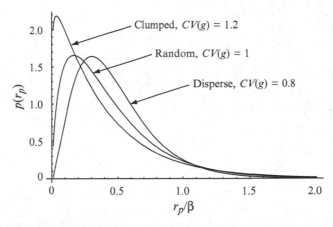

Fig. 9.11. Probability densities of pore radii at and near to the random case as given by equation (9.24).

so we observe that the expected pore radius is proportional to the expected polygon side length, or inter-crossing distance, and the constant of proportionality depends only on parameter α, hence on the coefficient of variation, §1.2, of inter-crossing distances, which can be considered a measure of the non-uniformity of the structure—in radiographs of real networks, increased coefficient of variation corresponds to increased contrast in the images. The coefficient of variation of equivalent pore radius is plotted against the parameter α in Figure 9.12.

Note also that for random networks, we have $\alpha = 1$ and,

$$\bar{r}_p^{random} = \frac{\sqrt{\pi}}{4}\,\mu \tag{9.28}$$

$$\sigma^2(r_p^{random}) = \left(\frac{1}{\pi} - \frac{\pi}{16}\right)\mu^2 \tag{9.29}$$

$$cv(r_p^{random}) = \frac{\sqrt{16 - \pi^2}}{\pi} \approx 0.788 \tag{9.30}$$

The probability density of pore radii, as given by equation (9.24) is plotted in Figure 9.11 for the random case and for networks with higher and lower degrees of uniformity as quantified by their coefficients of variation of rectangle side lengths; these are labelled 'clumped' and 'disperse' respectively. The influence of network uniformity on the mean and coefficient of variation of pore radii is shown in Figure 9.13.

9.5.1 Multiplanar Networks

In practice, real fibre networks are 3-dimensional though the hydrodynamics of filtering suspensions of fibres with relatively high aspect ratios at high speed yields layered structures—fibres tend to penetrate only a few fibre diameters

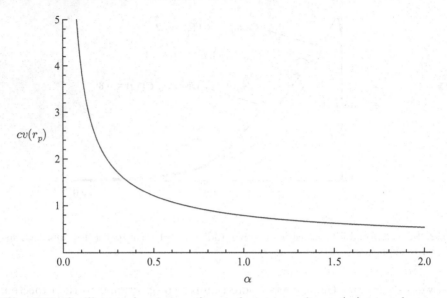

Fig. 9.12. Coefficient of variation of equivalent pore radius $cv(r_p)$, given by equation (9.27) as a function of gamma distribution parameter α. The random case has $\alpha = 1$.

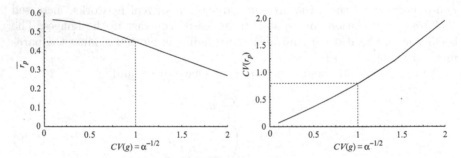

Fig. 9.13. Effect of network uniformity on mean (left) and coefficient of variation of pore radii (right) as given by equations (9.25) and (9.27) respectively.

up or down through the network as established by Radvan et al. [170]. However, sometimes the mean coverage of fibre networks is such that a significant proportion of fibres exist in planes several fibre diameters from the plane of support of the network. In such cases, the porous structure can be modelled by considering the superposition of several 2-dimensional layers.

As mentioned above, it turns out that the probability density for pore radii given by equation (9.24) is itself well approximated by a gamma distribution with the same mean and variance. Moreover, the distribution of a sum of independent gamma distributed random variables is itself a gamma distribution. Consider then a layered structure of circular voids with gamma distributed radii. The probability density function f and cumulative distribution function F, §1.2, for pore radii in a single layer are given by,

$$f(r_p) = \left(\frac{\alpha}{\mu}\right)^\alpha \frac{r_p^{\alpha-1}}{\Gamma(\alpha)} e^{-\alpha r_p/\mu}, \tag{9.31}$$

$$F(r_p) = 1 - \frac{\Gamma(\alpha, r_p\,\alpha/\mu)}{\Gamma(\alpha)}, \tag{9.32}$$

where $\Gamma(a, z)$ is the incomplete gamma function.

A second layer with an independent and identical distribution of pore radii is placed over the first layer such that the centres of pairs of voids in the two layers are aligned. For such a structure, we assign to each pair of pores the radius of the smaller pore such that we consider effectively the radii of the most constricted part of a path through the network. The probability density of these radii is given by,

$$f(r_p, 2) = 2\,(1 - F(r_p))\,f(r_p),$$
$$= 2\,\frac{\Gamma(\alpha, r_p\,\alpha/\mu)}{\Gamma(\alpha)}\,f(r_p). \tag{9.33}$$

The cumulative distribution function, §1.2, for two layers is

$$F(r_p, 2) = 1 - \left(\frac{\Gamma(\alpha, r_p\,\alpha/\mu)}{\Gamma(\alpha)}\right)^2. \tag{9.34}$$

Applying the same notation, the addition of further layers gives iteratively

$$f(r_p, 3) = 3\,(1 - F(r_p, 2))\,f(r_p) = 3\left(\frac{\Gamma(\alpha, r_p\,\alpha/\mu)}{\Gamma(\alpha)}\right)^2 f(r_p),$$

$$F(r_p, 3) = 1 - \left(\frac{\Gamma(\alpha, r_p\,\alpha/\mu)}{\Gamma(\alpha)}\right)^3,$$

$$f(r_p, 4) = 4\,(1 - F(r_p, 3))\,f(r_p) = 4\left(\frac{\Gamma(\alpha, r_p\,\alpha/\mu)}{\Gamma(\alpha)}\right)^3 f(r_p),$$

$$F(r_p, 4) = 1 - \left(\frac{\Gamma(\alpha, r_p\,\alpha/\mu)}{\Gamma(\alpha)}\right)^4, \quad etc.$$

So, for a structure composed of n layers we have the general expressions for the probability density and cumulative distribution functions of pore radii respectively:

$$f(r_p, n) = n\left(\frac{\Gamma(\alpha, r_p\,\alpha/\mu)}{\Gamma(\alpha)}\right)^{n-1} f(r_p), \tag{9.35}$$

$$F(r_p, n) = 1 - \left(\frac{\Gamma(\alpha, r_p\,\alpha/\mu)}{\Gamma(\alpha)}\right)^n. \tag{9.36}$$

To use equation (9.35) we need closed expressions giving the number of layers, n, and the expected pore radius μ of a monolayer network in terms of network and fibre properties.

Consider a network with mean coverage $\bar{c}$ and fractional void volume, or porosity, ε. An elemental plane of this network can be represented as a two-dimensional structure with fractional open area ε. From the Poisson statistics for coverage of this two-dimensional structure as given by equation (9.4) we have

$$\varepsilon = P(0) = e^{-\bar{c}_\varepsilon}, \tag{9.37}$$

where $\bar{c}_\varepsilon$ is the expected coverage of a two-dimensional network with fractional open area ε. It follows that

$$\bar{c}_\varepsilon = \log\left(1/\varepsilon\right), \tag{9.38}$$

so the number of layers n is given by

$$n = \frac{\bar{c}}{\bar{c}_\varepsilon} = \frac{\bar{c}}{\log\left(1/\varepsilon\right)}. \tag{9.39}$$

To determine the expected pore radius for a two-dimensional network with mean coverage $\bar{c}_\varepsilon$ we consider first the number of intersections between fibres occurring per unit area; for fibres of width ω this is given by Kallmes et al. [111] as

$$n_{\text{int}} = \frac{\bar{c}_\varepsilon^2}{\pi\,\omega^2}. \tag{9.40}$$

Inevitably, the number of polygons n_{poly} per unit area is approximately the same as the number of fibre intersections per unit area, and since $\bar{c}_\varepsilon = \log\left(1/\varepsilon\right)$ this is given by,

$$n_{\text{poly}} = \frac{\left(\log\left(1/\varepsilon\right)\right)^2}{\pi\,\omega^2}. \tag{9.41}$$

It follows that the expected area of a polygonal void is

$$\bar{a}_\varepsilon = \frac{\varepsilon}{n_{\text{poly}}}$$

$$= \frac{\pi\,\varepsilon\,\omega^2}{\left(\log\left(1/\varepsilon\right)\right)^2}. \tag{9.42}$$

Again, we define the pore radius as that of a circle with the same area

$$\bar{r}_{p,\varepsilon} = \sqrt{\frac{\bar{a}_\varepsilon}{\pi}}$$

$$= \frac{\omega\,\sqrt{\varepsilon}}{\log\left(1/\varepsilon\right)}. \tag{9.43}$$

Figure 9.14 shows the influence of mean coverage and porosity on the mean pore radius and standard deviation of pore radii as calculated from the

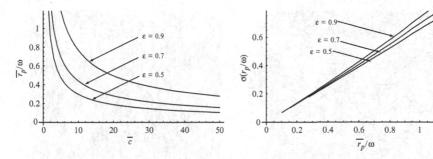

Fig. 9.14. Effect of coverage and porosity on pore radii as given by by equation (9.35) with equations (9.40) and (9.43) and $\alpha = \pi^2/(16 - \pi^2)$. Left: mean pore radius; right: standard deviation of pore radii plotted against the mean.

probability density given by equation (9.35) with equations (9.40) and (9.43) and with $\alpha = \pi^2/(16 - \pi^2)$ such that $cv(r_{p,\varepsilon})$ coincides with equation (9.30) for random networks. The mean pore radius decreases with mean coverage and porosity and for each case the standard deviation of pore radii is proportional to the mean. This latter property is important since it tells us that the coefficient of variation of pore radii is independent of the mean. There is only a weak dependence of the coefficient of variation on porosity and this is consistent with measurements reported in the literature.

9.6 Correlated Polygon Sides

There is a growing body of evidence confirming the suitability of the gamma distribution to describe the dimensions of voids in stochastic fibrous materials [40, 67, 181], and in general classes of stochastic porous materials [108, 109], but the influence on physical phenomena of the parameters characterising the distribution has yet to be satisfactorily explained. In part this arises because our models for pore areas and hence for pore radii so far assume that adjacent polygon sides have independent lengths; in practice they do not. The observation was made by Corte and Lloyd [48] that 'roundish' hence near-regular polygons are more frequent than irregular ones, and that 'slit-shaped' polygons are extremely rare [46]. See Miles [149] and Kovalenko [123] for proofs that this regularity is in fact a limiting property for random polygons as the area, perimeter or number of sides per polygon become large.

Graphical representations of two stochastic fibre networks are given in Figure 9.15; the histogram beneath each network shows the distribution of orientations of fibres in that network. Each network consists of 1500 fibres of unit length with their centres randomly positioned within a square of side 10 units. In the image on the left, the fibre orientation distribution is uniform; in the image on the right, the locations of the fibre centres are the same but fibre axes are preferentially oriented towards the vertical.

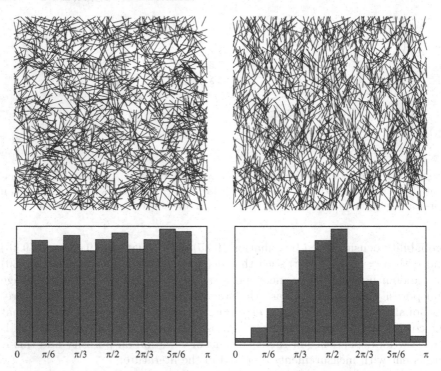

Fig. 9.15. Networks of 1500 unit length fibres with their centres randomly positioned within a square of side length 10 units. On the left, isotropy: fibre axes have a uniform orientation. On the right, anisotropy: fibre axes are preferentially oriented towards the vertical direction. The distributions of orientations are indicated in the histograms. The locations of fibre centres in each image are the same.

In the isotropic network on the left of Figure 9.15, inspection reveals that regions of high density have many short inter-crossing distances, and in regions of low density there are fewer but longer inter-crossing distances. So we have positive correlation between nearby polygonal side lengths and this tends to yield more regular polygons, simply from the random variations in the local density that arise from the underlying Poisson point process for fibre centres. This means that random isotropy has an inherent 'ground state' correlation of adjacent inter-crossing distances, which explains why pores seem mainly 'roundish' in real networks.

In the oriented network on the right we still see the effects of the local density of crossings on the lengths of nearby sides, but we observe fewer 'roundish' polygons and more polygons of appreciable aspect ratio. It is important to note however that even in the oriented example shown here, the correlation between pairs of adjacent polygon sides remains positive; so the effect of random variations in the local density overwhelms the superposed anisotropy. For a significant fraction of polygons to be 'slit-shaped', it would require pairs of adjacent polygon sides to consist of one short side and one long side. That would mean strong negative correlation, §1.3, and this is unlikely in stochastic fibre networks since the Poisson clustering of fibre centres inherent in such structures favours positive correlation. So we have a kind of meta-Theorem:

> A Poisson process of lines in a plane has an intrinsic ground state of positive correlation between adjacent polygon edges and this persists even when the process has substantial anisotropy.

This immediately begs two questions:

- What is the value of this ground state positive correlation for isotropic random networks?
- Is there a level of anisotropy that could reduce the ground state positive correlation to zero?

In [75] we used simulations of random line networks to show that the lengths of adjacent sides in random polygons are correlated with correlation coefficient $\rho \approx 0.616$.

Here we study the influence of the degree of correlation between the lengths of adjacent polygon sides on the statistics describing the distribution of pore sizes in thin fibre networks. We have developed computer routines that generate pairs of random numbers distributed according to the McKay bivariate gamma distribution, §4.1, which allows positive correlation $0 < \rho \leq 1$.

Pores are modelled as ellipses, each characterised by a pair (x, y) that represents its minor and major axes respectively. The eccentricity of each pore is given by,

$$e = \sqrt{1 - \left(\frac{x}{y}\right)^2} \tag{9.44}$$

and as a rule of thumb, we can consider pores to be 'roundish' if they have eccentricity less than about 0.6. The area and perimeter of an elliptical pore are given by

$$A = \pi\, x\, y, \quad P = 4\, y\, E(e^2) \tag{9.45}$$

where $E(e^2)$ is the complete elliptic integral.

The parameter of primary interest to us here is the equivalent radius of each pore and this is given by

$$r_{eq} = 2\,\frac{A}{P}. \tag{9.46}$$

For each generated pair of correlated (x, y) we obtain the eccentricity e, area A, perimeter P and hence the equivalent pore radius r_{eq}. In the next section we describe a simulator and use it to estimate the statistics describing the distribution of these parameters. First we need a source distribution for the pairs (x, y).

9.6.1 McKay Bivariate Gamma Distribution

The McKay bivariate gamma distribution for correlated $x < y$ has joint probability density,

$$m(x, y; \alpha_1, \sigma_{12}, \alpha_2) = \frac{\left(\frac{\alpha_1}{\sigma_{12}}\right)^{\frac{(\alpha_1 + \alpha_2)}{2}} x^{\alpha_1 - 1} (y - x)^{\alpha_2 - 1} e^{-\sqrt{\frac{\alpha_1}{\sigma_{12}}} y}}{\Gamma(\alpha_1)\Gamma(\alpha_2)}, \tag{9.47}$$

where $0 < x < y < \infty$ and $\alpha_1, \sigma_{12}, \alpha_2 > 0$. We may use it to model correlated polygon sides since the ordering of polygon sides is arbitrary.

The marginal probability densities of x and y are univariate gamma distributions with,

$$\bar{x} = \sqrt{\alpha_1 \sigma_{12}} \tag{9.48}$$

$$\bar{y} = (\alpha_1 + \alpha_2) \sqrt{\frac{\sigma_{12}}{\alpha_1}}. \tag{9.49}$$

The correlation coefficient between x and y is given by,

$$\rho = \sqrt{\frac{\alpha_1}{\alpha_1 + \alpha_2}}. \tag{9.50}$$

We need to keep the total fibre length constant, so we want $\bar{x} + \bar{y} = 2$. From equation (9.48) we have,

$$\bar{y} = 2 - \sqrt{\alpha_1 \sigma_{12}}. \tag{9.51}$$

Equating equations (9.51) and (9.49) yields, on manipulation,

$$\alpha_2 = 2 \left(\sqrt{\frac{\alpha_1}{\sigma_{12}}} - \alpha_1 \right), \tag{9.52}$$

and from equation (9.48) we have,

$$\alpha_1 = \frac{\bar{x}^2}{\sigma_{12}}. \tag{9.53}$$

On simplification, substitution of equations (9.52) and (9.53) in equation (9.50) yields

$$\rho = \sqrt{\frac{\bar{x}}{2 - \bar{x}}} = \sqrt{\frac{\bar{x}}{\bar{y}}}. \tag{9.54}$$

It follows directly that

$$\bar{x} = \frac{2\rho^2}{1 + \rho^2} \qquad \text{and} \qquad \bar{y} = \frac{2}{1 + \rho^2} \tag{9.55}$$

such that
$$\bar{x} + \bar{y} = 2.$$

The variances of x and y are given by,

$$\sigma^2(x) = \frac{\bar{x}}{\alpha_1} \tag{9.56}$$

$$\sigma^2(y) = \frac{\bar{y}}{\alpha_1 + \alpha_2}, \tag{9.57}$$

respectively, and from equation (9.50) we have

$$\alpha_2 = \alpha_1 \left(\frac{1}{\rho^2} - 1 \right). \tag{9.58}$$

Substitution of equations (9.55) and (9.58) in equations (9.56) and (9.57) respectively yields,

$$\sigma^2(x) = \frac{4\,\rho^4}{(1+\rho^2)^2\,\alpha_1} \tag{9.59}$$

$$\sigma^2(y) = \frac{4\,\rho^2}{(1+\rho^2)^2\,\alpha_1}, \tag{9.60}$$

such that
$$\sigma^2(x) = \rho^2\,\sigma^2(y).$$

By specifying the correlation coefficient ρ and the coefficient of variation, §1.2, of x :

$$cv(x) = \frac{\sigma_x}{\bar{x}} = \frac{1}{\sqrt{\alpha_1}}$$

we fully define the marginal and joint probability densities of x and y:

$$\alpha_1 = \frac{1}{cv(x)^2}$$

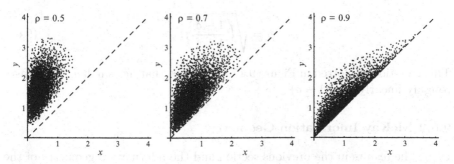

Fig. 9.16. Three outputs of the McKay simulator with differing values of correlation coefficient: $\rho = 0.5, 0.7, 0.9$. Each family of 5000 pairs (x, y) with $x < y$ has coefficient of variation $cv(x) = 0.6$.

$$\alpha_2 = \alpha_1 \left(\frac{1}{\rho^2} - 1 \right)$$

$$\bar{x} = \frac{2\rho^2}{1+\rho^2}$$

$$\bar{y} = \frac{2}{1+\rho^2}$$

$$\sigma_{12} = \frac{\bar{x}^2}{\alpha_1}.$$

It is easy to show that

$$cv(y) = \rho \, cv(x) \quad \text{hence} \quad cv(y) < cv(x) = \frac{1}{\sqrt{\alpha_1}}.$$

The McKay probability density functions do not include the situation when both marginal distributions are exponential; however, the two special cases when one of the marginals is exponential are of interest. These have respectively, $\alpha_1 = 1$ and $\alpha_1 + \alpha_2 = 1$ and then we can express the parameters in terms of the correlation coefficient ρ :

$$\alpha_1 = 1 \Rightarrow cv(x) = 1,$$

$$cv(y)^2 = \rho^2 = \frac{1}{1+\alpha_2}, \ \sigma_{12} = \left(\frac{2\rho^2}{1+\rho^2} \right)^2 \quad (9.61)$$

$$\alpha_1 + \alpha_2 = 1 \Rightarrow cv(x)^2 = \frac{1}{\rho^2},$$

$$cv(y)^2 = 1 + \rho^2 = \frac{2}{\sigma_{12}}, \ \alpha_1 = \rho^2. \quad (9.62)$$

Furthermore, in order to maintain constant total fibre length in our networks, we have controlled the mean values so that $\bar{x} + \bar{y} = 2$ so if $\bar{x} = 1 - \delta$ then $\bar{y} = 1 + \delta$. Then from (9.54) above we have

$$\rho = \sqrt{\left(\frac{1-\delta}{1+\delta} \right)}. \quad (9.63)$$

This equation is plotted in Figure 9.17 and we see that, for small δ, ρ decreases roughly linearly like $(1 - \delta)$.

9.6.2 McKay Information Geometry

Using the results in the previous section and the information geometry of the McKay distribution, we can illustrate how the variability of polygon side length $cv(x)$ and the correlation coefficient ρ for adjacent polygon sides influence the structure from an information geometric viewpoint. What we would like to

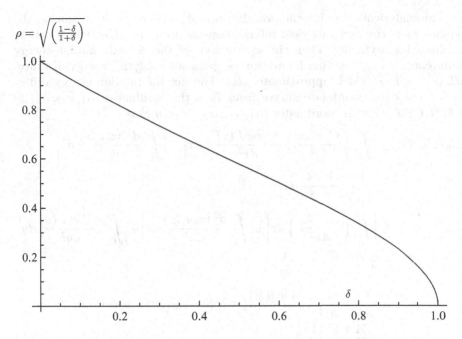

Fig. 9.17. McKay correlation coefficient ρ from equation (9.63) where $\bar{x} = 1 - \delta$ and $\bar{y} = 1 + \delta$.

show would be the information distance of a given random line structure from a reference structure; unfortunately, distances are hard to compute in Riemannian manifolds—the McKay manifold M is a 3-dimensional curved space.

Distance between a pair of points in a Riemannian manifold is defined as the infimum of arc lengths over all curves between the points. For sufficiently nearby pairs of points there will be a unique minimizing geodesic curve that realises the infimum arc length. In general, such curves are hard to find between more distant points. However, we can obtain an upper bound on distances between two points $T_0, T_1 \in M$ by taking the sum of arc lengths along coordinate curves that triangulate the pair of points with respect to the coordinate axes. We obtain the following upper bound on the information metric distance in M from T_0 with coordinates $(\alpha_1, \sigma_{12}, \alpha_2) = (A, B, C)$ to T_1 with coordinates $(\alpha_1, \sigma_{12}, \alpha_2) = (a, b, c)$

$$d_M(T_0, T_1) \leq \left| \int_A^a \sqrt{\left| \frac{C - 3x}{4x^2} + \frac{d^2 \log \Gamma(x)}{dx^2} \right|} \, dx \right| + \left| \int_C^c \sqrt{\left| \frac{d^2 \log \Gamma(y)}{dy^2} \right|} \, dy \right|$$

$$+ \left| \int_B^b \sqrt{\frac{A + C}{4 \, z^2}} \, dz \right|. \tag{9.64}$$

There is a further difficulty because of the presence of square roots arising from the norms of tangent vectors to coordinate curves so it is difficult to obtain analytically the information distance, d_M. However, by removing the square roots the integrals yield information-energy values E_M, which can be evaluated analytically. Then the square root of the net information-energy differences along the coordinate curves gives an analytic 'energy-distance' $dE_M = \sqrt{E_M}$, which approximates d_M. The net information-energy differences along the coordinate curves from T_0 with coordinates $(\alpha_1, \sigma_{12}, \alpha_2) = (A, B, C)$ to T_1 with coordinates $(\alpha_1, \sigma_{12}, \alpha_2) = (a, b, c)$ is

$$
E_M(T_0, T_1) \leq \left| \int_A^a \left(\frac{C - 3x}{4x^2} + \frac{d^2 \log \Gamma(x)}{dx^2} \right) dx \right| + \left| \int_C^c \frac{d^2 \log \Gamma(y)}{dy^2} dy \right|
$$

$$
+ \left| \int_B^b \frac{A + C}{4\,z^2} dz \right|
$$

$$
\leq \left| \int_A^a \left(\frac{C - 3x}{4x^2} \right) dx \right| + \left| \int_A^a \frac{d^2 \log \Gamma(x)}{dx^2} dx \right| + \left| \int_C^c \frac{d^2 \log \Gamma(y)}{dy^2} dy \right|
$$

$$
+ \frac{A + C}{4} \left| \frac{1}{b} - \frac{1}{B} \right|
$$

$$
= \left| \frac{C}{4\,a} - \frac{C}{4\,A} + \frac{3 \log(\frac{A}{a})}{4} \right| + |\psi(a) - \psi(A)| + |\psi(c) - \psi(C)|
$$

$$
+ \frac{A + C}{4} \left| \frac{1}{b} - \frac{1}{B} \right|. \tag{9.65}
$$

$$
dE_M = \sqrt{E_M}. \tag{9.66}
$$

where $\psi = \frac{\Gamma'}{\Gamma}$ is the digamma function. Now we take for T_0 the coordinate values corresponding to $cv(x) = 1$, so $\alpha_1 = 1$ giving the exponential distribution of minor axes of equivalent elliptical voids; doing this for arbitrary ρ gives:

$$
T_0 = \left(1, \frac{4\rho^4}{(\rho^2 + 1)^2}, \frac{1}{\rho^2} - 1 \right).
$$

For T_1 we allow $cv(x)$ to range through structures from dispersed ($cv(x) < 1$) through random ($cv(x) = 1$) to clustered ($cv(x) > 1$), for a range of ρ values:

$$
T_1 = \left(\frac{1}{cv(x)^2}, \frac{4\rho^4 cv(x)^2}{(\rho^2 + 1)^2}, \frac{\frac{1}{\rho^2} - 1}{cv(x)^2} \right).
$$

Making these substitutions in equation (9.65) we obtain for arbitrary

$$
E_M(cv(x), \rho)_{|[T_0 : \alpha_1 = 1]} = \frac{(\rho^2 + 1)^2}{16\rho^6} \left| \frac{1}{cv(x)^2} - 1 \right|
$$

$$
+ \frac{1}{4} \left| \left(1 - \frac{1}{cv(x)^2} \right) \left(1 - \frac{1}{\rho^2} \right) + 3 \log \left(cv(x)^2 \right) \right|
$$

$$
+ \left| \psi \left(\frac{1}{cv(x)^2} \frac{1}{\rho^2} - 1 \right) - \psi \left(\frac{1}{\rho^2} - 1 \right) \right|
$$

$$
+ \left| \psi \left(\frac{1}{cv(x)^2} \right) + \gamma \right| \tag{9.67}
$$

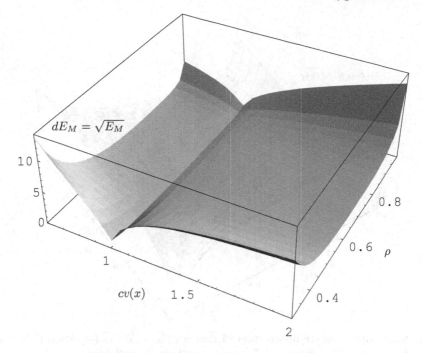

Fig. 9.18. Approximate information distances $dE_M = \sqrt{E_M}$ (equation (9.67)) in the McKay manifold, measured from distributions T_0 with exponential marginal distribution for x, so $\alpha_1 = 1$ and $cv(x) = 1$.

where γ is the Euler gamma constant, of numerical value about 0.577.

Figure 9.18 shows a plot of $dE_M = \sqrt{E_M(cv(x), \rho)}$ From equation (9.67). This is an approximation but we expect it represents the main features of the distance of arbitrary random line structures T_1 from T_0 with $\alpha_1 = 1$ and hence $cv(x) = 1$. The inherent variability of a Poisson process of lines seems to yield a ground state of positive correlation between adjacent polygon edges and this plot suggests that the information distance increases more rapidly away from the line $cv(x) = 1$ when ρ becomes very small or very large.

Repeating the above procedure for the case when T_0 has $\alpha_1 + \alpha_2 = 1$, we obtain

$$E_M(\alpha_1, \alpha_2)_{|[T_0:\alpha_1+\alpha_2=1]} = |\psi(\alpha_2) - \psi(1 - \alpha_1)|$$

$$+ \frac{1}{4}\left|\frac{(2\alpha_1 + \alpha_2)^2}{4\alpha_1} + \frac{1}{2}(-\alpha_1 - 1)\right|. \quad (9.68)$$

This is plotted in Figure 9.19.

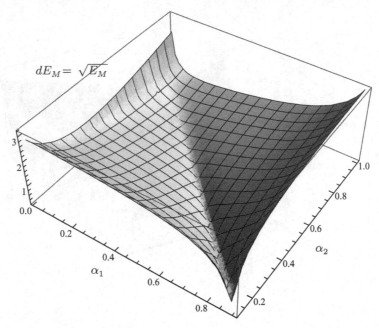

$$dE_M = \sqrt{E_M}$$

Fig. 9.19. Approximate information distances $dE_M = \sqrt{E_M}$ (equation (9.68)) in the McKay manifold, measured from distributions T_0 with exponential marginal distribution for y, so $\alpha_1 + \alpha_2 = 1$.

9.6.3 McKay Information Entropy

We recall from Figure 1.4 that among the gamma distributions, the exponential distribution has maximum information entropy. The entropy of the McKay bivariate gamma densities is

$$S(\alpha_1, \sigma_{12}, \alpha_2) = \alpha_1^{-\frac{\alpha_1}{2}} \, \sigma_{12}^{\frac{1}{2}(\alpha_1+1)} \, L \tag{9.69}$$

$$\text{where } L = \log\left(\frac{\alpha_1}{\sigma_{12}\Gamma(\alpha_1)\Gamma(\alpha_2)}\right)$$
$$+ \psi(\alpha_1)(\alpha_1 - 1) - \alpha_1 + \psi(\alpha_2)(\alpha_2 - 1) - \alpha_2.$$

In fact, using

$$\sigma_{12} = \frac{1}{\alpha_1}\left(\frac{2\rho^2}{1+\rho^2}\right)^2$$

we can re-express this in terms of α_1, α_2, ρ in the form

$$S\left(\alpha_1, \frac{1}{\alpha_1}\left(\frac{2\rho^2}{1+\rho^2}\right)^2, \alpha_2\right) = 2^{\alpha_1+1}\left(\frac{\rho^2}{\rho^2+1}\right)^{\alpha_1+1} \alpha_1^{-\frac{\alpha_1}{2}} K \tag{9.70}$$

where $K = \log \left(\dfrac{(\rho^2 + 1)^2 \alpha_1}{4\rho^4 \Gamma(\alpha_1) \Gamma(\alpha_2)} \right)$

$+ \psi(\alpha_1)(\alpha_1 - 1) - \alpha_1 + \psi(\alpha_2)(\alpha_2 - 1) - \alpha_2.$

Then, at $\alpha_1 = 1$, we have in terms of α_2 and ρ

$$S\left(1, \left(\frac{2\rho^2}{1 + \rho^2} \right)^2, \alpha_2 \right) = \frac{4\rho^4}{(\rho^2 + 1)^2} \left(\log \left(\frac{4\rho^4 \Gamma(\alpha_2)}{(\rho^2 + 1)^2} \right) \right.$$
$$\left. + \left(\psi^{(0)}(\alpha_2) - 1 \right) (\alpha_2 - 1) \right).$$

$$(9.71)$$

Figure 9.20 shows the information entropy S from equation (9.71) for McKay probablity density functions, in terms of α_2, ρ when $\alpha_1 = 1$ and $cv(x) = 1$. So here, the minor axis has an exponential distribution. Making the substitution

$$\alpha_2 = \alpha_1 \left(\frac{1}{\rho^2} - 1 \right)$$

from the previous section we obtain the entropy in terms of ρ alone. It turns out that there is a shallow maximum local entropy near $\rho = 0.24$ and a pronounced local minimum near $\rho = 0.75$, Figure 9.21. In Figure 9.17 we can see that $\rho = 0.24$ corresponds to $\delta \approx 0.89$. In Figure 9.22 we plot $\frac{dS}{d\rho}$, the total derivative of entropy with respect to ρ for the interval $0.1 < \rho < 0.76$ and we

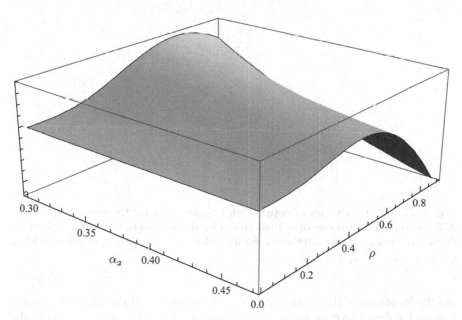

Fig. 9.20. Information entropy S from equation (9.71) for McKay probablity density functions, in terms of α_2, ρ when $\alpha_1 = 1$ and $cv(x) = 1$.

$S(\rho)$ for $\alpha_1 = 1$

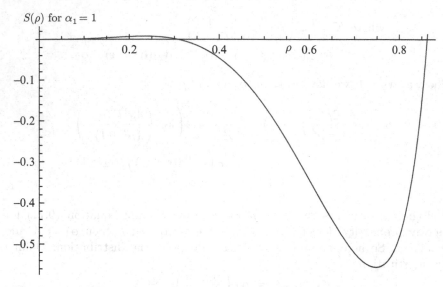

Fig. 9.21. McKay information entropy S in terms of ρ when $cv(x) = 1 = \alpha_1$.

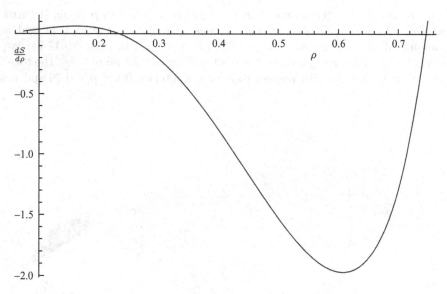

Fig. 9.22. Total derivative of entropy with respect to ρ for the interval $0.1 < \rho < 0.76$ showing the locations of critical points by the intersections with the abscissa. From equation (9.71) for McKay probability density functions, in terms of ρ when $\alpha_2 = \alpha_1 \left(\frac{1}{\rho^2} - 1 \right)$ and $cv(x) = 1 = \alpha_1$.

see the locations of these critical points by the intersections with the absisca. Figure 4.4 shows geodesics with $\alpha_1 = 1$ passing through $(c, \alpha_2) = (1, 1)$ in the McKay submanifold M_1 where

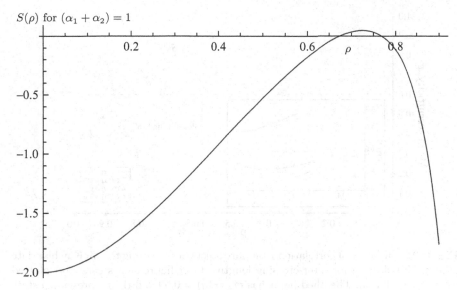

Fig. 9.23. McKay information entropy in terms of ρ when $(\alpha_1 + \alpha_2) = 1$.

$$c = \sqrt{\frac{\alpha_1}{\sigma_{12}}} = \sqrt{\frac{1}{\sigma_{12}}} = \frac{1}{\overline{x}}$$

and so also $cv(x) = 1$.

Figure 9.23 shows a plot of the information entropy as a function of ρ when $(\alpha_1 + \alpha_2) = 1$; there is a maximum near $\rho = 0.72$, cf. also Figure 9.17 from which we see that this network has $\delta \approx 0.32$.

9.6.4 Simulation Results

Our simulator generates pairs (x, y) distributed according to the McKay bivariate gamma distribution as given by equation (9.47) and we use these pairs to compute the structural characteristics of voids as discussed previously for the isotropic case. The inputs to the routine are the target values of the correlation coefficient, ρ, and the coefficient of variation, §1.2, of x which controls the overall variability in the network—$cv(x) = 1$ corresponds to a random network.

To generate the McKay distributed pairs, first a sufficiently large part of the x-y plane is partitioned into square regions labelled by their lower left corners. We compute the total McKay probability for pairs in the $(i, j)^{th}$ region of the plane as:

$$P_{ij} = \int_{x_i}^{x_i + \Delta x} \int_{y_j}^{y_j + \Delta y} m(x_i, y_j)\, dy\, dx \qquad (9.72)$$

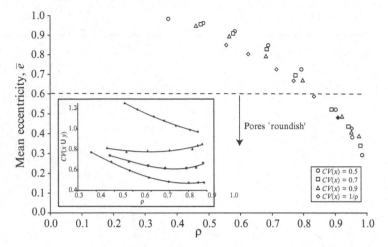

Fig. 9.24. Influence of correlation ρ on pore eccentricity e using the McKay bivariate gamma distribution for inter-crossing lengths. Inset figure shows $cv(x \cup y)$ plotted against correlation. The filled point has $cv(x \cup y) = 0.99$ and thus represents a state close to a random network.

This integral is evaluated numerically for each (i, j) and these probabilities multiplied by the required total number of pairs (x, y) define the simulated number of points in each region. The algorithm then generates this number of pairs of uniformly distributed random numbers in the interval $(x_i + \Delta x, y_j + \Delta y)$ for each (i, j). In the simulations presented here using $0 < x < y < 4$, intervals of $\Delta x = \Delta y = 0.1$ have been sufficient to yield McKay distributed pairs (x, y) with ρ within 3% of the target. Examples of the outputs of the generator are shown in Figure 9.16 where each plot shows 5000 points each representing an (x, y) pair.

The simulator has been used to compute the statistics of elliptical voids for a range of correlation coefficients and a range of $cv(x)$. Each time the simulator was run, 5000 pairs (x, y) and the coefficient of variation of all x and y, $cv(x \cup y)$ were computed from the resulting data.

The mean eccentricity is plotted against the correlation coefficient in Figure 9.24. In this, and all figures arising from our analysis of the McKay bivariate gamma distribution, we present data where $\bar{e} \leq 0.95$; these being associated with correlation coefficients greater than about 0.5. The data form a curve rather like

$$\bar{e}^2 = 1 - \rho^2$$

This gives us another meta-Theorem:

Modelling pores in a random network of lines by equivalent ellipses, the mean eccentricity essentially measures the variation in major axis y not due to positive linear correlation with minor axis x.

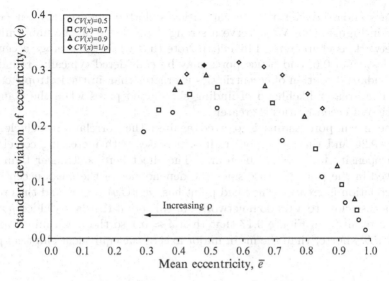

Fig. 9.25. Standard deviation of eccentricity plotted against mean eccentricity using the McKay bivariate gamma distribution for free-fibre-lengths. The filled point has $cv(x \cup y) = 0.99$ and thus represents a case close to a random network.

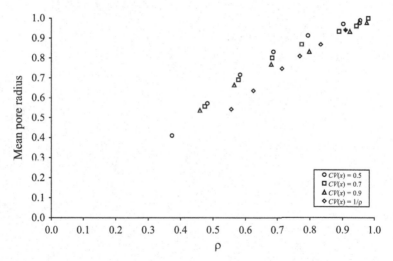

Fig. 9.26. Influence of correlation on the mean pore radius using McKay bivariate gamma distribution for inter-crossing lengths. The filled point has $cv(x \cup y) = 0.99$ and thus represents a case close to a random network.

Network uniformity is characterised by the grand union coefficient of variation $cv(x \cup y)$ taken over all x, y. This latter is a rather weak function of ρ and $cv(x)$, as illustrated in the inset plot, but the ranking of $cv(x \cup y)$ is the same as that of $cv(x)$ so this may still be used as an approximate measure of uniformity.

The standard deviation of eccentricities is plotted against the mean eccentricity in Figure 9.25. We observe a strong dependence on the uniformity of the network as characterised by $cv(x)$. Note that when the average eccentricity is less than 0.6, and hence pores may be considered typically 'roundish', the standard deviation of eccentricities is greater than in the isotropic case—there is a greater likelihood of finding slit-shaped pores when the standard deviation of eccentricities is greater.

The mean pore radius is plotted against the correlation coefficient in Figure 9.26 and the mean pore radius increases with increasing correlation and propensity for pores to be round. The effect here is stronger than that observed in the isotropic case since the dependence of the mean eccentricity on correlation is greater. The filled point has $cv(x \cup y) = 0.99$ and thus represents a case close to a random network; for this point the data yielded $\rho \approx 0.9$ and we recall from Figure 9.17 that then $\delta \approx 0.1$ so the mean minor axis is about 10% shorter than the mean major axis of the equivalent elliptical pore.

10

Stochastic Porous Media and Hydrology

With J. Scharcanski and S. Felipussi

Stochastic porous media arise naturally in many situations; the common feature is a spatial statistical process of extended objects, such as voids distributed in a solid or a connected matrix of distributed solids in air. We have modelled real examples above of cosmological voids among stochastic galactic clusters and at the other extreme of scale are the inter-fibre voids in stochastic fibrous networks. The main context in the present chapter is that of voids in agricultural soils.

10.1 Hydrological Modelling

Yue et al. [216] reviewed various bivariate distributions that are constructed from gamma marginals and concluded that such bigamma distribution models will be useful in hydrology. Here we study the application of the McKay bivariate gamma distribution, which has positive covariance, to model the joint probability distribution of adjacent void and capillary sizes in soils. In this context we compare the discriminating power of an information theoretic metric with two classical distance functions in the space of probability distributions. We believe that similar methods may be applicable elsewhere in hydrology, to characterize stochastic structures of porous media and to model correlated flow variables. Phien [166] considered the distribution of the storage capacity of reservoirs with gamma inflows that are either independent or first-order autoregressive and our methods may have relevance in modelling and quantifying correlated inflow processes. Govindaraju and Kavvas [98] used gamma or Gaussian distributions to model rill depth and width at different spatial locations and again an information geometric approach using a bivariate gamma or Gaussian model may be useful in further probing the joint behavior of these rill geometry variables.

K. Arwini, C.T.J. Dodson, *Information Geometry.*
Lecture Notes in Mathematics 1953,
© Springer-Verlag Berlin Heidelberg 2008

10.2 Univariate Gamma Distributions and Randomness

The family of gamma distributions, §1.4.1, has event space $\Omega = \mathbb{R}^+$ and probability density functions given by

$$S = \{f(x; \gamma, \alpha) | \gamma, \alpha \in \mathbb{R}^+\}$$

so here $M \equiv \mathbb{R}^+ \times \mathbb{R}^+$ and the random variable is $x \in \Omega = \mathbb{R}^+$ with

$$f(x; \gamma, \alpha) = \left(\frac{\alpha}{\gamma}\right)^\alpha \frac{x^{\alpha-1}}{\Gamma(\alpha)} e^{-x\alpha/\gamma}.$$

It is an exponential family, §3.2 with mean γ and it includes as a special case ($\alpha = 1$) the exponential distribution itself, which complements the Poisson process on a line, §1.1.3, §1.2.2. It is pertinent to our interests that the property of having sample standard deviation proportional to the mean actually characterizes gamma distributions, as shown recently [106], cf. Theorem 1.1. Of course, the exponential distribution has unit coefficient of variation, §1.2.

The univariate gamma distribution is widely used to model processes involving a continuous positive random variable, for example, in hydrology the inflows of reservoirs [166] and the depth and width of rills [98]. As we have seen above, the information geometry of gamma distributions has been applied recently to represent and metrize departures from Poisson randomness of, for example, the processes that allocate gaps between occurrences of each amino acid along a protein chain within the *Saccharomyces cerevisiae* genome [34], Chapter 7, clustering of galaxies and communications [63, 61, 64], Chapter 6, and control [78]. We have made precise and proved the statement that around every Poisson random process on the real line there is a neighborhood of processes governed by the gamma distribution, so gamma distributions can approximate any small enough departure from Poisson randomness, Chapter 5 [14]. Such results are, by their topological nature, stable under small perturbations of a process, which is important in real applications. This, and their uniqueness property [106] Theorem 1.1, gives confidence in the use of gamma distributions to model near-Poisson processes. Moreover, the information-theoretic heritage of the metric for the neighborhoods lends significance to the result.

10.3 Mckay Bivariate Gamma 3-Manifold

It is logical next to consider bivariate processes, §1.3, which may depart from independence and from Poisson randomness. Natural choices arise for marginal distributions, §1.3: Gaussian, §1.2.3, or log-Normal distributions, §3.6, and gamma distributions, §1.4.1. For example, recently in hydrology, bivariate gamma distributions have been reviewed [216], and from [98] we may expect that rill depth and width admit bivariate gamma or bivariate Gaussian models with positive covariance. Here we concentrate on the case when the marginals are gamma and the covariance is positive, §1.3, which has application to the modelling of void and capillary size in porous media like soils.

Positive covariance and gamma marginals gives rise to one of the earliest forms of the bivariate gamma distribution, due to Mckay [146] which we discussed in §4.1, defined by the density function

$$f_M(x,y) = \frac{c^{(\alpha_1+\alpha_2)} x^{\alpha_1-1}(y-x)^{\alpha_2-1}e^{-cy}}{\Gamma(\alpha_1)\Gamma(\alpha_2)} \quad \text{defined on } y > x > 0\,,\ \alpha_1, c, \alpha_2 > 0$$

(10.1)

One way to view this is that $f_M(x,y)$ is the probability density for the two random variables X and $Y = X + Z$ where X and Z both have gamma distributions. The marginal distributions of X and Y are gamma with shape parameters α_1 and $\alpha_1 + \alpha_2$, respectively. The covariance Cov and correlation coefficient ρ_M of X and Y are given by :

$$Cov(X,Y) = \frac{\alpha_1}{c^2} = \sigma_{12}$$

(10.2)

$$\rho_M(X,Y) = \sqrt{\frac{\alpha_1}{\alpha_1 + \alpha_2}}.$$

(10.3)

Observe that in this bivariate distribution the covariance, and hence correlation, tends to zero only as α_1 tends to zero.

In the coordinates $(\alpha_1, \sigma_{12}, \alpha_2)$ the McKay densities (10.1) are given by:

$$f(x,y;\alpha_1,\sigma_{12},\alpha_2) = \frac{\left(\frac{\alpha_1}{\sigma_{12}}\right)^{\frac{(\alpha_1+\alpha_2)}{2}} x^{\alpha_1-1}(y-x)^{\alpha_2-1}e^{-\sqrt{\frac{\alpha_1}{\sigma_{12}}}y}}{\Gamma(\alpha_1)\Gamma(\alpha_2)}\,,$$

(10.4)

defined on $0 < x < y < \infty$ with parameters $\alpha_1, \sigma_{12}, \alpha_2 > 0$. Where σ_{12} is the covariance of X and Y. The correlation coefficient and marginal functions, of X and Y re given by :

$$\rho_M(X,Y) = \sqrt{\frac{\alpha_1}{\alpha_1 + \alpha_2}}$$

(10.5)

$$f_X(x) = \frac{\left(\frac{\alpha_1}{\sigma_{12}}\right)^{\frac{\alpha_1}{2}} x^{\alpha_1-1}e^{-\sqrt{\frac{\alpha_1}{\sigma_{12}}}x}}{\Gamma(\alpha_1)}\,,\quad x > 0$$

(10.6)

$$f_Y(y) = \frac{\left(\frac{\alpha_1}{\sigma_{12}}\right)^{\frac{(\alpha_1+\alpha_2)}{2}} y^{(\alpha_1+\alpha_2)-1}e^{-\sqrt{\frac{\alpha_1}{\sigma_{12}}}y}}{\Gamma(\alpha_1+\alpha_2)}\,,\quad y > 0$$

(10.7)

We consider the Mckay bivariate gamma model as a 3-manifold, §4.1, equipped with Fisher information as Riemannian metric, §4.1 from equation (4.5):

$$[g_{ij}] = \begin{bmatrix} \frac{-3\,\alpha_1 + \alpha_2}{4\,\alpha_1{}^2} + \psi'(\alpha_1) & \frac{\alpha_1 - \alpha_2}{4\,\alpha_1\,\sigma_{12}} & -\frac{1}{2\,\alpha_1} \\ \frac{\alpha_1 - \alpha_2}{4\,\alpha_1\,\sigma_{12}} & \frac{\alpha_1 + \alpha_2}{4\,\sigma_{12}{}^2} & \frac{1}{2\,\sigma_{12}} \\ -\frac{1}{2\,\alpha_1} & \frac{1}{2\,\sigma_{12}} & \psi'(\alpha_2) \end{bmatrix} \qquad (10.8)$$

where $\psi(\alpha_i) = \frac{\Gamma'(\alpha_i)}{\Gamma(\alpha_i)}$ $(i = 1, 2)$.

10.4 Distance Approximations in the McKay Manifold

Distance between a pair of points in a Riemannian manifold is defined as the infimum of arc lengths, §2.1, over all curves between the points. For sufficiently nearby pairs of points there will be a unique minimizing geodesic curve that realises the infimum arc length, §2.1. In general, such curves are hard to find between more distant points. However, we can obtain an upper bound on distances between two points $T_0, T_1 \in M$ by taking the sum of arc lengths along coordinate curves that triangulate the pair of points with respect to the coordinate axes. We adopted similar methods in the gamma manifold for univariate processes [34, 76]. Here, following the methodology in §9.6.2, we use the metric for the McKay manifold (10.8) §4.1, and obtain the following upper bound on the information distance in M from T_0 with coordinates $(\alpha_1, \sigma_{12}, \alpha_2) = (A, B, C)$ to T_1 with coordinates $(\alpha_1, \sigma_{12}, \alpha_2) = (a, b, c)$

$$d_M(T_0, T_1) \le \left| \int_A^a \sqrt{\left| \frac{C - 3x}{4x^2} + \frac{d^2 \log \Gamma(x)}{dx^2} \right|}\, dx \right| + \left| \int_C^c \sqrt{\left| \frac{d^2 \log \Gamma(y)}{dy^2} \right|}\, dy \right|$$
$$+ \left| \int_B^b \sqrt{\frac{A + C}{4\,z^2}}\, dz \right| . \qquad (10.9)$$

The square roots arise from the norms of tangent vectors, §2.0.5, to coordinate curves and it is difficult to obtain the closed form solution for information distance, d_M. However, by removing the square roots the integrals yield information-energy values E_M, which can be evaluated analytically. Then the square root of the net information-energy differences along the coordinate curves gives a closed analytic 'energy-distance' $dE_M = \sqrt{E_M}$, which we can compare with d_M. The net information-energy differences along the coordinate curves from T_0 with coordinates $(\alpha_1, \sigma_{12}, \alpha_2) = (A, B, C)$ to T_1 with coordinates $(\alpha_1, \sigma_{12}, \alpha_2) = (a, b, c)$ is

$$E_M(T_0, T_1) \le \left| \int_A^a \left(\frac{C - 3x}{4x^2} + \frac{d^2 \log \Gamma(x)}{dx^2} \right) dx \right| + \left| \int_C^c \frac{d^2 \log \Gamma(y)}{dy^2}\, dy \right|$$
$$+ \left| \int_B^b \frac{A + C}{4\,z^2}\, dz \right|$$

$$\leq \left| \int_A^a \left(\frac{C - 3x}{4x^2} \right) dx \right| + \left| \int_A^a \frac{d^2 \log \Gamma(x)}{dx^2} dx \right| + \left| \int_C^c \frac{d^2 \log \Gamma(y)}{dy^2} dy \right|$$

$$+ \frac{A + C}{4} \left| \frac{1}{b} - \frac{1}{B} \right|$$

$$= \left| \frac{C}{4a} - \frac{C}{4A} + \frac{3 \log(\frac{A}{a})}{4} \right| + |\psi(a) - \psi(A)| + |\psi(c) - \psi(C)|$$

$$+ \frac{A + C}{4} \left| \frac{1}{b} - \frac{1}{B} \right| . \tag{10.10}$$

$$dE_M = \sqrt{E_M}. \tag{10.11}$$

where $\psi = \frac{\Gamma'}{\Gamma}$ is the digamma function.

Next we compare distances between bivariate gamma distributions obtained using this information metric upper bound (10.9) in the McKay manifold metric (4.5) with the classical Bhattacharyya distance [27] between the distributions. Some further discussion of classical distance measures can be found in Chapter 3 of Fukunga [90]; the Bhattacharyya distance is actually a special case of the Chernoff distance [90].

The Bhattacharyya distance from T_0 to T_1 defined on $0 < x < y < \infty$ is given by

$$d_B(T_0, T_1) = -\log \int_{y=0}^\infty \int_{x=0}^y \sqrt{T_0 T_1} \, dx \, dy$$

$$= -\log \left(\frac{W}{\sqrt{\Gamma(A) \Gamma(C) \Gamma(a) \Gamma(c)}} \right) \tag{10.12}$$

where

$$W = \Gamma \left(\frac{A + a}{2} \right) \Gamma \left(\frac{C + c}{2} \right) \left(\frac{A}{B} \right)^{\frac{A+C}{4}} \left(\frac{a}{b} \right)^{\frac{a+c}{4}} \left(\frac{\sqrt{a}}{2\sqrt{b}} + \frac{\sqrt{A}}{2\sqrt{B}} \right)^{\frac{-(A+C+a+c)}{2}} \tag{10.13}$$

The Kullback-Leibler 'distance' [126] or 'relative entropy' from T_0 to T_1 defined on $0 < x < y < \infty$ is given by

$$KL(T_0, T_1) = \int_{y=0}^\infty \int_{x=0}^y T_0 \log \frac{T_0}{T_1} \, dx \, dy$$

$$= -A + \psi(A) (A - a) - C + \psi(C) (C - c) + \log \left(\frac{\Gamma(a) \Gamma(c)}{\Gamma(A) \Gamma(C)} \right)$$

$$+ \frac{(a + c)}{2} \log \left(\frac{A b}{a B} \right) + (A + C) \sqrt{\frac{a B}{A b}} . \tag{10.14}$$

and we symmetrize this to give a true distance

$$d_K(T_0, T_1) = \frac{KL(T_0, T_1) + KL(T_1, T_0)}{2}.$$ (10.15)

10.5 Modelling Stochastic Porous Media

Structural characterization and classification of stochastic porous materials has attracted the attention of researchers in different application areas, because of its great economic importance. For example, problems related to mass transfer and retention of solids in multi-phase fluid flow through stochastic porous materials are ubiquitous in different areas of chemical engineering. One application of gamma distributed voids to stochastic porous media has admitted a direct statistical geometric representation of stochastic fibrous networks [72], §9.5, and their fluid transfer properties [74]. Agricultural engineering is one of the sectors that has received attention recently, mostly because of the changing practices in agriculture in developing countries, and in developed countries, with great environmental impact [206, 207].

Phenomenologically, mass transfer in porous media depends strongly on the morphological aspects of the media—such as the shape and size of pores, and depends also on the topological attributes of these media, such as the pore network connectivity [74].

Several approaches have been presented in the literature for structural characterization of porous media, involving morphological and topological aspects. [12] proposed the analysis of porous media sections for their pore shape and size distributions. In their work, images are obtained using a scanning electron microscope for micro-structural characterization, and an optical microscope for macro-structural characterization. However, the acquisition of samples for the analysis is destructive, and it is necessary to slice the porous media so that sections can be obtained, and then to introduce epoxy resin for contrast. These procedures influence the structure of the media solid phase and consequently the morphology and topology of the porous phase, which implies that the three-dimensional reconstruction is less reliable for soil samples. In order to overcome similar difficulties, a few years earlier, the non-destructive testing in soil samples using tomographic images was proposed [28], but their goal was the evaluation of the swelling and shrinkage properties of loamy soil. Also, other researchers have concentrated on the 'fingering' phenomenon occurring during fluid flow in soils [159]. More recently, researchers have proposed geometrical and statistical approaches for porous media characterization. A skeletonization method based on the Voronoi diagram [54] was introduced to estimate the distributions of local pore sizes in porous media, see also [116].

The statistical characterization and classification of stochastic porous media, is essential for the simulation and/or prediction of the mass transfer properties of a particular stochastic medium. Much work has been done on the characterization of porous media but the discrimination among different models from observed data still remains a challenging issue. This is particularly true considering the tomographic images of porous media, often used in soil analysis; for two recent studies see [3] and [116].

We apply our distance measures to experimental porous media data obtained from tomographic images of soil, to data from model porous media and to simulation data drawn by computer from bivariate correlated gamma processes. It turns out that tomographic images of soil structure reveal a bivariate stochastic structure of sizes of voids and their interconnecting capillaries. The information geometry of the Riemannian 3-manifold of McKay bivariate gamma distributions, §4.1, provides a useful mechanism for discriminating between treatments of soil. This method is more sensitive than that using the classical Bhattacharyya distance between the bivariate gamma distributions and in most cases better than the Kullback-Leibler measure for distances between such distributions.

10.5.1 Adaptive Tomographic Image Segmentation

Image analysis is an important tool for the structural characterization of porous materials, and its applications can be found in several areas, such as in oil reservoir modeling, and in estimates of soil permeability or bone density [199]. The structural properties of porous media can be represented by statistical and morphological aspects, such as distributions of pore sizes and pore shapes; or by topological properties, such as pore network inter-connectivity.

Here we describe a method for statistical characterization of porous media based on tomographic images, with some examples. Three-dimensional tomographic images obtained from porous media samples are represented based on statistical and topological features (i.e. pore and throat sizes distributions, as well as pore connectivity and flow tortuosity). Based on selected geometrical and statistical features, soil sample classification is performed. Our experimental results enable the analysis of the soil compaction resulting from different soil preparation techniques commonly used nowadays.

We outline next some morphological concepts and then our adaptive image segmentation approach will be described, as well as our proposed feature extraction scheme.

Pores, Grains and Throats

In general, porous media are constituted by two types of elements: grains and pore bodies. Let S be a cross-section of a porous medium, given by a 2D binary representation like the one shown in Figure 10.1, where pores (i.e. the void fraction) are represented by white regions, and grains (i.e. the solid fraction) by black regions. The phase function Ω is defined as in [83]:

$$\Omega(\Xi) = \begin{cases} 1, & \text{when } \Xi \text{ belongs to the pore phase;} \\ 0, & \text{otherwise.} \end{cases} \tag{10.16}$$

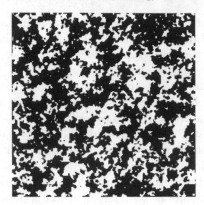

Fig. 10.1. A 2D section representing the pores (clear regions) and grains (dark regions).

where Ξ denotes the vector specifying a position in S. In fact, Ξ denotes the set of integer pairs, (χ, υ), being multiples of the measuring unit. In our case, S is represented by a 2D image I of size $Nh \times Nv$, and (χ, υ) labels the pixel coordinates within this image. Each pixel represents the local density of the porous medium. Within an image, we may identify sets of connected pixels (i.e. pixels belonging to the same region). A path from pixel U to pixel V is a set of pixels forming the sequence $A_1, A_2, ..., A_n$, where $A_1 = U$, $A_n = V$, and A_{w+1} is a neighbor of A_{w+1}, $w = 2, \cdots, n - 1$. A region is defined as a set of pixels in which there is at least one path between each pixel pair of the region [192]; so it is a pixel-path connected set.

The images obtained by computerized tomography are slices (cross-sections) of the 3D objects, and this imaging technique has been applied recently to the analysis of porous media [199]. In a slice z, the pore bodies (i.e. pores) are disjoint regions P_i, with $i = 1, ..., N$ and N is the number of pores in z :

$$P_i(z) = \{I(\chi, \upsilon) | (\chi, \upsilon) \text{ connected}\}. \tag{10.17}$$

Let $P_i(z)$ be a pore at slice z, and $P_j(z + 1)$ a pore in an adjacent slice z+1, with $j = 1, \cdots, M$ (where M is the number of pore bodies at slice z+1). If there is at least one pixel in common between $P_i(z)$ and $P_j(z + 1)$, then there exists a throat connecting these pores, defined as follows:

$$T(P_i(z), P_j(z + 1)) = \cap(P_i(z), P_j(z + 1)). \tag{10.18}$$

Figure 10.2 shows a pore $P_i(z)$ in slice z, and a pore $P_j(z + 1)$ in slice z+1 (darker regions). The regions in common are depicted as clearer areas and represent the throats for better visualization.

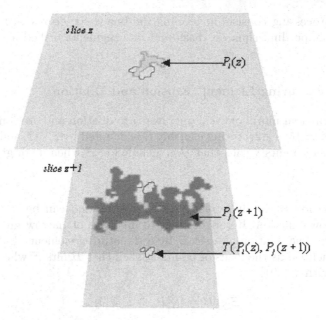

Fig. 10.2. An illustration representing the intersection between two regions of two slices and the resultant throat.

Tortuosity in Porous Media

In hydrological surveys, tortuosity is one of the most informative parameters. However, it has different definitions in the literature. Prasher et al. [169] introduced the concept of tortuosity as the square of the ratio of the effective average path in the porous medium (L_c) to the shortest distance measured along the direction of the pore (L). Here, as the analysis is based on sequences of images of thin slices, the coefficient of tortuosity is evaluated based on three consecutive slices. Considering three interconnecting pores in slices n, $n + 1$ and $n + 2$, the local tortuosity coefficient is estimated as the Euclidean distance between the centroids of their throats $T(P_i(n), P_j(n + 1))$ and $T(P_i(n + 1), P_j(n + 2))$:

$$\text{Tort} = \sqrt{(\bar{\chi} - \chi)^2 + (\bar{v} - v)^2},\qquad(10.19)$$

where, the centroids (χ, v) and $(\bar{\chi}, \bar{v})$ correspond to the throats between pores at slices $(n) - (n + 1)$ and $(n + 1)(n + 2)$, respectively.

10.5.2 Mathematical Morphology Concepts

We shall use morphological operators for extracting the image components needed in the pore and throat size representation. The morphological approach to image processing consists in probing the image structures with a pattern of known shape, link squares, disks and line segments, called a structuring element B.

Planar Structuring Element, Erosion and Dilation

The two principal morphological operators are dilation and erosion. The erosion of a set Ξ by a structuring element B is denoted by $\Xi \ominus \Upsilon$ and is defined as the locus of points χ such that B is included in Ξ when its origin is placed at χ [191]:

$$\Xi \ominus B = \{\chi | B_\chi \subseteq \Xi\}. \tag{10.20}$$

The dilation is the dual operation of erosion, and can be thought of as a fill or grow function. It can be used to fill holes or narrow gulfs between objects [29]. The dilation of a set Ξ by a structuring element B, denoted by $\Xi \oplus B$, is defined as the locus of points χ, such that B hits Ξ when its origin coincides with χ [191]:

$$\Xi \oplus B = \{\chi | B_\chi \cap \Xi \neq \emptyset\}. \tag{10.21}$$

Geodesic Dilation and Reconstruction

The so called 'geodesic methods' are morphological transformations that can operate only on some parts of an image [192]. To perform geodesic operations, we only need the definition of a geodesic distance. The simplest geodesic distance is the one which is built from a set Ξ. The distance of two points p and q belonging to Ξ is the length of the shortest path (if any) included in Ξ and joining p and q (see Figure 10.3) [205].

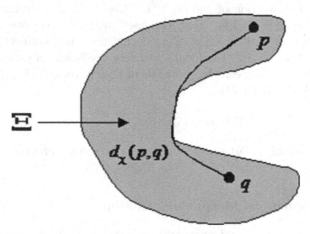

Fig. 10.3. Geodesic distance $d_\chi(p, q)$ within a set Ξ.

The geodesic transformations are employed in the reconstruction of a set Ξ marked by another reference set Υ, non-empty, denominated marker. In geodesic dilation two images are considered: a marker image and a mask image. These images must have the same definition; domain and marker image must be lower or equal to that of the mask image. The geodesic dilation uses a structuring element, and the resulting image is then forced to remain restricted to the mask image. Formally, the marker image Υ is any set included in the mask image Ξ. We can compute the set of all points of Ξ that are at a finite geodesic distance $n \geq 0$ from Υ:

$$\delta_{\Xi}^{(n)}(\Upsilon) = \{p \in \Xi | d_{\Xi}(p, \Upsilon) \leq n\}. \tag{10.22}$$

The elementary geodesic dilation of size 1 $(\delta_{\Xi}^{(1)})$ of a set Υ inside Ξ is obtained as the intersection of the unit-size dilation of Υ (with respect to the structuring element B) with the set Ξ [192]:

$$\delta_{\Xi}^{(1)}(\Upsilon) = (\Upsilon \oplus B) \cap \Xi. \tag{10.23}$$

The geodesic dilations of a given size n can be obtained by iterating n elementary geodesic dilations [205, 192]:

$$\delta_{\Xi}^{(n)}(\Upsilon) = \delta_{\Xi}^{(1)}(\delta_{\Xi}^{(i)}(\Upsilon)), \tag{10.24}$$

where, $n \geq 2$, $i = 1, \cdots, n - 1$, and $\delta_{\Xi}^{(n)}(\Upsilon)$ is the Ξ reconstructed set by the marker set Υ. It is constituted of all the connected components of Ξ that are marked by Υ. This transformation can be achieved by iterating elementary geodesic dilations until idempotence (see Figure 10.4), when $\delta_{\Xi}^{(n)}(\Upsilon) = \delta_{\Xi}^{(n+1)}(\Upsilon)$ (i.e. until no further modification occurs). This operation is called reconstruction and is denoted by $\rho_{\Xi}(\Upsilon)$. Formally [192],

$$\rho_{\Xi}(\Upsilon) = \lim_{n \to \infty} \delta_{\Xi}^{(n)}(\Upsilon). \tag{10.25}$$

The extension of geodesic transformation to greyscale images is based on the fact that, at least in the discrete case, any increasing transformation defined for binary images can be extended to greyscale images [205, 192]. By increasing, we mean a transformation Ψ such that:

$$\Upsilon \subseteq \Xi \Rightarrow \Psi(\Upsilon) \subseteq \Psi(\Xi), \ \forall \Xi, \Upsilon \subset Z^2. \tag{10.26}$$

The transformation Ψ can be generalized by viewing a greyscale image I as a stack of binary images obtained by successive thresholding. Let D_I be the domain of the image I, and the grey values of image I be in $0, 1, \cdots, N - 1$. The thresholded images $T_k(I)$ are [205, 192]:

$$T_k(I) = \{p \in D | I(p) \geq k\}. \tag{10.27}$$

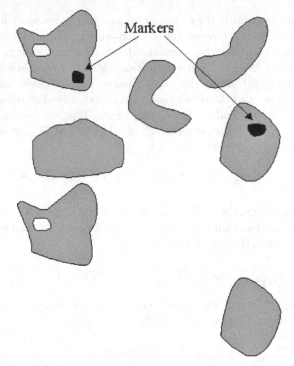

Fig. 10.4. Reconstruction of Ξ (shown in light grey) from markers Υ (black). The reconstructed result is shown below.

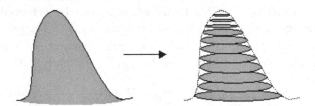

Fig. 10.5. Threshold decomposition of a greylevel image.

The idea of threshold decomposition is illustrated in Figure 10.5. The threshold-decomposed images $T_k(I)$ satisfy the inclusion relationship:

$$T_k(I) \subseteq T_{k-1}(I) \ , \ \forall k \in [1, N-1] \tag{10.28}$$

Consider the increasing transformation Ψ applied in each threshold-decomposed image; then their inclusion relationships are preserved. Thus the transformation Ψ can be extended to greyscale images using the following threshold decomposition principle [192]:

$$\forall p \in D, \Psi(I)(p) = \max\left\{k \in [0, N-1] | p \in \Psi(T_k(I))\right\}. \tag{10.29}$$

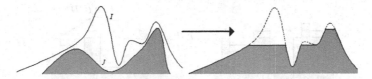

Fig. 10.6. Grayscale reconstruction of mask I from markers J.

Since the binary geodesic is an increasing transformation, it satisfies:

$$\Upsilon_1 \subseteq \Upsilon_2, \ \Xi_1 \subseteq \Xi_2, \ \Upsilon_1 \subseteq \Xi_1, \ \Upsilon_2 \subseteq \Xi_2, \ \Rightarrow$$
$$\rho_{\Xi_1}(\Upsilon_1) \subseteq \rho_{\Xi_2}(\Upsilon_2). \tag{10.30}$$

Therefore, using the threshold decomposition principle (see Equation 10.29), we can generalize the binary reconstruction to greyscale reconstruction [205, 192]. Let J and I be two greyscale images defined on the same domain D, with grey-level values from the discrete interval $[0, 1, \cdots, N-1]$. If for each pixel $p \in D$, $J(p) \leq I(p)$, the greyscale reconstruction $\rho_I(J)$ of the mask image I from the marker image J is given by:

$$\forall p \in D, \ \rho_I(J)(p) = \max \left\{ k \in [0, N-1] \, | \, p \in \rho_{T_k(I)}(T_k(J)) \right\} \tag{10.31}$$

This transformation is illustrated in Figure 10.6, where the greyscale reconstruction extracts the peaks of the mask-image I, which is marked by the marker-image J.

Regional Maxima and Minima

Reconstruction turns out to provide a very efficient method to extract *regional maxima* and *minima* from greyscale images [205]. It is important not to confuse these concepts with local maxima and local minima. Being a local maximum is not a regional property, but a property of a pixel. A pixel is a local maximum if its value is greater than or equal to any pixel in its neighborhood. By definition, a regional maximum M (minimum) of a greyscale image I is a connected set of pixels with an associated value h (plateau at height h), such that every neighboring pixel of M has strictly lower (higher) value than h [191].

Regional maxima and minima often mark relevant image objects, and image peaks or valleys can be used to create marker images for the morphological reconstruction. In order to obtain the set of regional maxima, an image I can be reconstructed from $I - 1$, and the result subtracted from I. Therefore, the set of regional maxima of a greyscale image I, denoted by $RMAX(I)$, is defined by [205] (see Figure 10.7):

$$RMAX(I) = \{ p \in D / (I - \rho_I(I-1))(p) \}. \tag{10.32}$$

By duality, the set of regional minima can be computed replacing I by its complement I^C:

$$RMIN(I) = RMAX(I^C). \tag{10.33}$$

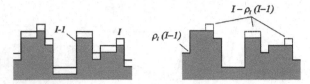

Fig. 10.7. Maximum detection by reconstruction subtracting I from $\rho_I(I-1)$. At the left, a 1D profile of I and marker image $I-1$ are shown. The figure at the right shows the reconstruction of mask image I from marker image $I-1$, and the difference $I-\rho_I(I-1)$ keeps only the "domes" (set of regional maxima) and removes the background.

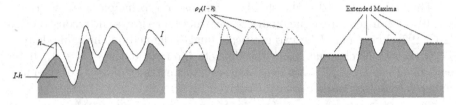

Fig. 10.8. Left: 1D profile of I and $I-h$. Center: the result of the h-maxima transform. Right: the result of extended maxima transform (dashed).

h-Extrema and Extended Extrema Transformations

The h-maxima transform allows the reconstruction of I from $I-h$, where h represents an arbitrary greylevel constant. In this way, the h-maxima transform provide a tool to filter the image using a contrast criterion.

By definition, the h-maxima transform suppresses all maxima whose height is smaller than a given threshold level h. Therefore, the h-maxima detects peaks using a contrast parameter, without involving any size or shape criterion, as it is imposed by other techniques like openings and closings (see Figure 10.8. Formally, we have:

$$HMAX_h(I) = \rho_I(I-h). \tag{10.34}$$

Analogously, the h-minima transform reconstructs I from $I+h$, where h represents an arbitrary grey-level. By definition, the h-minima transform suppresses all minima whose depth is less than a given threshold level h [205]:

$$HMIN_h(I) = \rho_I(I+h). \tag{10.35}$$

The extended-maxima $EMAX$ (extended-minima $EMIN$) are the regional maxima (minima) of the corresponding h-extrema transformation [191]:

$$EMAX_h(I) = RMAX(HMAX_h(I)), \tag{10.36}$$

and,

$$EMIN_h(I) = RMIN(HMIN_h(I)). \tag{10.37}$$

10.5.3 Adaptive Image Segmentation and Representation

Image segmentation is a very important step in porous media modelling, but it is one of the most difficult tasks in image analysis. Our goal is to obtain an image where the pore and solid phases of the original image are discriminated. In the literature, several authors have proposed the segmentation based on a global threshold as an early step in the porous media characterization [204, 96, 103]. Basically, the central question in these methods is the selection of a threshold to discriminate between pores and the solid phase based on the image histogram. This is typically done interactively, by visual inspection of the results. Global thresholding generally results in serious segmentation flaws if the object of interest and background are not uniform (as in our case).

Next, we discuss a locally adaptive image segmentation procedure to discriminate between pores and the solid phase, using extended minima or maxima. The choice of the parameter h is now the central issue, because the h-minima/maxima transform suppresses all minima/maxima regions whose depth/height is less than h (i.e. considering a greylevel image relief). Comparing the extended minima in Figure 10.9 we can verify that when h is increased, the area of some objects increase, and some other objects disappear. In our case, the images are noisy and present low contrast, making the choice of the parameter h critical. Therefore, let us consider the union of the pixel sets of the regions obtained by the extended minima/maxima when h varies between 1 and k (see Figure 10.10). Formally, we have:

$$T_{MIN}^{k}(I) = \bigcup_{h=1}^{k} EMIN_h(I),\qquad(10.38)$$

and,

$$T_{MAX}^{k} = \bigcup_{h=1}^{k} EMAX_h(I).\qquad(10.39)$$

If the parameter k is small, we obtain small regions centered on regional minima/maxima of the image. If k increases, the regions grow and can be merged. Some important issues arising in the analysis of soil tomographies

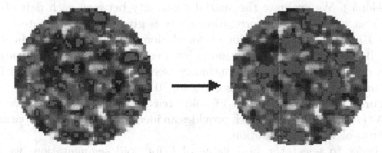

Fig. 10.9. Left: the extended minima with $h = 10$ was applied to segment the dark regions. Right the segmented image with $h = 50$; notice that some regions disappeared and others grew with increased h.

Fig. 10.10. The result of increasing h in the interval $[1, \cdots, 50]$, the regions with low contrast are preserved.

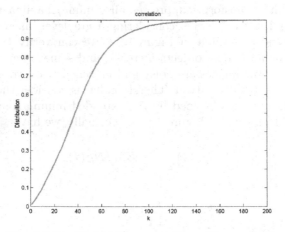

Fig. 10.11. Plot of the correlation with varying k.

are that such images tend to be noisy and present low contrast, affecting the spatial continuity in three-dimensions (since these structures are three-dimensional). We evaluate the spatial continuity between each pair of slices in the image stack, using correlation analysis given an appropriate k value. Figure 10.11 illustrates the plot of correlation between adjacent slices as k varies. If k is small, the correlation is low because the pixels of the adjacent images do not present spatial continuity (see Figure 10.12, top), and if k is large, the correlation is higher (see Figure 10.12, bottom).

Figure 10.13 shows the plot of void areas sum for the slice stack, given $TMIN^k(I)$ (i.e. $1 - area$), which provides an idea of the effect of the parameter k in terms of void segmentation.

In order to select the best value of k for void segmentation, we use a criterion expressed by the following function:

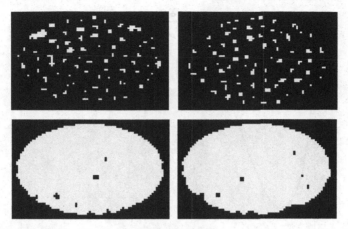

Fig. 10.12. Top left: segmentation of first slice with $k = 1$. Top right: second slice of the stack also with $k = 1$. Bottom left and right: first and second slice with $k = 100$, respectively. Notice that the correlation is high for k large and low if k is small.

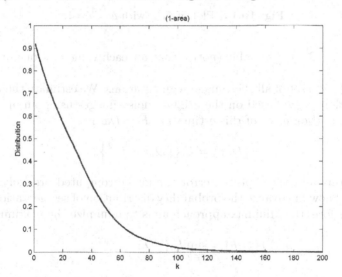

Fig. 10.13. Plot of (1-area) for the slice stack.

$$f(k) = a(1 - area) + b(correlation), \quad \forall a, b \in [0, 1], \qquad (10.40)$$

which provides a relationship $(1 - area) \times (correlation)$, and is shown in Figure 10.14.

To find proper values for the parameters k, a and b given the image data we use an optimization algorithm based on the Minimax approach [138]. In this case, the Minimax approach is used to estimate the values of k, a and b as those leading to the smallest risk/error in void segmentation, considering

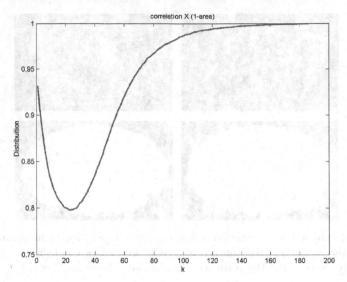

Fig. 10.14. Plot of $f(k)$ with $a = b = 1$.

all operators $D(k)$ obtainable (notice that for each value of k one operator is defined).

Let θ be the set of all slice image segmentations. We estimate the best void segmentation $f \in \theta$ based on the original noisy images using an operator D. The segmentation error of this estimation $\bar{F} = D\Xi$ is:

$$r(D, f) = E\left\{\|D\Xi - f\|^2\right\}. \tag{10.41}$$

Unfortunately, the expected error cannot be computed precisely, because we do not know in advance the probability distribution of segmentation images in θ. Therefore, the Minimax approach aims to minimize the maximum error:

$$r(D, \theta) = \sup_{f \in \theta} E\left\{\|D\Xi - f\|^2\right\}, \tag{10.42}$$

and the Minimax error is the lower bound computed over all operators D:

$$r_n(\theta) = \inf_{D \in O_n} r(D, \theta), \text{ with } O_n \text{ the set of all operators.} \tag{10.43}$$

In practice, we must find a decision operator D that is simple to implement (small k), such that $r(D, \theta)$ is close to Minimax error $r_n(\theta)$, and we have:

$$r_n(\theta) = \inf_{D \in O_n} \left\{\sup_{f \in \theta} E\left\{\|D\Xi - f\|^2\right\}\right\}. \tag{10.44}$$

In our case, we must approximate the error function based on problem constraints, i.e. we wish to have slice continuity along the slice stack, and at the

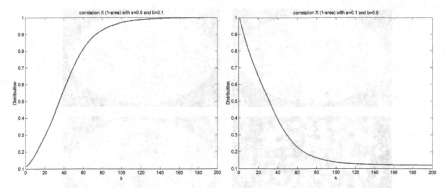

Fig. 10.15. Left: function f(k) that prioritizes feature continuity (correlation) with $a = 0.9$ and $b = 0.1$. Right: complement of the areas is prioritized using $a = 0.1$ and $b = 0.9$, in both cases $r_n(\Theta)$ is small, but the resulting operator D not is adequate for segmentation.

same time we want to segment the largest (maximum) number of voids. These are obviously conflicting constraints because by nature such tomographic images have low contrast and are noisy and, at the same time the porous media structure may be nearly Poisson random. Therefore, we approximate the error function by the measurable data described by equation (10.40), with the additional constraint $|a - b| \leq \delta$, with $\delta \in [0, 1]$. If this constraint is not considered, the function f prioritizes just one of the constraints, and returns an inadequate result (see Figure 10.15).

An overview of the segmentation algorithm is outlined below:

1. Compute the image $\rho_I \left\{ \bigcup_{h=1}^{k} EMIN_h(I) \right\}$ performing the greyscale reconstruction with the union of the sets of the extended minima to all $k \in [1, \cdots, N]$;
2. Compute for each k, the correlation between each pair of slices of the stack of images;
3. Compute $T_{MIN}^k(I)$ for all slices of the image stack;
4. Compute the function in equation (10.40) for all k values in the interval;
5. Using the Minimax principle, find a decision operator $D(k)$ leading to the smallest maximum error (i.e. deepest valley in the plot).

The image $\rho_I \left\{ \bigcup_{h=1}^{k} EMIN_h(I) \right\}$ is our void image segmentation. The void data used in our experiments was obtained using the segmentation algorithm just described. Section 10.5.4 provides a detailed discussion based on experimental data obtained with the application of the segmentation algorithm to 465 images of soil samples (see some examples in Figure 10.17), and 58 images of glass spheres immersed in water (see some examples in Figure 10.16).

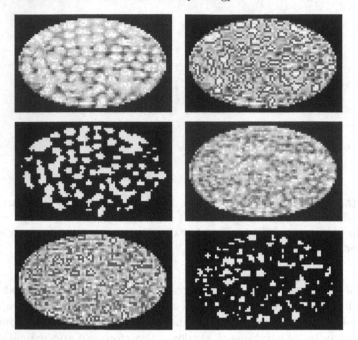

Fig. 10.16. Top row (left): original soil sample. Middle row: boundaries of the segmented regions. Third row: segmented image.

10.5.4 Soil Tomographic Data

Three-dimensional tomographic images were obtained from thin slices of soil and model samples and the new algorithms described in § 10.5.3 were used to reveal features of the pore size distribution, and pore connectivity. These features are relevant for the quantitative analysis of samples in several applications of economic importance, such as the effect of conventional and direct planting methods on soil properties, as compared to untreated soil (i.e. natural forest soil). The three-dimensional profiles for these soils are shown in Figure 10.18.

The methodology is applicable generally to stochastic porous media but here we focus on the analysis of soil samples, in terms of the soil compaction resulting from different soil preparation techniques. The interconnectivity of the pore network is analyzed through a fast algorithm that simulates flow. The image analysis methods employed to extract features from these images are beyond our present scope and will be discussed elsewhere.

The two variables $0 < x < y < \infty$ correspond as follows: y represents the cross-section area of a pore in the soil and x represents the corresponding cross-sectional area of the throats or capillaries that connect it to neighbouring voids. It turns out that these two variables have a positive covariance and can be fitted quite well to the McKay bivariate gamma distribution (10.4),

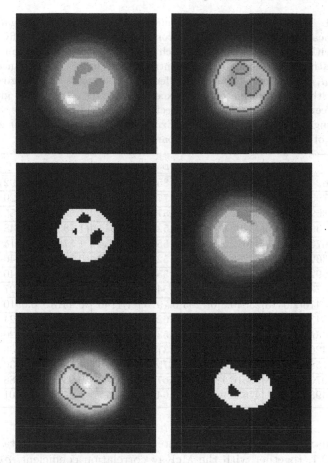

Fig. 10.17. Top row (left): original soil sample. Middle row: boundaries of the segmented regions. Third row: segmented image.

Fig. 10.18. Profile three-dimensional reconstructions of soil samples. Left: Forest; center: Conventional Planting; right: Direct Planting.

Table 10.1. Maximum likelihood parameters of McKay bivariate gamma fitted to hydrological survey data extracted from tomograghic images of five soil samples, each with three treatments. For each soil, the natural state is untreated forest; two treatments are compared: conventional and direct. The distance functions d_M, dE_M, d_B, d_K are used to measure effects of treatments—values given are distances from the forest untreated case, except that values in brackets give the distances between the conventional and direct treatments. In most cases, d_M is most discriminating and dE_M is second best. Except for Sample A, all distances agree on the ranking of effects of treatments compared with untreated soils.

Sample	α_1	α_2	σ_{12}	ρ_M	ρ_{Data}	d_M	dE_M	d_B	d_K
A forest	7.6249	3.581	19199.4	0.8249	0.8555	(4.1904)	(1.4368)	(0.1292)	(0.5775)
A conv	4.7931	7.4816	22631.2	0.6249	0.7725	2.8424	1.2643	0.7920	3.2275
A direct	1.8692	3.4911	33442.6	0.5905	0.5791	3.181	1.5336	0.5855	2.7886
B forest	1.2396	2.2965	41402.8	0.5920	0.5245	(4.1611)	(1.7835)	(0.3948)	(1.9502)
B conv	5.6754	4.8053	34612.2	0.7359	0.5500	3.7034	1.8784	0.3214	1.6816
B direct	1.3622	2.0074	30283.5	0.6358	0.5215	0.6322	0.5820	0.0346	0.1390
C forest	1.6920	2.6801	37538.3	0.6221	0.5582	(2.7858)	(1.8896)	(0.2140)	(1.0462)
C conv	0.7736	1.2488	30697.5	0.6185	0.5466	2.3931	1.5372	0.1727	0.7403
C direct	2.8476	2.6413	25840.8	0.7203	0.7975	1.1146	0.9518	0.0910	0.3741
D forest	1.8439	1.7499	15818.7	0.7163	0.6237	(6.1671)	(2.2155)	(1.6124)	(8.6557)
D conv	0.963	1.2533	26929.7	0.6592	0.5324	1.7529	1.3028	0.0526	0.2221
D direct	3.4262	9.9762	39626.4	0.5056	0.3777	5.5669	1.7660	2.3136	10.3993
E forest	2.7587	1.4647	26501.5	0.8082	0.7952	(1.4423)	(1.1962)	(0.0830)	(0.3519)
E conv	3.0761	2.4388	38516.1	0.7468	0.6799	1.3283	0.9314	0.1388	0.5609
E direct	1.4987	2.107	47630.9	0.6447	0.6400	1.8977	1.2728	0.2401	0.9932

§4.1. The maximum likelihood parameters $(\alpha_1, \sigma_{12}, \alpha_2)$ for the data are shown in Table 10.1, together with the McKay correlation coefficient, ρ_M and the measured correlation ρ_{Data}.

In these experiments, we used tomographic images of soil samples, and packings of spheres of different sizes as phantoms. The soil samples were selected from untreated (i.e. forest soil type), and treated (i.e. conventional mechanized cultivated soil, and direct plantation cultivated soil). The image analysis methods employed to measure the pore and throat size distributions in these images are out of the scope of this paper, and will be discussed elsewhere. Typical scatterplots of the throat area and pore area data are shown in Figure 10.19 for the untreated soil forest A, which shows strong correlation, and in Figure 10.20 for the model structure 1 made from beds of spheres, which shows weak correlation.

We see from Table 10.1 that the theoretical McKay correlation agrees well with that found experimentally. The four distance functions d_M, dE_M, d_B, d_K are given for the four soil treatments. The information metric d_M is the most discriminating and the Bhattacharyya distance d_B is the least discriminating. Over all treatments the grand means for distances from untreated (forest) soils

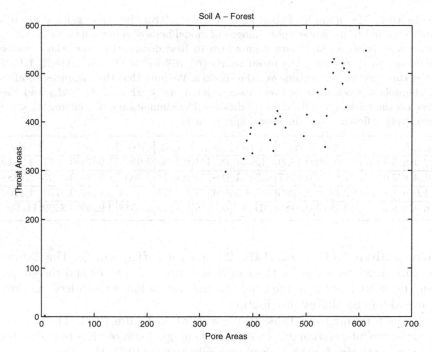

Fig. 10.19. Scatterplot of the data for untreated forest A soil sample, Table 10.1.

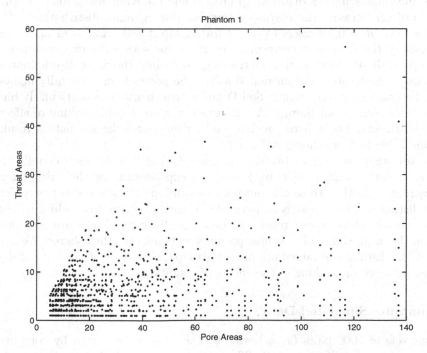

Fig. 10.20. Scatterplot of the data for model sphere bed structure Sample 1 of Table 10.2.

Table 10.2. Maximum likelihood parameters of McKay bivariate gamma fitted to data extracted from tomographic images of model beds of spheres and a simulation. There is no reference structure from which to base distances so here the distances shown are measured from the mean values $(\overline{\alpha_1}, \overline{\sigma_{12}}, \overline{\alpha_2}) = (1.6151, 19.9672, 1.1661)$ of the three parameters taken over the spheres. We note that the distances obtained for the sphere models are in every case ordered: $d_B < dE_M < d_K < d_M$ and they agree on the ranking of effects of conditions. The simulation data, having $\alpha_1 << 1$, seems very different from the model sphere beds.

Sample	α_1	α_2	σ_{12}	ρ_M	ρ_{Data}	d_M	dE_M	d_B	d_K
1(2.4 a 3.3 mm)	1.0249	0.1469	54.8050	0.9341	0.3033	3.6369	1.9071	0.4477	2.9319
2(1.4 a 2.0 mm)	1.6789	0.5117	4.3863	0.8755	0.1714	2.3057	1.5185	0.4195	1.7888
3(1.0 a 1.7 mm)	2.1416	2.8396	0.7103	0.6557	0.1275	4.5013	2.1216	0.6751	4.1202
Simulation	0.1514	0.3185	4137	0.5676	0.1118	8.5318	2.9209	0.8539	11.4460

are respectively 2.423, 1.302, 0.474, 2.113, for d_M, dE_M, d_B, d_K. The distance measures are found also for the model structures of spheres and the simulation, Table 10.2, but here the experimental correlation is much less than that expected for the McKay distribution.

The soil results from Table 10.1 are shown in Figure 10.21. The first two plots use the information distance d_M and energy-distance dE_M bounds (10.9, 10.11 respectively) for the McKay manifold metric (10.8), the other two plots use the Bhattacharyya distance d_B (10.12) and the Kullback-Leibler distance d_K (10.12) between the corresponding bivariate gamma distributions. The base plane $d = 0$ represents the natural or forest soil; data is in pairs, two points of the same size correspond to the same soils with two treatments. Importantly, the information metric, d_M, is mainly the most discriminating between the treated and untreated soils—the points being generally highest in the graphic for d_M, though Soil D direct treatment has a particularly high d_K value. Except for Sample A, all distances agree on the ranking of effects of treatments. The sphere packing model results and the simulation results from Table 10.2 are shown in Figure 10.22.

Note that the McKay bivariate gamma distribution does not contain the case of both marginal distributions being exponential nor the case of zero covariance—both of these are limiting cases and so cannot serve as base points for distances. Thus, there is no natural reference structure from which to base distances in these model results so here the distances shown are measured from the mean values of the three parameters taken over the spheres. We note that the distances obtained are in each case ordered: $d_B < dE_M < d_K < d_M$; they all agree on ranking of the effects of conditions.

Computer Simulated Data

Four sets of 5000 pairs (x_i, y_i) with $0 < x_i < y_i$ were drawn by computer from gamma distributions, with different parameters and with varying positive

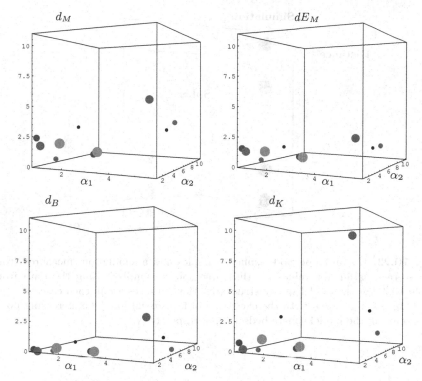

Fig. 10.21. Distances of 5 pairs of treated soil samples from the untreated forest soil, in terms of their porous structure from tomographic images, using the data from Table 10.1. Clockwise from the top left the plots use: the information theoretic bounds (10.9) d_M and root energy dE_M for the McKay manifold metric (4.5), the Bhattacharyya distance (10.12) d_B and the Kullback-Leibler distance (10.12) d_K between the corresponding bivariate gamma distributions. The plane $d = 0$ represents the natural or forest soil; data is in pairs, two points of the same size correspond to the same soils with two treatments. The information metric, top left, is most discriminating.

covariance between the two variables, Figure 10.23. This data was analyzed and from it maximum likelihood fits were made of McKay bivariate gamma distributions.

Table 10.3 summarizes the parameters, and the distances measured between the corresponding points are shown in Tables 10.4, 10.5, 10.6 and 10.7. These experiments confirm that data sets 1 and 2 are more similar to each other, than to the data sets 3 and 4, which is verified by visual inspection of the scatterplots shown in Figure 10.23.

If we consider that data sets 1 and 2 form one cluster, and that data sets 3 and 4 form another cluster, it is important to verify how the distance measures we are comparing perform in terms of data discrimination. Table 10.8 shows the ratios between the mean inter and intra cluster distances, indicating that

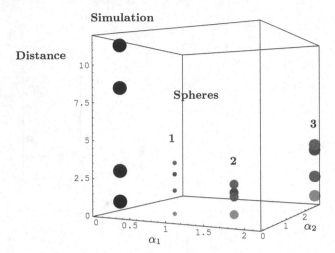

Fig. 10.22. Distances of model sphere samples and a simulation, measured from the average parameter values for the three sphere samples, using the data from Table 10.2. For the model sphere structures, the distances are in each case ordered: $d_B < dE_M < d_K < d_M$, with the exception of the simulation. The increasing point sizes refer to: the model sphere beds 1, 2, 3, respectively.

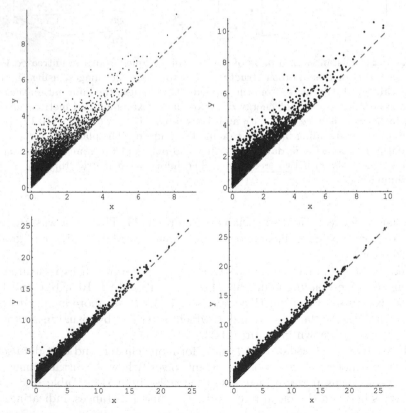

Fig. 10.23. Scatterplot of the data for computer simulations of the four positively correlated gamma processes with $x < y$: top row # 1,2, second row # 3,4. Maximum likelihood McKay parameters are given in Table 10.3.

Table 10.3. Maximum likelihood McKay parameters for the four simulated bivariate gamma data sets 1,2,3,4.

#	$\bar{x}$	$\bar{y}$	α_1	α_2	σ_{12}	ρ_M	ρ_{data}
1	0.9931	1.5288	1.0383	0.9174	0.9607	0.7286	0.9017
2	1.1924	1.7275	1.0176	0.7934	1.4117	0.7443	0.9304
3	2.9793	3.5151	1.0383	0.3598	8.5813	0.8618	0.9873
4	3.2072	3.7429	1.0165	0.3432	10.1006	0.8646	0.9892

Table 10.4. Pairwise McKay information energy distances dE_M^{sym} for the four simulated bivariate gamma data sets 1,2,3,4.

dE_M^{sym}	sample set 2	sample set 3	sample set 4
sample set 1	1.2505	2.5489	2.2506
sample set 2	0.0000	2.1854	2.2988
sample set 3	0.0000	0.0000	0.2067

Table 10.5. Pairwise McKay information distances d_M^{sym} for the four simulated bivariate gamma data sets 1,2,3,4.

d_M^{sym}	sample set 2	sample set 3	sample set 4
sample set 1	1.2625	2.4836	2.6511
sample set 2	0.0000	2.0532	2.1788
sample set 3	0.0000	0.0000	0.1653

Table 10.6. Pairwise Kullback-Leibler distances d_K for the four simulated bivariate gamma data sets 1,2,3,4.

d_K	sample set 2	sample set 3	sample set 4
sample set 1	0.6804	1.0368	1.1793
sample set 2	0.0000	0.7013	0.1828
sample set 3	0.0000	0.0000	0.0040

Table 10.7. Pairwise Bhattacharyya distances d_B for the four simulated bivariate gamma data sets 1,2,3,4.

d_B	sample set 2	sample set 3	sample set 4
sample set 1	0.1661	0.2254	0.2506
sample set 2	0.0000	0.1596	0.1815
sample set 3	0.0000	0.0000	0.0010

Table 10.8. Expected ratio between inter/intra cluster distances for the four simulated bivariate gamma data sets 1,2,3,4.

dE_M^{sym}	d_M^{sym}	d_K	d_B
1.6734	1.6401	1.2225	1.1323

the best data separability is obtained by dE_M^{sym} and d_M^{sym}, using the symmetrization

$$d^{sym} = \frac{1}{2}(d(A, B) + d(B, A)).$$

In most cases, the information geometry, which uses a maximum likelihood metric, is more discriminating than the classical Bhattacharyya distance, or the Kullback-Leibler divergence, between pairs of bivariate gamma distributions. We have also available the information geometry of bivariate Gaussian and bivariate exponential distributions and we expect that our methodology may have other applications in the modelling of bivariate statistical processes in hydrology.

11

Quantum Chaology

This chapter, based on Dodson [66], is somewhat speculative in that it is clear
that gamma distributions do not precisely model the analytic systems discussed here, but some features may be useful in studies of qualitative generic
properties in applications to data from real systems which manifestly seem
to exhibit behaviour reminiscent of near-random processes. Quantum counter-
parts of certain simple classical systems can exhibit chaotic behaviour through
the statistics of their energy levels and the irregular spectra of chaotic systems are modelled by eigenvalues of infinite random matrices. We use known
bounds on the distribution function for eigenvalue spacings for the Gaussian
orthogonal ensemble (GOE) of infinite random real symmetric matrices and
show that gamma distributions, which have the important uniqueness property Theorem 1.1, can yield an approximation to the GOE distribution. This
has the advantage that then both chaotic and non chaotic cases fit in the
information geometric framework of the manifold of gamma distributions.
Additionally, gamma distributions give approximations, to eigenvalue spacings for the Gaussian unitary ensemble (GUE) of infinite random hermitian
matrices and for the Gaussian symplectic ensemble (GSE) of infinite random
hermitian matrices with real quaternionic elements. Interestingly, the spacing
distribution between zeros of the Riemann zeta function is approximated by
the GUE distribution, and we investigate the stationarity of the coefficient
of variation of the numerical data with respect to location and sample size.
The review by Deift [52] illustrates how random matrix theory has significant
links to a wide range of mathematical problems in the theory of functions as
well as to mathematical physics.

11.1 Introduction

Berry introduced the term quantum chaology in his 1987 Bakerian Lecture [24]
as the study of semiclassical but non-classical behaviour of systems whose classical motion exhibits chaos. He illustrated it with the statistics of energy levels,

K. Arwini, C.T.J. Dodson, *Information Geometry.*   223
Lecture Notes in Mathematics 1953,
© Springer-Verlag Berlin Heidelberg 2008

following his earlier work with Tabor [25] and with related developments from the study of a range of systems. In the regular spectrum of a bound system with $n \geq 2$ degrees of freedom and n constants of motion, the energy levels are labelled by n quantum numbers, but the quantum numbers of nearby energy levels may be very different. In the case of an irregular spectrum, such as for an ergodic system where only energy is conserved, we cannot use quantum number labelling. This prompted the use of energy level spacing distributions to allow comparisons among different spectra [25]. It was known, eg from the work of Porter [168], that the spacings between energy levels of complex nuclei and atoms are modelled by the spacings of eigenvalues of random matrices and that the Wigner distribution [214] gives a very good fit. It turns out that the spacing distributions for generic regular systems are negative exponential, that is Poisson random, §1.1.3; but for irregular systems the distributions are skew and unimodal, at the scale of the mean spacing. Mehta [145] provides a detailed discussion of the numerical experiments and functional approximations to the energy level spacing statistics, Alt et al [5] compare eigenvalues from numerical analysis and from microwave resonator experiments, also eg. Bohigas et al [30] and Soshnikov [193] confirm certain universality properties. Miller [151] provides much detail on a range of related number theoretic properties, including random matrix theory links with L-functions. Forrester's online book [88] gives a wealth of analytic detail on the mathematics and physics of eigenvalues of infinite random matrices for the three ensembles of particular interest: Gaussian orthogonal (GOE), unitary (GUE) and symplectic (GSE), being real, complex and quaternionic, respectively.

Definition 11.1. *There are three cases of interest [88]*

GOE: A random real symmetric $n \times n$ matrix belongs to the Gaussian orthogonal ensemble (GOE) if the diagonal and upper triangular elements are independent random variables with Gaussian distributions of zero mean and standard deviation 1 for the diagonal and $\frac{1}{\sqrt{2}}$ for the upper triangular elements.

GUE: A random hermitian $n \times n$ matrix belongs to the Gaussian unitary ensemble (GUE) if the diagonal elements m_{jj} (which must be real) and the upper triangular elements $m_{jk} = u_{jk} + iv_{jk}$ are independent random variables with Gaussian distributions of zero mean and standard deviation $\frac{1}{\sqrt{2}}$ for the m_{jj} and $\frac{1}{2}$ for each of the u_{jk} and v_{jk}.

GSE: A random hermitian $n \times n$ matrix with real quaternionic elements belongs to the Gaussian symplectic ensemble (GSE) if the diagonal elements z_{jj} (which must be real) are independent with Gaussian distribution of zero mean and standard deviation $\frac{1}{2}$ and the upper triangular elements $z_{jk} = u_{jk} + iv_{jk}$ and $w_{jk} = u'_{jk} + iv'_{jk}$ are independent random variables with Gaussian distributions of zero mean and standard deviation $\frac{1}{2\sqrt{2}}$ for each of the u_{jk}, u'_{jk}, v'_{jk} and v_{jk}.

Then the matrices in these ensembles are respectively invariant under the appropriate orthogonal, unitary and symplectic transformation groups, and moreover in each case the joint density function of all independent elements is controlled by the trace of the matrices and is of form [88]

$$p(X) = A_n\, e^{-\frac{1}{2}TrX^2} \tag{11.1}$$

where A_n is a normalizing factor. Barndorff-Nielsen et al [19] give some background mathematical statistics on the more general problem of quantum information and quantum statistical inference, including reference to random matrices.

Here we show that gamma distributions, §1.4.1, provide approximations to eigenvalue spacing distributions for the GOE distribution comparable to the Wigner distribution at the scale of the mean and for the GUE and GSE distributions, except near the origin. That may be useful because the gamma distribution has a well-understood and tractable information geometry [14, 65] as well as the important uniqueness property of Hwang and Hu [106] Theorem 1.1 above, which says that: *For independent positive random variables with a common probability density function f, having independence of the sample mean and the sample coefficient of variation is equivalent to f being the gamma distribution.*

It is noteworthy also that the non-chaotic case has an exponential distribution of spacings between energy levels and that the sum of n independent identical exponential random variables follows a gamma distribution and the sum of n independent identical gamma random variables follows a gamma distribution, §1.4; moreover, the product of independent gamma distributions are well-approximated by gamma distributions, cf. eg. §9.5 §9.5.1.

From a different standpoint, Berry and Robnik [26] gave a statistical model using a mixture of energy level spacing sequences from exponential and Wigner distributions. Monte Carlo methods were used by Caër et al. [33] to investigate such a mixture. Caër et al. established also the best fit of GOE, GUE and GSE unit mean distributions, for spacing $s > 0$, using the generalized gamma density which we can put in the form

$$g(s; \beta, \omega) = a(\beta, \omega)\, s^\beta\, e^{-b(\beta,\omega)s^\omega} \quad \text{for } \beta, \omega > 0 \tag{11.2}$$

$$\text{where} \quad a(\beta, \omega) = \frac{\omega\left[\Gamma((2+\beta)/\omega)\right]^{\beta+1}}{\left[\Gamma((1+\beta)/\omega)\right]^{\beta+2}} \quad \text{and} \quad b(\beta, \omega) = \left[\frac{\Gamma((2+\beta)/\omega)}{\Gamma((1+\beta)/\omega)}\right]^\omega .$$

Then the best fits of (11.2) had the parameter values [33]

Ensemble	β	ω	Variance
Exponential	0	1	1
GOE	1	1.886	0.2856
GUE	2	1.973	0.1868
GSE	4	2.007	0.1100

and were accurate to within $\sim$ 0.1% of the true distributions from Forrester [88]. Observe that the exponential distribution is recovered by the choice $g(s; 0, 1) = e^{-s}$. These distributions are shown in Figure 11.3. Götze and Kösters [97] show that the asymptotic results for the second-order correlation function of the characteristic polynomial of a random matrix from the Gaussian Unitary Ensemble essentially continue to hold for a general Hermitian Wigner matrix.

11.2 Eigenvalues of Random Matrices

The two classes of spectra are illustrated in two dimensions by bouncing geodesics in plane billiard tables: eg in the de-symmetrized 'stadium of Bunimovich' with ergodic chaotic behaviour and irregular spectrum on the one hand, and on the other hand in the symmetric annular region between concentric circles with non-chaotic behaviour, regular spectrum and random energy spacings [25, 30, 145, 24].

It turns out that the mean spacing between eigenvalues of infinite symmetric real random matrices—the Gaussian Orthogonal Ensemble (GOE)—is bounded and therefore it is convenient to normalize the distribution to have unit mean; also, in fact, the same is true for the GUE and GSE cases. Barnett [21] provides a numerical tabulation of the first 1,276,900 GOE eigenvalues. In fact, Wigner [212, 213, 214] had already surmised that the cumulative probability distribution function, §1.2, for the spacing $s > 0$ at the scale of the mean spacing should be of the form:

$$W(s) = 1 - e^{-\frac{\pi s^2}{4}}. \qquad (11.3)$$

This has unit mean and variance $\frac{4-\pi}{\pi} \approx 0.273$ with probability density function

$$w(s) = \frac{\pi}{2} s \, e^{-\frac{\pi s^2}{4}}. \qquad (11.4)$$

Note that the particular case $a(1, 2) = \frac{\pi}{2}$, $b(1, 2) = \frac{\pi}{4}$ reduces (11.2) to (11.4) [214]. Remarkably, Wigner's surmise gave an extremely good fit with numerical computation of the true GOE distribution, cf. Mehta [145] Appendix A.15, and with a variety of observed data from atomic and nuclear systems [214, 25, 24, 145]. Caër et al. [33] showed that the generalized gamma (11.2) was within $\sim$ 0.1% of the true GOE distribution from Forrester [88], rather better even than the original Wigner [214] surmise (11.4).

From Mehta [145] p 171, we have bounds on P, the cumulative probability distribution function, §1.2, for the spacings between eigenvalues of infinite symmetric real random matrices:

$$L(s) = 1 - e^{-\frac{1}{16}\pi^2 s^2} \leq P(s) \leq U(s) = 1 - e^{-\frac{1}{16}\pi^2 s^2} \left(1 - \frac{\pi^2 s^2}{48}\right). \qquad (11.5)$$

Here the lower bound L has mean $\frac{2}{\sqrt{\pi}} \approx 1.13$ and variance $\frac{4(4-\pi)}{\pi^2} \approx 0.348$, and the upper bound U has mean $\frac{5}{3\sqrt{5}} \approx 0.940$ and variance $\frac{96-25\pi}{9\pi^2} \approx 0.197$.

The family of probability density functions for gamma distributions, §1.4.1, with dispersion parameter $\kappa > 0$ and mean $\kappa/\nu > 0$ for positive random variable s is given by

$$f(s; \nu, \kappa) = \nu^\kappa \frac{s^{\kappa-1}}{\Gamma(\kappa)} e^{-s\nu}, \quad \text{for } \nu, \kappa > 0 \tag{11.6}$$

with variance $\frac{\kappa}{\nu^2}$. Then the subset having unit mean is given by

$$f(s; \kappa, \kappa) = \kappa^\kappa \frac{s^{\kappa-1}}{\Gamma(\kappa)} e^{-s\kappa}, \quad \text{for } \kappa > 0 \tag{11.7}$$

with variance $\frac{1}{\kappa}$. These parameters ν, κ are called natural coordinates, §3.3, because they admit presentation of the family (11.6) as an exponential family [11], §3.2, and thereby provide an associated natural affine immersion in $\mathbb{R}^3$ [68], §3.4

$$h : \mathbb{R}^+ \times \mathbb{R}^+ \to \mathbb{R}^3 : \begin{pmatrix} \nu \\ \kappa \end{pmatrix} \mapsto \begin{pmatrix} \nu \\ \kappa \\ \log \Gamma(\kappa) - \kappa \log \nu \end{pmatrix}. \tag{11.8}$$

The generalized gamma distributions (11.2) do not constitute an exponential family, except in the case $\omega = 1$, so they do not admit an affine immersion. The gamma family affine immersion, (11.8), was used [14] to present tubular neighbourhoods of the 1-dimensional subspace consisting of exponential distributions ($\kappa = 1$), so giving neighbourhoods of random processes, §5.1. The maximum entropy case has $\kappa = 1$, the exponential distribution, which corresponds to an underlying Poisson random event process, §1.4.1, and so models spacings in the spectra for non-chaotic systems; for $\kappa > 1$ the distributions are skew unimodular. The gamma unit mean distribution fit to the true GOE distribution from Mehta [145] has variance ≈ 0.379 and hence $\kappa \approx 2.42$.

In fact, κ is a geodesic coordinate, §2.1, in the Riemannian 2-manifold of gamma distributions with Fisher information metric, §3.5; arc length, §2.0.5, along this coordinate from $\kappa = a$ to $\kappa = b$ is given by

$$\left| \int_a^b \sqrt{\frac{d^2 \log(\Gamma(\kappa))}{d\kappa^2} - \frac{1}{\kappa}} \, d\kappa \right|. \tag{11.9}$$

Plotted in Figure 11.1 are the cumulative distributions for the bounds (11.5) (dashed), the gamma distribution (thick solid) fit to the true GOE distribution with unit mean, §1.4.1, and the Wigner surmised distribution (11.3) (thin solid).

The corresponding probability density functions are in Figure 11.2: gamma distribution fit (thick solid) to the true GOE distribution from Mehta [145]

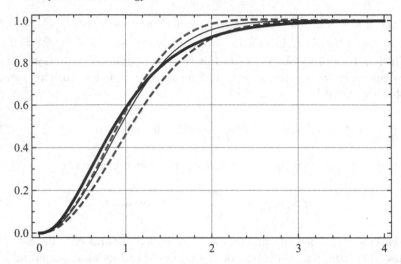

Fig. 11.1. The bounds on the normalized cumulative distribution function of eigenvalue spacings for the GOE of random matrices (11.5) (dashed), the Wigner surmise (11.3) (thin solid) and the unit mean gamma distribution fit to the true GOE distribution from Mehta [145] Appendix A.15 (thick solid).

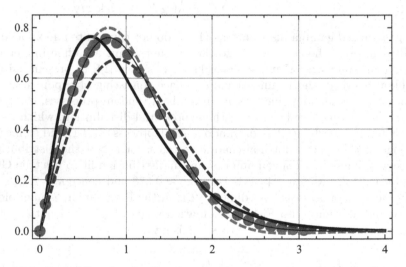

Fig. 11.2. Probabilty density function for the unit mean gamma distribution fit (thick solid) to the true GOE distribution from Mehta [145] Appendix A.15 (points), the Wigner surmised density (11.4) (thin solid) and the probability densities for the bounds (11.10) (dashed) for the distribution of normalized spacings between eigenvalues for infinite symmetric real random matrices.

Appendix A.15 (points), the Wigner surmised density (11.4) (thin solid) and the probability densities for the bounds (11.10) (dashed), respectively,

$$l(s) = \frac{\pi s}{2} e^{-\frac{\pi s^2}{4}}, \quad u(s) = \frac{\pi^2 s(64 - \pi^2 s^2)}{384} e^{-\frac{1}{16}\pi^2 s^2}. \tag{11.10}$$

11.3 Deviations

Berry has pointed out [23] that the behaviour of the eigenvalue spacing probability density near the origin is an important feature of the ensemble statistics of these matrices; in particular it is linear for the GOE case and for the Wigner approximation. Moreover, for the unitary ensemble (GUE) of complex hermitian matrices, near the origin, the behaviour is $\sim s^2$ and for the symplectic ensemble (GSM, representing half-integer spin particles with time-reversal symmetric interactions) it is $\sim s^4$.

From (11.7) we see that at unit mean the gamma density behaves like $s^{\kappa-1}$ near the origin, so linear behaviour would require $\kappa = 2$ which gives a variance of $\frac{1}{\kappa} = \frac{1}{2}$ whereas the GOE fitted gamma distribution has $\kappa \approx 2.42$ and hence variance ≈ 0.379. This may be compared with variances for the lower bound l, $\frac{4(4-\pi)}{\pi^2} \approx 0.348$, the upper bound u, $\frac{96-25\pi}{9\pi^2} \approx 0.197$, and the Wigner distribution w, $\frac{4-\pi}{\pi} \approx 0.273$. The gamma distributions fitted to the lower and upper bounding distributions have, respectively, $\kappa_L = \frac{\pi}{4-\pi} \approx 3.660$ and $\kappa_U = \frac{5\pi^2}{96-25\pi} \approx 2.826$. Figure 11.3 shows the probability density functions for the

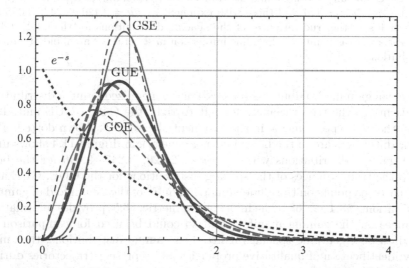

Fig. 11.3. Probability density functions for the unit mean gamma distributions (dashed) and generalized gamma distribution (solid) fits to the true variances for left to right the GOE , GUE and GSE cases. The two types coincide in the exponential case, e^{-s}, shown dotted.

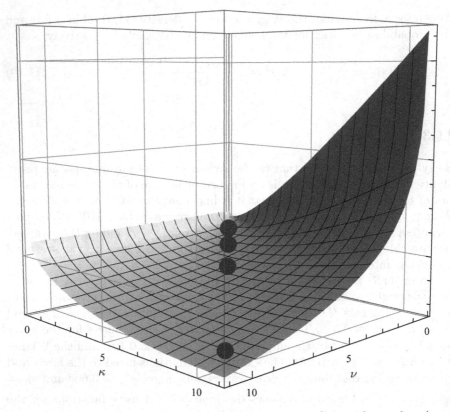

Fig. 11.4. The unit mean gamma distributions corresponding to the random (non-chaotic) case, $\kappa = \nu = 1$ and those with exponent $\kappa = \nu = 2.420$, 4.247, 9.606 for the best fits to the true variances of the spacing distributions for the GOE, GUE and GSE cases, as points on the affine immersion in $\mathbb{R}^3$ of the 2-manifold of gamma distributions.

unit mean gamma distributions (dashed) and generalized gamma distribution (solid) fits to the true variances for left to right the GOE , GUE and GSE cases; the two types coincide in the exponential case, e^{-s}, shown dotted. The major differences are in the behaviour near the origin. Figure 11.4 shows unit mean gamma distributions with $\kappa = \nu = 2.420$, 4.247, 9.606 for the best fits to the true variances of the spacing distributions for the GOE, GUE and GSE cases, as points on the affine immersion in $\mathbb{R}^3$ of the 2-manifold of gamma distributions, §3.4, cf. [68]. The information metric, §3.5, provides information distances on the gamma manifold and so could be used for comparison of real data on eigenvalue spacings if fitted to gamma distributions; that may allow identification of qualitative properties and represent trajectories during structural changes of systems.

The authors are indebted to Rudnick [178] for pointing out that the GUE eigenvalue spacing distribution is rather closely followed by the distribution

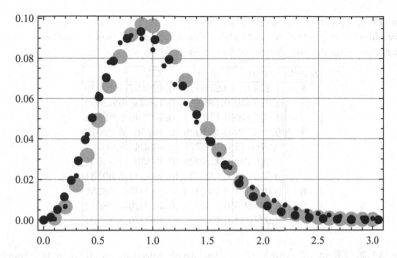

Fig. 11.5. Probability plot with unit mean for the spacings between the first 2,001,052 zeros of the Riemann zeta function from the tabulation of Odlyzko [157] (large points), that for the true GUE distribution from the tabulation of Mehta [145] Appendix A.15 (medium points) and the gamma fit to the true GUE (small points).

of zeros for the Riemann zeta function; actually, Hilbert had conjectured this, as mentioned along with a variety of other probabilistic aspects of number theory by Schroeder [184]. This can be seen in Figure 11.5 which shows with unit mean the probability distribution for spacings among the first 2,001,052 zeros from the tabulation of Odlyzko [157] (large points), that for the true GUE distribution from the tabulation of Mehta [145] Appendix A.15 (medium points) and the gamma fit to the true GUE (small points), which has $\kappa \approx 4.247$. The grand mean spacing between zeros from the data was ≈ 0.566, the coefficient of variation ≈ 0.422 and variance ≈ 0.178.

Table 11.1 shows the effect of location on the statistical data for spacings in the first ten consecutive blocks of 200,000 zeros of the Riemann zeta function normalized with unit grand mean; Table 11.2 shows the effect of sample size. For gamma distributions we expect the coefficient of variation to be independent of both sample size and location, by Theorem 1.1.

Remark 11.2. The gamma distribution provides approximations to the true distributions for the spacings between eigenvalues of infinite random matrices for the GOE, GUE and the GSE cases. However, it is clear that gamma distributions do not precisely model the analytic systems discussed here, and do not give correct asymptotic behaviour at the origin, as is evident from the results of Caër et al. [33] who obtained excellent approximations for GOE, GUE and GSE distributions using the generalized gamma distribution (11.2) §11.1. The differences may be seen in Figure 11.3 which shows the unit mean distributions for gamma (dashed) and generalized gamma [33] (solid) fits to the true variances for the Poisson, GOE, GUE and GSE ensembles.

Table 11.1. Effect of location: Statistical data for spacings in the first ten consecutive blocks of 200,000 zeros of the Riemann zeta function normalized with unit grand mean from the tabulation of Odlyzko [157].

Block	Mean	Variance	CV	κ
1	1.232360	0.276512	0.426697	5.49239
2	1.072330	0.189859	0.406338	6.05654
3	1.025210	0.174313	0.407240	6.02974
4	0.996739	0.165026	0.407563	6.02019
5	0.976537	0.158777	0.408042	6.00607
6	0.960995	0.154008	0.408367	5.99651
7	0.948424	0.150136	0.408544	5.99131
8	0.937914	0.147043	0.408845	5.98250
9	0.928896	0.144285	0.408926	5.98014
10	0.921034	0.142097	0.409276	5.96991

Table 11.2. Effect of sample size: Statistical data for spacings in ten blocks of increasing size $200,000m$, $m = 1, 2, \ldots, 10$, for the first 2,000,000 zeros of the Riemann zeta function, normalized with unit grand mean, from the tabulation of Odlyzko [157].

m	Mean	Variance	CV	κ
1	1.23236	0.276511	0.426696	5.49242
2	1.15234	0.239586	0.424765	5.54246
3	1.10997	0.221420	0.423934	5.56421
4	1.08166	0.209725	0.423384	5.57869
5	1.06064	0.201303	0.423018	5.58833
6	1.04403	0.194799	0.422748	5.59548
7	1.03037	0.189538	0.422527	5.60133
8	1.01881	0.185161	0.422357	5.60584
9	1.00882	0.181418	0.422207	5.60983
10	1.00004	0.178180	0.422094	5.61282

Unfortunately, from our present perspective, the generalized gamma distributions do not have a tractable information geometry and so some features of the gamma distribution approximations may be useful in studies of qualitative generic properties in applications to data from real systems. It would be interesting to investigate the extent to which data from real atomic and nuclear systems has generally the qualitative property that the sample coefficient of variation is independent of the mean. That, by Theorem 1.1, is an information-theoretic distinguishing property of the gamma distribution.

It would be interesting to know if there is a number-theoretic property that corresponds to the apparently similar qualitative behaviour of the spacings of zeros of the Riemann zeta function, Tables 11.1, 11.2. Since the non-chaotic case has an exponential distribution of spacings between energy levels and the sum of n independent identical exponential random variables follows a gamma distribution and moreover the sum of n independent identical gamma random variables follows a gamma distribution, a further analytic development would be to calculate the eigenvalue distributions for gamma or loggamma-distributed matrix ensembles. Information geometrically, the Riemannian manifolds of gamma and loggamma families are isometric, §3.6 but the loggamma random variables have bounded domain and their distributions contain the uniform distribution, §3.6 and §5.2, which may be important in modelling some real physical processes.

References

1. I. Akyildiz. Mobility management in current and future communication networks. *IEEE Network Mag.* 12, 4 (1998) 39-49.
2. I. Akyildiz. Performance modeling of next generation wireless systems, Keynote Address, Conference on Simulation Methods and Applications, 1-3 November 1998, Orlando, Florida.
3. R.I. Al-Raoush and C. S. Willson. Extraction of physically realistic pore network properties from three-dimensional synchrotron X-ray microtomography images of unconsolidated porous media systems. Moving through scales of flow and transport in soil. *Journal of Hydrology*, 300, 1-4, (2005) 44-64.
4. P.L. Alger, Editor. **Life and times of Gabriel Kron**, Mohawk, New York, 1969. Cf. C.T.J Dodson, Diakoptics Past and Future, pp 288-9 ibid.
5. H. Alt, C. Dembrowski, H.D. Graf, R. Hofferbert, H. Rehfield, A. Richter and C. Schmit. Experimental versus numerical eigenvalues of a Bunimovich stadium billiard: A comparison. *Phys. Rev. E* 60, 3 (1999) 2851-2857.
6. S-I. Amari. **Diakoptics of Information Spaces** Doctoral Thesis, University of Tokyo, 1963.
7. S-I. Amari. Theory of Information Spaces—A Geometrical Foundation of the Analysis of Communication Systems. *Research Association of Applied Geometry Memoirs* 4 (1968) 171-216.
8. S-I. Amari. **Differential Geometrical Methods in Statistics** Springer Lecture Notes in Statistics 28, Springer-Verlag, Berlin 1985.
9. S-I. Amari, O.E. Barndorff-Nielsen, R.E. Kass, S.L. Lauritzen and C.R. Rao. **Differential Geometry in Statistical Inference**. Lecture Notes Monograph Series, Institute of Mathematical Statistics, Volume 10, Hayward California, 1987.
10. S-I. Amari. Dual Connections on the Hilbert Bundles of Statistical Models. In **Proc. Workshop on Geometrization of Statistical Theory** 28-31 October 1987. Ed. C.T.J. Dodson, ULDM Publications, University of Lancaster, 1987, pp 123-151.
11. S-I. Amari and H. Nagaoka. **Methods of Information Geometry**, American Mathematical Society, Oxford University Press, 2000.
12. Flavio S. Anselmetti, Stefan Luthi and Gregor P. Eberli. Quantitative Characterization of Carbonate Pore Systems by Digital Image Analysis. *AAPG Bulletin*, 82, 10, (1998) 1815-1836.

13. Khadiga Arwini. **Differential geometry in neighbourhoods of randomness and independence**. PhD thesis, Department of Mathematics, University of Manchester Institute of Science and Technology (2004).
14. Khadiga Arwini and C.T.J. Dodson. Information geometric neighbourhoods of randomness and geometry of the McKay bivariate gamma 3-manifold. *Sankhya: Indian Journal of Statistics* 66, 2 (2004) 211-231.
15. Khadiga Arwini and C.T.J. Dodson. Neighbourhoods of independence and associated geometry in manifolds of bivariate Gaussians and Freund distributions. *Central European J. Mathematics* 5, 1 (2007) 50-83.
16. Khadiga Arwini, L. Del Riego and C.T.J. Dodson. Universal connection and curvature for statistical manifold geometry. *Houston Journal of Mathematics* 33, 1 (2007) 145-161.
17. Khadiga Arwini and C.T.J. Dodson. Alpha-geometry of the Weibull manifold. Second Basic Science Conference, 4-8 November 2007, Al-Fatah University, Tripoli, Libya.
18. C. Baccigalupi, L. Amendola and F. Occhionero. Imprints of primordial voids on the cosmic microwave background *Mon. Not. R. Astr. Soc.* 288, 2 (1997) 387-96.
19. O.E. Barndorff-Nielsen, R.D. Gill and P.E. Jupp. On quantum statistical inference. *J. Roy. Statist. Soc.* B 65 (2003) 775-816.
20. O.E. Barndorff-Nielsen and D.R. Cox. **Inference and Asymptotics**. Monographs on Statistics and Applied Probability, 52. Chapman & Hall, London, 1994
21. A.H. Barnett. http://math.dartmouth.edu/~ahb/pubs.html
22. A.J. Benson, F. Hoyle, F. Torres and M.J, Vogeley. LGalaxy voids in cold dark matter universes. *Mon. Not. R. Astr. Soc.* 340 (2003) 160-174.
23. M.V. Berry. Private communication. 2008.
24. M.V. Berry. Quantum Chaology. *Proc. Roy. Soc. London A* 413, (1987) 183-198.
25. M.V. Berry and M. Tabor. Level clustering in the regular spectrum. *Proc. Roy. Soc. London A* 356, (1977) 373-394.
26. M.V. Berry and M. Robnik. Semiclassical level spacings when regular and chaotic orbits coexist. *J. Phys. A Math. General* 17, (1984) 2413-2421.
27. A. Bhattacharyya. On a measure of divergence between two statistical populations defined by their distributions. *Bull. Calcutta Math. Soc.* 35 (1943) 99-110.
28. Marcelo Biassusi. **Estudo da Deformao de um Vertissolo Atravs da Tomografia Computadorizada de Dupla Energia Simultnea**. *Phd Thesis*, UFRGS - Federal University of Rio Grande do Sul, Porto Alegre, Brazil. February 1996.
29. D. Bloomberg. Basic Definitions in Mathematical Morphology. www.leptonica.com/papers, April 2003.
30. O. Bohigas, M.J. Giannoni and C. Schmit. Characterization of Chaotic Quantum Spectra and Universality of Level Fluctuation Laws. *Phys. Rev. Lett.* 52, 1 (1984) 1-4.
31. K. Borovkov. **Elements of Stochastic Modelling**, World Scientific and Imperial College Press, Singapore and London, 2003.
32. U. Boudriot, R. Dersch, A. Greiner, and J.H. Wendorf. Electrospinning approaches toward scaffold engineering–A brief overview. *Artificial Organs* 10 (2006) 785-792.

33. G. Le Caër, C. Male and R. Delannay. Nearest-neighbour spacing distributions of the β-Hermite ensemble of random matrices. *Physica A* (2007) 190-208. Cf. also their Erratum: *Physica A* 387 (2008) 1713.

34. Y. Cai, C.T.J. Dodson, O. Wolkenhauer and A.J. Doig. Gamma Distribution Analysis of Protein Sequences shows that Amino Acids Self Cluster. *J. Theoretical Biology* 218, 4 (2002) 409-418.

35. M. Calvo and J.M. Oller. An explicit solution of information geodesic equations for the multivariate normal model. *Statistics & Decisions* 9, (1990) 119-138.

36. D.Canarutto and C.T.J.Dodson. On the bundle of principal connections and the stability of b-incompleteness of manifolds. *Math. Proc. Camb. Phil. Soc.* 98, (1985) 51-59.

37. B. Canvel. **Timing Tags for Exponentiations for RSA** MSc Thesis, Department of Mathematics, University of Manchester Institute of Science and Technology, 1999.

38. B. Canvel and C.T.J. Dodson. Public Key Cryptosystem Timing Analysis. *Rump Session*, **CRYPTO 2000**, Santa Barbara, 20-24 August 2000.
    http://www.maths.manchester.ac.uk/~kd/PREPRINTS/rsatim.ps

39. A. Cappi, S. Maurogordato and M. Lachiéze-Rey A scaling law in the distribution of galaxy clusters. *Astron. Astrophys.* 243, 1 (1991) 28-32.

40. J. Castro and M. Ostoja-Starzewski. Particle sieving in a random fiber network. *Appl. Math. Modelling* 24, 8-9, (2000) 523-534.

41. S. Chari, C.S. Jutla, J.R. Rao and P. Rohatgi. Towards sound approaches to counteract power-analysis attacks. In **Advances in Cryptology-CRYPTO '99**, Ed. M. Wiener, Lecture Notes in Computer Science 1666, Springer, Berlin 1999 pp 398-412.

42. P. Coles. Understanding recent observations of the large-scale structure of the universe. *Nature* 346 (1990) 446.

43. L.A. Cordero, C.T.J. Dodson and M. deLeon. **Differential Geometry of Frame Bundles.** Kluwer, Dordrecht, 1989.

44. L.A. Cordero, C.T.J. Dodson and P.E. Parker. Connections on principal S^1-bundles over compacta. *Rev. Real Acad. Galega de Ciencias* XIII (1994) 141-149.

45. H. Corte. Statistical geometry of random fibre networks. In **Structure, Solid Mechanics and Engineering Design** (M. Te'eni, ed.), Proc. Southampton Civil Engineering Materials Conference, 1969. pp. 341-355. Wiley Interscience, London, 1971.

46. H. Corte. Statistical geometry of random fibre networks. In **Structure, Solid Mechanics and Engineering Design**, Proc. Southampton 1969 Civil Engineering Materials Conference, vol. 1, (ed. M. Te'eni) pp341-355. Wiley-Interscience, London, 1971.

47. H. Corte and C.T.J. Dodson. Über die Verteilung der Massendichte in Papier. Erster Teil: Theoretische Grundlagen **Das Papier**, 23, 7, (1969) 381-393.

48. II. Corte and E.H. Lloyd. Fluid flow through paper and sheet structure. In **Consolidation of the Paper Web** *Trans. IIIrd Fund. Res. Symp. 1965* (F. Bolam, ed.), pp 981-1009, BPBMA, London 1966.

49. D.J. Croton et al. (The 2dFGRS Team).The 2dF Galaxy Redshift Survey: Higher order galaxy correlation functions. Preprint, arXiv:astro-ph/0401434 v2 23 Aug 2004.

238   References

50. D.J. Croton et al. (The 2dFGRS Team). The 2dF Galaxy Redshift Survey: Voids and hierarchical scaling models. Preprint, arXiv:astro-ph/0401406 v2 23 Aug 2004.

51. R. Dawkins. **The Selfish Gene** Oxford University Press, Oxford 1976—cf. also the enlarged 1989 edition.

52. P. Deift. Some open problems in random matrix theory and the theory of integrable systems. Preprint, arXiv:arXiv:0712.0849v1 6 December 2007.

53. L. Del Riego and C.T.J. Dodson. Sprays, universality and stability. *Math. Proc. Camb. Phil. Soc.* 103(1988), 515-534.

54. J.F. Delrue, E. Perrier, Z.Y. Yu and B. Velde. New Algorithms in 3D Image Analysis and Their Application to the Measurement of a Spatialized Pore Size Distribution in Soils. *Phys. Chem. Earth*, 24, 7, (1999) 639-644.

55. M. Deng. **Differential Geometry in Statistical Inference** PhD thesis, Department of Statistics, Pennsylvania State University, 1990.

56. M. Deng and C.T.J. Dodson. **Paper: An Engineered Stochastic Structure**. Tappi Press, Atlanta (1994).

57. G. Di Crescenzo and R. Ostrovsky. On concurrent zero-knowledge with preprocessing. In **Advances in Cryptology-CRYPTO '99** Ed. M. Wiener, Lecture Notes in Computer Science 1666, Springer, Berlin 1999 pp 485-502.

58. C.T.J. Dodson. Spatial variability and the theory of sampling in random fibrous networks. *J. Royal Statist. Soc.* 33, 1, (1971) 88-94.

59. C.T.J. Dodson. Systems of connections for parametric models. In **Proc. Workshop on Geometrization of Statistical Theory** 28-31 October 1987. Ed. C.T.J. Dodson, ULDM Publications, University of Lancaster, 1987, pp 153-170.

60. C.T.J. Dodson. Gamma manifolds and stochastic geometry. In: **Proceedings of the Workshop on Recent Topics in Differential Geometry**, Santiago de Compostela 16-19 July 1997. *Public. Depto. Geometría y Topología* 89 (1998) 85-92.

61. C.T.J. Dodson. Information geodesics for communication clustering. *J. Statistical Computation and Simulation* 65, (2000) 133-146.

62. C.T.J. Dodson. Evolution of the void probability function. Presented at **Workshop on Statistics of Cosmological Data Sets**, 8-13 August 1999, Isaac Newton Institute, Cambridge.
http://www.maths.manchester.ac.uk/kd/PREPRINTS/vpf.ps . Cf. also [65].

63. C.T.J. Dodson. Spatial statistics and information geometry for parametric statistical models of galaxy clustering. *Int. J. Theor. Phys.*, 38, 10, (1999) 2585-2597.

64. C.T.J. Dodson. Geometry for stochastically inhomogeneous spacetimes. *Nonlinear Analysis*, 47 (2001) 2951-2958.

65. C.T.J. Dodson. Quantifying galactic clustering and departures from randomness of the inter-galactic void probablity function using information geometry. http://arxiv.org/abs/astro-ph/0608511 (2006).

66. C.T.J. Dodson. A note on quantum chaology and gamma manifold approximations to eigenvalue spacings for infinite random matrices. Proceedings CHAOS 2008, Charnia Crete 3-6 June 2008. http://arxiv.org/abs/math-ph/0802.2251

67. C.T.J. Dodson, A.G. Handley, Y. Oba and W.W. Sampson. The pore radius distribution in paper. Part I: The effect of formation and grammage. *Appita Journal* 56, 4 (2003) 275-280.

68. C.T.J. Dodson and Hiroshi Matsuzoe. An affine embedding of the gamma manifold. *InterStat*, January 2002, 2 (2002) 1-6.
69. C.T.J. Dodson and M. Modugno. Connections over connections and universal calculus. In Proc. **VI Convegno Nazionale di Relativita General a Fisic Della Gravitazione** Florence, 10-13 Octobar 1984, Eds. R. Fabbri and M. Modugno, pp. 89-97, Pitagora Editrice, Bologna, 1986.
70. C.T.J. Dodson and T. Poston. **Tensor Geometry** Graduate Texts in Mathematics 130, Second edition, Springer-Verlag, New York, 1991.
71. C.T.J. Dodson and W.W. Sampson. The effect of paper formation and grammage on its pore size distribution. *J. Pulp Pap. Sci.* 22(5) (1996) J165-J169.
72. C.T.J. Dodson and W.W. Sampson. Modeling a class of stochastic porous media. *App. Math. Lett.* 10, 2 (1997) 87-89.
73. C.T.J. Dodson and W.W. Sampson. Spatial statistics of stochastic fibre networks. *J. Statist. Phys.* 96, 1/2 (1999) 447-458.
74. C.T.J. Dodson and W.W. Sampson. Flow simulation in stochastic porous media. *Simulation*, 74:6, (2000) 351-358.
75. C.T.J. Dodson and W.W. Sampson. Planar line processes for void and density statistics in thin stochastic fibre networks. *J. Statist. Phys.* 129 (2007) 311-322.
76. C.T.J. Dodson and J. Scharcanski. Information Geometric Similarity Measurement for Near-Random Stochastic Processes. *IEEE Transactions on Systems, Man and Cybernetics - Part A*, 33, 4, (2003) 435-440.
77. C.T.J. Dodson and S.M. Thompson. A metric space of test distributions for DPA and SZK proofs. *Poster Session*, **Eurocrypt 2000**, Bruges, 14-19 May 2000. http://www.maths.manchester.ac.uk/kd/PREPRINTS/mstd.pdf.
78. C.T.J. Dodson and H. Wang. Iterative approximation of statistical distributions and relation to information geometry. *J. Statistical Inference for Stochastic Processes* 147, (2001) 307-318.
79. A.G. Doroshkevich, D.L. Tucker, A. Oemler, R.P. Kirshner, H. Lin, S.A. Shectman, S.D. Landy and R. Fong. Large- and Superlarge-scale Structure in the Las Campanas Redshift Survey. *Mon. Not. R. Astr. Soc.* 283 4 (1996) 1281-1310.
80. F. Downton. Bivariate exponential distributions in reliability theory. *J. Royal Statist. Soc. Series B* 32 (1970) 408-417.
81. G. Efstathiou. Counts-in-cells comparisons of redshift surveys. *Mon. Not. R. Astr. Soc.* 276, 4 (1995) 1425-1434.
82. A.P. Fairall. **Large-scale structure in the universe** Wiley-Praxis, Chichester 1998.
83. Fernandes, C.P., Magnani, F.S. 1996. Multiscale Geometrical Reconstruction of Porous Structures. *Physical Review E*, 54, 1734-1741.
84. W. Feller. **An Introduction to Probability Theory and its Applications.** Volume 1, John Wiley, Chichester 1968.
85. W. Feller. **An Introduction to Probability Theory and its Applications.** Volume 2, John Wiley, Chichester 1971.
86. R.A. Fisher. Theory of statistical estimation. *Proc. Camb. Phil. Soc.* 122 (1925) 700-725.
87. M. Fisz. **Probability Theory and Mathematical Statistics.** 3rd edition, John Wiley, Chichester 1963.
88. P. J. Forrester, Log-Gases and Random Matrices, Chapter 1 Gaussian matrix ensembles. Online book manuscript http://www.ms.unimelb.edu.au/~matpjf/matpjf.html, 2007.

89. R.J. Freund. A bivariate extension of the exponential distribution. *Journal of the American Statistical*, 56, (1961) 971-977.
90. K. Fukunga. *Introduction to Statistical Pattern Recognition*, 2nd Edition, Academic Press, Boston 1991.
91. B. Ghosh. Random distances within a rectangle and between two rectangles. *Calcutta Math. Soc.* 43, 1 (1951) 17-24.
92. S. Ghigna, S. Borgani, M. Tucci, S.A. Bonometto, A. Klypin and J.R. Primack. Statistical tests for CHDM and Lambda CDM cosmologies. *Astrophys. J.* 479, 2, 1 (1997) 580-91.
93. J. Gleick. **CHAOS: Making a New Science**. Heinemann, London 1988.
94. O. Goldreich, A. Sahai and S. Vadham. Can Statistical Zero-Knowledge be made non-interactive? Or, on the relationship of SZK and NISZK. In **Advances in Cryptology-CRYPTO '99**, Ed. M. Wiener, Lecture Notes in Computer Science 1666, Springer, Berlin 1999 pp 467-484.
95. A. Goffeau, B.G. Barrell, H. Bussey, R.W. Davis, B. Dujon, H. Feldmann, F. Galibert, J.D. Hoheisel, C. Jacq, M. Johnston, E.J. Louis, H.W. Mewes, Y. Murakami, P. Philippsen, H. Tettelin and S.G. Oliver. Life with 6000 genes. *Science* 274, 546, (1996) 563-567.
96. R. Gosine,X. Zhao and S. Davis. Automated Image Analysis for Applications in Reservoir Characterization. In **International Conference on Knowledge-Based Intelligent Engineering Systems and Allied Technologies**, September 2000, Brighton, UK.
97. F. Götze and H. Kösters. On the Second-Order Correlation Function of the Characteristic Polynomial of a Hermitian Wigner Matrix. http://arxiv.org/abs/math-ph/0803.0926 (2008).
98. Rao S. Govindaraju and M. Levent Kavvas. Characterization of the rill geometry over straight hillslopes through spatial scales. *Journal of Hydrology*, 130, 1, (1992) 339-365.
99. A. Gray **Modern Differential Geometry of Curves and Surfaces** 2nd Edition, CRC Press, Boca Raton 1998.
100. R.C. Griffiths. The canonical correlation coefficients of bivariate gamma distributions. *Annals Math. Statist.* 40, 4 (1969) 1401-1408.
101. P. Grzegorzewski and R. Wieczorkowski. Entropy-based goodness-of-fit test for exponentiality. *Commun. Statist. Theory Meth.* 28, 5 (1999) 1183-1202.
102. F.A. Haight. **Handbook of the Poisson Distribution** J. Wiley, New York, 1967.
103. A.W.J. Heijs, J. Lange, J.F. Schoute and J. Bouma. Computed Tomography as a Tool for Non-destructive Analysis of Flow Patterns in Macroporous Clay Soils. *Geoderma*, 64, (1995) 183-196.
104. F. Hoyle and M.S. Vogeley. Voids in the 2dF Galaxy Redshift Survey. *Astrophys. J.* 607 (2004) 751-764.
105. T.P. Hutchinson and C.D. Lai. **Continuous Multivariate Distributions, Emphasising Applications**, Rumsby Scientific Publishing, Adelaide 1990.
106. T-Y. Hwang and C-Y. Hu. On a characterization of the gamma distribution: The independence of the sample mean and the sample coefficient of variation. *Annals Inst. Statist. Math.* 51, 4 (1999) 749-753.
107. E.T. Jaynes. Information theory and statistical inference. *The Physical Review* 106 (1957) 620-630 and 108 (1957) 171-190. Cf. also the collection E.T. Jaynes, **Papers on probability, statistics and statistical physics** Ed. R. D. Rosenkrantz, Synthese Library, 158. D. Reidel Publishing Co., Dordrecht, 1983.

108. P.R. Johnston. The most probable pore size distribution in fluid filter media. *J. Testing and Evaluation* 11, 2 (1983) 117-121.

109. P.R. Johnston. Revisiting the most probable pore size distribution in filter media. The gamma distribution. *Filtration and Separation.* 35, 3 (1998) 287-292.

110. A.M. Kagan, Y.V. Linnik and C.R. Rao. **Characterization Problems in Mathematical Statistics** John Wiley, New York, 1973.

111. O. Kallmes and H. Corte. The structure of paper, I. The statistical geometry of an ideal two dimensional fiber network. *Tappi J.* 43, 9 (1960) 737-752. Cf. also: Errata 44, 6 (1961) 448.

112. O. Kallmes, H. Corte and G. Bernier. The structure of paper, V. The bonding states of fibres in randomly formed papers. *Tappi Journal* 46, 8, (1963) 493-502.

113. R.E. Kass and P.W. Vos. **Geometrical Foundations of Asymptotic Inference.** Wiley Series in Probability and Statistics: Probability and Statistics. A Wiley-Interscience Publication. John Wiley & Sons, Inc., New York, 1997.

114. G. Kauffmann and A.P. Fairall. Voids in the distribution of galaxies: an assessment of their significance and derivation of a void spectrum. *Mon. Not. R. Astr. Soc.* 248 (1990) 313-324.

115. M.D. Kaytor and S.T. Warren. Aberrant protein deposition and neurological disease. *J. Biological Chemistry* 53, (1999) 37507-37510.

116. R.A. Ketcham and Gerardo J. Iturrino. Nondestructive high-resolution visualization and measurement of anisotropic effective porosity in complex lithologies using high-resolution X-ray computed tomography. *Journal of Hydrology*, 302, 1-4, (2005) 92-106.

117. M. Kendall and A. Stuart. **The Advanced Theory of Statistics, Volume 2 Inference and Relationship** 4th Edition. Charles Griffin, London, 1979.

118. W.F. Kibble. A two variate gamma-type distribution. *Sankhyā* 5 (1941) 137-150.

119. P. Kocher, J. Jaffe and B. Jun. Differential Power Analysis. In **Advances in Cryptology-CRYPTO '99**, Ed. M. Wiener, Lecture Notes in Computer Science 1666, Springer, Berlin 1999 pp 388-397.

120. S. Kokoska and C. Nevison. **Statistical Tables and Formulae** Springer Texts in Statistics, Springer-Verlag, New York 1989.

121. K. Kondo, Editor. **Research Association of Applied Geometry Memoirs** Volume IV, Tokyo 1968.

122. S. Kotz, N.Balakrishnan and N.Johnson. **Continuous Multivariate Distributions** 2nd Edition, Volume 1 (2000).

123. I. Kovalenko. A simplified proof of a conjecture of D.G. Kendall concerning shapes of random polygons. *J. Appl. Math. Stochastic Anal.* 12, 4 (1999) 301-310.

124. G. Kron. Diakoptics—The Science of Tearing, Tensors and Topological Models. *RAAG Memoirs* Volume II, (1958) 343-368.

125. G. Kron. **Diakoptics—The Piecewise Solution of Large-Scale Systems**. MacDonald, London 1963.

126. S. Kullback. **Information and Statistics**, J. Wiley, New York, 1959.

127. T. Kurose. On the divergences of 1-conformally flat statistical manifolds. *Tôhoku Math. J.*, 46 (1994) 427-433.

128. F. Sylos Labini, A. Gabrielli, M. Montuori and L. Pietronero. Finite size effects on the galaxy number counts: Evidence for fractal behavior up to the deepest scale. *Physica A.* 226, 3-411 (1996) 195-242.

129. F. Sylos Labini, M. Montuori and L. Pietronero. Scale Invariance of galaxy clustering. *Physics Reports* 293 (1998)61-226.

130. M. Lachiéze-Rey, L.N. Da-Costa and S. Maurogordato. Void probability function in the Southern Sky Redshift Survey. *Astrophys. J.* 399 (1992) 10-15.

131. C.D. Lai. Constructions of bivariate distributions by a generalized trivariate reduction technique. *Statistics and Probability Letters* 25, 3 (1995) 265-270.

132. W.H. Landschulz, P.F. Johnson and S.L. McKnight. The Leucine Zipper - A hypothetical structure common to a new class of DNA-binding proteins. *Science* 240, (1988) 1759-1764.

133. S.D. Landy, S.A. Shectman, H. Lin, R.P. Kirshner, A.A. Oemler and D. Tucker. Two-dimensional power spectrum of the Las Campanas redshift survey: Detection of excess power on $100\, h^{-1} Mpc$ scales. *Astroph. J.* 456, 1, 2 (1996) L1-7.

134. S.L. Lauritzen. Statistical Manifolds. In **Differential Geometry in Statistical Inference**, Institute of Mathematical Statistics Lecture Notes, Volume 10, Berkeley 1987, pp 163-218.

135. S. Leurgans, T.W-Y. Tsai and J. Crowley. Freund's bivariate exponential distribution and censoring, in **Survival Analysis** (R. A. Johnson, eds.), IMS Lecture Notes, Hayward, California: Institute of Mathematical Statistics, 1982.

136. H. Lin, R.P. Kirshner, S.A. Schectman, S.D. Landy, A. Oemler, D.L. Tucker and P.L. Schechter. The power spectrum of galaxy clustering in the Las Campanas Redshift Survey. *Astroph. J.* 471, 2, 1 (1996) 617-635.

137. A. Lupas. Coiled coils: New structures and new functions. *Trends Biochem. Sci.* 21, 10 (1996) 375-382.

138. S. Mallat. **A Wavelet Tour of Signal Processing**. Academic Press, San Diego, 1998.

139. R.E. Mark. Structure and structural anisotropy. Ch. 24 in **Handbook of Physical and Mechanical Testing of Paper and Paperboard**. (R.E. Mark, ed.). Marcel Dekker, New York, 1984.

140. L. Mangiarotti and M. Modugno. Fibred spaces, jet spaces and connections for field theories. In Proc. International Meeting on **Geometry and Physics**, Florence, 12-15 October 1982, ed. M.Modugno, Pitagora Editrice, Bologna, 1983 pp 135-165.

141. K.V. Mardia. **Families of Bivariate Distributions**. Griffin, London 1970.

142. A.W. Marshall and I. Olkin. A generalized bivariate exponential distribution. *J. Appl. Prob.* 4 (1967) 291-302.

143. H. Matsuzoe. On realization of conformally-projectively flat statistical manifolds and the divergences. *Hokkaido Math. J.*, 27 (1998) 409-421.

144. H. Matsuzoe. Geometry of contrast functions and conformal geometry. *Hiroshima Math. J.*, 29 (1999) 175-191.

145. Madan Lal Mehta. **Random Matrices** 3rd Edition, Academic Press, London 2004.

146. A.T. McKay. Sampling from batches. *J. Royal Statist. Soc.* 2 (1934) 207-216.

147. R.E. Miles. Random polygons determined by random lines in a plane. *Proc. Nat. Acad. Sci. USA* 52, (1964) 901-907,1157-1160.

148. R.E. Miles. The various aggregates of random polygons determined by random lines in a plane. *Advances in Math.* 10, (1973) 256-290.

149. R.E. Miles. A heuristic proof of a long-standing conjecture of D.G. Kendall concerning the shapes of certain large random polygons. *Adv. in Appl. Probab.* 27, 2 (1995) 397-417.

150. G.K. Miller and U.N. Bhat. Estimation for renewal processes with unobservable gamma or Erlang interarrival times. *J. Statistical Planning and Inference* 61, 2 (1997) 355-372.

151. Steven J. Miller and Ramin Takloo-Bighash. **An Invitation to Modern Number Theory** Princeton University Press, Princeton 2006. Cf. also the seminar notes:
    - Steven J. Miller: Random Matrix Theory, Random Graphs, and L-Functions: How the Manhatten Project helped us understand primes. Ohio State University Colloquium 2003.
      `http://www.math.brown.edu/~sjmiller/math/talks/colloquium7.pdf` .
    - Steven J. Miller. Random Matrix Theory Models for zeros of L-functions near the central point (and applications to elliptic curves). Brown University Algebra Seminar 2004.
    `http://www.math.brown.edu/~sjmiller/math/talks/RMTandNTportrait.pdf`

152. M. Modugno. Systems of vector valued forms on a fibred manifold and applications to gauge theories. In Proc. Conference **Differential Geomeometric Methods in Mathematical Physics**, Salamanca 1985, Lecture Notes in Mathematics 1251, Springer-Verlag, Berlin 1987, pp. 238-264.

153. M.K. Murray and J.W. Rice. **Differential Geometry and Statistics**. Monographs on Statistics and Applied Probability, 48. Chapman & Hall, London, 1993.

154. K. Nomizu and T. Sasaki. **Affine differential geometry: Geometry of Affine Immersions**. Cambridge University Press, Cambridge, 1994.

155. B. Norman. Overview of the physics of forming. In **Fundamentals of Papermaking**, *Trans. IXth Fund. Res. Symp.*, (C.F. Baker, ed.), Vol III, pp. 73149, Mechanical Engineering Publications, London, 1989.

156. Y. Oba. **Z-directional structural development and density variation in paper**. Ph.D. Thesis, Department of Paper Science, University of Manchester Institute of Science and Technology, 1999.

157. A. Odlyzko. Tables of zeros of the Riemann zeta function.
    `http://www.dtc.umn.edu:80/~odlyzko/zeta_tables/index.html`.

158. S.H. Ong. Computation of bivariate-gamma and inverted-beta distribution functions. *J. Statistal Computation and Simulation* 51, 2-4 (1995) 153-163.

159. R.N. Onody, A.N.D. Posadas and S. Crestana. Experimental Studies of the Fingering Phenomena in Two Dimensions and Simulation Using a Modified Invasion Percolation Model. *Journal of Applied Physics*, 78, 5, (1995) 2970-2976.

160. E.K. O'Shea,R. Rutkowski and P.S. Kim. Evidence that the leucine zipper is a coiled coil. *Science* 243, (1989) 538-542.

161. A. Papoulis. **Probability, Random Variables and Stochastic Processes** 3rd edition, McGraw-Hill, New York 1991.

162. P.J.E. Peebles. **Large Scale Structure of the Universe** Princeton University Press, Princeton 1980.

163. S. Penel, R.G. Morrison, R.J. Mortishire-Smith and A.J. Doig. Periodicity in a-helix lengths and C-capping preferences. *J. Mol. Biol.* 293, (1999) 1211-1219.

164. R. Penrose. **The Emperor's New Mind** Oxford University Press, Oxford 1989.

165. Q.P. Pham, U. Sharma and A.G. Mikos. Characterization of scaffolds and measurement of cellular infiltration. *Biomacromolecules* 7, 10 (2006) 2796-2805.

166. Huynh Ngoc Phien. Reservoir storage capacity with gamma inflows. *Journal of Hydrology*, 146, 1, (1993) 383-389.

167. T. Piran, M.Lecar, D.S. Goldwirth, L. Nicolaci da Costa and G.R. Blumenthal. Limits on the primordial fluctuation spectrum: void sizes and anisotropy of the cosmic microwave background radiation. *Mon. Not. R. Astr. Soc.* 265, 3 (1993) 681-8.

168. C.F. Porter. **Statistical Theory of Spectra: Fluctuations** Edition, Academic Press, London 1965.

169. S.O. Prasher, J. Perret,A. Kantzas and C. Langford. Three-Dimensional Quantification of Macropore Networks in Undisturbed Soil Cores. *Soil Sci. Soc. Am. Journal,* 63 (1999) 1530-1543.

170. B. Radvan, C.T.J. Dodson and C.G. Skold. Detection and cause of the layered structure of paper. In **Consolidation of the Paper Web** *Trans. IIIrd Fund. Res. Symp. 1965* (F. Bolam, ed.), pp 189-214, BPBMA, London 1966.

171. C.R. Rao. Information and accuracy attainable in the estimation of statistical parameters. *Bull. Calcutta Math. Soc.* 37, (1945) 81-91.

172. S.A. Riboldi, M. Sampaolesi, P. Neuenschwander, G. Cossub and S. Mantero. Electrospun degradable polyesterurethane membranes: potential scaffolds for skeletal muscle tissue engineering. *Biomaterials* 26, 22 (2005) 4606-4615.

173. B.D. Ripley. **Statistical Inference for Spatial Processes**. Cambridge University Press, Cambridge 1988.

174. R.L. Rivest, A. Shamir and L.M. Adleman. A method for obtaining digital key signatures and public-key cryptosystems. *Communications of the ACM* 21 (1978) 120-126.

175. S. Roman. **Coding and Information Theory**. Graduate Texts in Mathematics, 134 Springer-Verlag, New York, 1992.

176. S. Roman. **Introduction to Coding and Information Theory**. Undergraduate Texts in Mathematics. Springer-Verlag, New York, 1997.

177. Colin Rose and Murray D.Smith. **Mathematical Statistics with *Mathematica*** Springer texts in statistics, Springer-Verlag, Berlin 2002.

178. Z. Rudnick. Private communication. 2008. Cf. also Z. Rudnick. What is Quantum Chaos? *Notices A.M.S.* 55, 1 (2008) 33-35.

179. A. Rushkin, J. Soto et al. **A Statistical Test Suite for Random and Pseudorandom Number Generators for Cryptographic Applications**. *National Institute of Standards & Technology,* Gaithersburg, MD USA, 2001.

180. B.Ya. Ryabko and V.A. Monarev. Using information theory approach to randomness testing. Preprint: *arXiv:CS.IT/0504006 v1,* 3 April 2005.

181. W.W. Sampson. Comments on the pore radius distribution in near-planar stochastic fibre networks. *J. Mater. Sci.* 36, 21 (2001) 5131-5135.

182. W.W. Sampson. The structure and structural characterisation of fibre networks in papermaking processes.

183. Y. Sato, K. Sugawa and M. Kawaguchi. The geometrical structure of the parameter space of the two-dimensional normal distribution. Division of information engineering, Hokkaido University, Sapporo, Japan (1977).

184. M.R. Schroeder. **Number Theory in Science and Communication. With Applications in Cryptography, Physics, Digital Information, Computing, and Self-Similarity**. Springer Series in Information Science, 3rd edition, Springer, Berlin 1999.

185. K. Schulgasser. Fiber orientation in machine made paper. *J. Mater. Sci.* 20, 3 (1985) 859-866.

186. C.E. Shannon. A mathematical theory of communication. *Bell Syst. Tech. J.* 27, (1948) 379-423 and 623-656.

187. D. Shi and C.D. Lai. Fisher information for Downton's bivariate exponential distribution. *J. Statistical Computation and Simulation* 60, 2 (1998) 123-127.
188. S.D. Silvey. **Statistical Inference** Chapman and Hall, Cambridge 1975.
189. L.T. Skovgaard. A Riemannian geometry of the multivariate normal model. *Scand. J. Statist.* 11 (1984) 211-223.
190. D. Slepian, ed. **Key papers in the development of information theory**, IEEE Press, New York, 1974.
191. P. Soille. **Morphological Image Analysis: Principles and Applications**. Springer-Verlag, Heidelberg 1999.
192. M. Sonka and H. Hlavac. **Image Processing, Analysis, and Machine Vision**, 2nd. Ed. *PWS Publishing Co.*, 1999.
193. A. Soshnikov. Universality at the edge of the spectrum in Wigner random matrices. *Commun. Math. Phys.* 207 (1999) 697-733.
194. M. Spivak. **Calculus on Manifolds**. W.A. Benjamin, New York 1965.
195. M. Spivak. **A Comprehensive Introduction to Differential Geometry, Vols. 1-5,** 2nd edn. Publish or Perish, Wilmington 1979.
196. D. Stoyan, W.S. Kendall and J. Mecke. **Stochastic Geometry and its Applications** 2nd Edition, John Wiley, Chichester, 1995.
197. I. Szapudi, A. Meiksin and R.C. Nichol. Higher order statistics from the Edinburgh Durham Southern Galaxy Catalogue Survey. 1. Counts in cells. *Astroph. J.* 473, 1, 1 (1996) 15-21.
198. J.C. Tanner. The proportion of quadrilaterals formed by random lines in a plane. *J. Appl. Probab.* 20, 2 (1983) 400-404.
199. Taud H., Martinez-Angeles T. et al. 2005. Porosity Estimation Method by X-ray Computed Tomography. *Journal of Petroleum Science and Engineering*, 47, 209-217.
200. M. Tribus. **Thermostatics and Thermodynamics** D. Van Nostrand and Co., Princeton N.J., 1961.
201. M. Tribus, R. Evans and G. Crellin. The use of entropy in hypothesis testing. In Proc. **Tenth National Symposium on Reliability and Quality Control** 7-9 January 1964.
202. R. van der Weygaert. Quasi-periodicity in deep redshift surveys. *Mon. Not. R. Astr. Soc.* 249 (1991) 159.
203. R. van der Weygaert and V. Icke. Fragmenting the universe II. Voronoi vertices as Abell clusters. *Astron. Astrophys.* 213 (1989) 1-9.
204. B. Velde, E. Moreau and F. Terribile. Pore Networks in an Italian Vertisol: Quantitative Characterization by Two Dimensional Image Analysis. *Geoderma*, 72, (1996) 271-285.
205. L. Vincent. Morphological Grayscale Reconstruction in Image Analysis: Applications and Efficient Algorithms. *IEEE Transactions of Image Processing*, 2 (1993) 176-201.
206. H.J. Vogel and A. Kretzchmar. *Topological Characterization of Pore Space in Soil-Sample Preparation and Digital Image-Processing. Geoderma*, 73, (1996) 23-18.
207. H.J. Vogel and K. Roth. Moving through scales of flow and transport in soil. *Journal of Hydrology*, 272, 1-4, (2003) 95-106.
208. M.S. Vogeley, M.J. Geller, C. Park and J.P. Huchra. Voids and constraints on nonlinear clustering of galaxies. *Astron. J.* 108, 3 (1994) 745-58.
209. H. Weyl. **Space Time Matter** Dover, New York 1950.

210. S.D.M. White. The hierarchy of correlation functions and its relation to other measures of galaxy clustering. *Mon. Not. R. Astr. Soc.* 186, (1979) 145-154.
211. H. Whitney. Differentiable manifolds, *Annals of Math.* 41 (1940) 645-680.
212. E.P. Wigner. Characteristic vectors of bordered matrices with infinite dimensions. *Annals of Mathematics* 62, 3 (1955) 548-564.
213. E.P. Wigner. On the distribution of the roots of certain symmetric matrices. *Annals of Mathematics* 67, 2 (1958) 325-327.
214. E.P. Wigner. Random matrices in physics. *SIAM Review* 9, 1 (1967) 1-23.
215. S. Wolfram. **The Mathematica Book** 3rd edition, Cambridge University Press, Cambridge, 1996.
216. S. Yue, T.B.M.J. Ouarda and B. Bobée. A review of bivariate gamma distributions for hydrological application. *Journal of Hydrology*, 246, 1-4, (2001) 1-18.

Index

Lecture Notes in Mathematics

For information about earlier volumes
please contact your bookseller or Springer
LNM Online archive: springerlink.com

Vol. 1867: J. Sneyd (Ed.), Tutorials in Mathematical Biosciences II. Mathematical Modeling of Calcium Dynamics and Signal Transduction. (2005)

Vol. 1868: J. Jorgenson, S. Lang, $Pos_n(R)$ and Eisenstein Series. (2005)

Vol. 1869: A. Dembo, T. Funaki, Lectures on Probability Theory and Statistics. Ecole d'Eté de Probabilités de Saint-Flour XXXIII-2003. Editor: J. Picard (2005)

Vol. 1870: V.I. Gurariy, W. Lusky, Geometry of Müntz Spaces and Related Questions. (2005)

Vol. 1871: P. Constantin, G. Gallavotti, A.V. Kazhikhov, Y. Meyer, S. Ukai, Mathematical Foundation of Turbulent Viscous Flows, Martina Franca, Italy, 2003. Editors: M. Cannone, T. Miyakawa (2006)

Vol. 1872: A. Friedman (Ed.), Tutorials in Mathematical Biosciences III. Cell Cycle, Proliferation, and Cancer (2006)

Vol. 1873: R. Mansuy, M. Yor, Random Times and Enlargements of Filtrations in a Brownian Setting (2006)

Vol. 1874: M. Yor, M. Émery (Eds.), In Memoriam Paul-André Meyer - Séminaire de Probabilités XXXIX (2006)

Vol. 1875: J. Pitman, Combinatorial Stochastic Processes. Ecole d'Eté de Probabilités de Saint-Flour XXXII-2002. Editor: J. Picard (2006)

Vol. 1876: H. Herrlich, Axiom of Choice (2006)

Vol. 1877: J. Steuding, Value Distributions of L-Functions (2007)

Vol. 1878: R. Cerf, The Wulff Crystal in Ising and Percolation Models, Ecole d'Eté de Probabilités de Saint-Flour XXXIV-2004. Editor: Jean Picard (2006)

Vol. 1879: G. Slade, The Lace Expansion and its Applications, Ecole d'Eté de Probabilités de Saint-Flour XXXIV-2004. Editor: Jean Picard (2006)

Vol. 1880: S. Attal, A. Joye, C.-A. Pillet, Open Quantum Systems I, The Hamiltonian Approach (2006)

Vol. 1881: S. Attal, A. Joye, C.-A. Pillet, Open Quantum Systems II, The Markovian Approach (2006)

Vol. 1882: S. Attal, A. Joye, C.-A. Pillet, Open Quantum Systems III, Recent Developments (2006)

Vol. 1883: W. Van Assche, F. Marcellàn (Eds.), Orthogonal Polynomials and Special Functions, Computation and Application (2006)

Vol. 1884: N. Hayashi, E.I. Kaikina, P.I. Naumkin, I.A. Shishmarev, Asymptotics for Dissipative Nonlinear Equations (2006)

Vol. 1885: A. Telcs, The Art of Random Walks (2006)

Vol. 1886: S. Takamura, Splitting Deformations of Degenerations of Complex Curves (2006)

Vol. 1887: K. Habermann, L. Habermann, Introduction to Symplectic Dirac Operators (2006)

Vol. 1888: J. van der Hoeven, Transseries and Real Differential Algebra (2006)

Vol. 1889: G. Osipenko, Dynamical Systems, Graphs, and Algorithms (2006)

Vol. 1890: M. Bunge, J. Funk, Singular Coverings of Toposes (2006)

Vol. 1891: J.B. Friedlander, D.R. Heath-Brown, H. Iwaniec, J. Kaczorowski, Analytic Number Theory, Cetraro, Italy, 2002. Editors: A. Perelli, C. Viola (2006)

Vol. 1892: A. Baddeley, I. Bárány, R. Schneider, W. Weil, Stochastic Geometry, Martina Franca, Italy, 2004. Editor: W. Weil (2007)

Vol. 1893: H. Hanßmann, Local and Semi-Local Bifurcations in Hamiltonian Dynamical Systems, Results and Examples (2007)

Vol. 1894: C.W. Groetsch, Stable Approximate Evaluation of Unbounded Operators (2007)

Vol. 1895: L. Molnár, Selected Preserver Problems on Algebraic Structures of Linear Operators and on Function Spaces (2007)

Vol. 1896: P. Massart, Concentration Inequalities and Model Selection, Ecole d'Été de Probabilités de Saint-Flour XXXIII-2003. Editor: J. Picard (2007)

Vol. 1897: R. Doney, Fluctuation Theory for Lévy Processes, Ecole d'Été de Probabilités de Saint-Flour XXXV-2005. Editor: J. Picard (2007)

Vol. 1898: H.R. Beyer, Beyond Partial Differential Equations, On linear and Quasi-Linear Abstract Hyperbolic Evolution Equations (2007)

Vol. 1899: Séminaire de Probabilités XL. Editors: C. Donati-Martin, M. Émery, A. Rouault, C. Stricker (2007)

Vol. 1900: E. Bolthausen, A. Bovier (Eds.), Spin Glasses (2007)

Vol. 1901: O. Wittenberg, Intersections de deux quadriques et pinceaux de courbes de genre 1, Intersections of Two Quadrics and Pencils of Curves of Genus 1 (2007)

Vol. 1902: A. Isaev, Lectures on the Automorphism Groups of Kobayashi-Hyperbolic Manifolds (2007)

Vol. 1903: G. Kresin, V. Maz'ya, Sharp Real-Part Theorems (2007)

Vol. 1904: P. Giesl, Construction of Global Lyapunov Functions Using Radial Basis Functions (2007)

Vol. 1905: C. Prévôt, M. Röckner, A Concise Course on Stochastic Partial Differential Equations (2007)

Vol. 1906: T. Schuster, The Method of Approximate Inverse: Theory and Applications (2007)

Vol. 1907: M. Rasmussen, Attractivity and Bifurcation for Nonautonomous Dynamical Systems (2007)

Vol. 1908: T.J. Lyons, M. Caruana, T. Lévy, Differential Equations Driven by Rough Paths, Ecole d'Été de Probabilités de Saint-Flour XXXIV-2004 (2007)

Vol. 1909: H. Akiyoshi, M. Sakuma, M. Wada, Y. Yamashita, Punctured Torus Groups and 2-Bridge Knot Groups (I) (2007)

Vol. 1910: V.D. Milman, G. Schechtman (Eds.), Geometric Aspects of Functional Analysis. Israel Seminar 2004-2005 (2007)

Vol. 1911: A. Bressan, D. Serre, M. Williams, K. Zumbrun, Hyperbolic Systems of Balance Laws. Cetraro, Italy 2003. Editor: P. Marcati (2007)

Vol. 1912: V. Berinde, Iterative Approximation of Fixed Points (2007)

Vol. 1913: J.E. Marsden, G. Misiołek, J.-P. Ortega, M. Perlmutter, T.S. Ratiu, Hamiltonian Reduction by Stages (2007)

Vol. 1914: G. Kutyniok, Affine Density in Wavelet Analysis (2007)

Vol. 1915: T. Bıyıkoğlu, J. Leydold, P.F. Stadler, Laplacian Eigenvectors of Graphs. Perron-Frobenius and Faber-Krahn Type Theorems (2007)

Vol. 1916: C. Villani, F. Rezakhanlou, Entropy Methods for the Boltzmann Equation. Editors: F. Golse, S. Olla (2008)

Vol. 1917: I. Veselić, Existence and Regularity Properties of the Integrated Density of States of Random Schrödinger (2008)

Vol. 1918: B. Roberts, R. Schmidt, Local Newforms for GSp(4) (2007)

Vol. 1919: R.A. Carmona, I. Ekeland, A. Kohatsu-Higa, J.-M. Lasry, P.-L. Lions, H. Pham, E. Taflin, Paris-Princeton Lectures on Mathematical Finance 2004.

Editors: R.A. Carmona, E. Çinlar, I. Ekeland, E. Jouini, J.A. Scheinkman, N. Touzi (2007)

Vol. 1920: S.N. Evans, Probability and Real Trees. Ecole d'Été de Probabilités de Saint-Flour XXXV-2005 (2008)

Vol. 1921: J.P. Tian, Evolution Algebras and their Applications (2008)

Vol. 1922: A. Friedman (Ed.), Tutorials in Mathematical BioSciences IV. Evolution and Ecology (2008)

Vol. 1923: J.P.N. Bishwal, Parameter Estimation in Stochastic Differential Equations (2008)

Vol. 1924: M. Wilson, Littlewood-Paley Theory and Exponential-Square Integrability (2008)

Vol. 1925: M. du Sautoy, L. Woodward, Zeta Functions of Groups and Rings (2008)

Vol. 1926: L. Barreira, V. Claudia, Stability of Nonautonomous Differential Equations (2008)

Vol. 1927: L. Ambrosio, L. Caffarelli, M.G. Crandall, L.C. Evans, N. Fusco, Calculus of Variations and Non-Linear Partial Differential Equations. Cetraro, Italy 2005. Editors: B. Dacorogna, P. Marcellini (2008)

Vol. 1928: J. Jonsson, Simplicial Complexes of Graphs (2008)

Vol. 1929: Y. Mishura, Stochastic Calculus for Fractional Brownian Motion and Related Processes (2008)

Vol. 1930: J.M. Urbano, The Method of Intrinsic Scaling. A Systematic Approach to Regularity for Degenerate and Singular PDEs (2008)

Vol. 1931: M. Cowling, E. Frenkel, M. Kashiwara, A. Valette, D.A. Vogan, Jr., N.R. Wallach, Representation Theory and Complex Analysis. Venice, Italy 2004. Editors: E.C. Tarabusi, A. D'Agnolo, M. Picardello (2008)

Vol. 1932: A.A. Agrachev, A.S. Morse, E.D. Sontag, H.J. Sussmann, V.I. Utkin, Nonlinear and Optimal Control Theory. Cetraro, Italy 2004. Editors: P. Nistri, G. Stefani (2008)

Vol. 1933: M. Petkovic, Point Estimation of Root Finding Methods (2008)

Vol. 1934: C. Donati-Martin, M. Émery, A. Rouault, C. Stricker (Eds.), Séminaire de Probabilités XLI (2008)

Vol. 1935: A. Unterberger, Alternative Pseudodifferential Analysis (2008)

Vol. 1936: P. Magal, S. Ruan (Eds.), Structured Population Models in Biology and Epidemiology (2008)

Vol. 1937: G. Capriz, P. Giovine, P.M. Mariano (Eds.), Mathematical Models of Granular Matter (2008)

Vol. 1938: D. Auroux, F. Catanese, M. Manetti, P. Seidel, B. Siebert, I. Smith, G. Tian, Symplectic 4-Manifolds and Algebraic Surfaces. Cetraro, Italy 2003. Editors: F. Catanese, G. Tian (2008)

Vol. 1939: D. Boffi, F. Brezzi, L. Demkowicz, R.G. Durán, R.S. Falk, M. Fortin, Mixed Finite Elements, Compatibility Conditions, and Applications. Cetraro, Italy 2006. Editors: D. Boffi, L. Gastaldi (2008)

Vol. 1940: J. Banasiak, V. Capasso, M.A.J. Chaplain, M. Lachowicz, J. Miękisz, Multiscale Problems in the Life Sciences. From Microscopic to Macroscopic. Będlewo, Poland 2006. Editors: V. Capasso, M. Lachowicz (2008)

Vol. 1941: S.M.J. Haran, Arithmetical Investigations. Representation Theory, Orthogonal Polynomials, and Quantum Interpolations (2008)

Vol. 1942: S. Albeverio, F. Flandoli, Y.G. Sinai, SPDE in Hydrodynamic. Recent Progress and Prospects. Cetraro, Italy 2005. Editors: G. Da Prato, M. Röckner (2008)

Vol. 1943: L.L. Bonilla (Ed.), Inverse Problems and Imaging. Martina Franca, Italy 2002 (2008)

Vol. 1944: A. Di Bartolo, G. Falcone, P. Plaumann, K. Strambach, Algebraic Groups and Lie Groups with Few Factors (2008)

Vol. 1945: F. Brauer, P. van den Driessche, J. Wu (Eds.), Mathematical Epidemiology (2008)

Vol. 1946: G. Allaire, A. Arnold, P. Degond, T.Y. Hou, Quantum Transport. Modelling, Analysis and Asymptotics. Cetraro, Italy 2006. Editors: N.B. Abdallah, G. Frosali (2008)

Vol. 1947: D. Abramovich, M. Mariño, M. Thaddeus, R. Vakil, Enumerative Invariants in Algebraic Geometry and String Theory. Cetraro, Italy 2005. Editors: K. Behrend, M. Manetti (2008)

Vol. 1948: F. Cao, J-L. Lisani, J-M. Morel, P. Musé, F. Sur, A Theory of Shape Identification (2008)

Vol. 1949: H.G. Feichtinger, B. Helffer, M.P. Lamoureux, N. Lerner, J. Toft, Pseudo-Differential Operators. Quantization and Signals. Cetraro, Italy 2006. Editors: L. Rodino, M.W. Wong (2008)

Vol. 1950: M. Bramson, Stability of Queueing Networks, Ecole d'Eté de Probabilités de Saint-Flour XXXVI-2006 (2008)

Vol. 1951: A. Moltó, J. Orihuela, S. Troyanski, M. Valdivia, A Non Linear Transfer Technique for Renorming (2008)

Vol. 1952: R. Mikhailov, I.B.S. Passi, Lower Central and Dimension Series of Groups (2008)

Vol. 1953: K. Arwini, C.T.J. Dodson, Information Geometry (2008)

Vol. 1954: P. Biane, L. Bouten, F. Cipriani, N. Konno, N. Privault, Q. Xu, Quantum Potential Theory. Editors: U. Franz, M. Schuermann (2008)

Vol. 1955: M. Bernot, V. Caselles, J.-M. Morel, Optimal transportation networks (2008)

Vol. 1956: C.H. Chu, Matrix Convolution Operators on Groups (2008)

Vol. 1957: A. Guionnet, On Random Matrices: Macroscopic Asymptotics, Ecole d'Eté de Probabilités de Saint-Flour XXXVI-2006 (2008)

Vol. 1958: M.C. Olsson, Compactifying Moduli Spaces for Abelian Varieties (2008)

Recent Reprints and New Editions

Vol. 1702: J. Ma, J. Yong, Forward-Backward Stochastic Differential Equations and their Applications. 1999 – Corr. 3rd printing (2007)

Vol. 830: J.A. Green, Polynomial Representations of GL_n, with an Appendix on Schensted Correspondence and Littelmann Paths by K. Erdmann, J.A. Green and M. Schoker 1980 – 2nd corr. and augmented edition (2007)

Vol. 1693: S. Simons, From Hahn-Banach to Monotonicity (Minimax and Monotonicity 1998) – 2nd exp. edition (2008)

Vol. 470: R.E. Bowen, Equilibrium States and the Ergodic Theory of Anosov Diffeomorphisms. With a preface by D. Ruelle. Edited by J.-R. Chazottes. 1975 – 2nd rev. edition (2008)

Vol. 523: S.A. Albeverio, R.J. Høegh-Krohn, S. Mazzucchi, Mathematical Theory of Feynman Path Integral. 1976 – 2nd corr. and enlarged edition (2008)

Vol. 1764: A. Cannas da Silva, Lectures on Symplectic Geometry 2001 – Corr. 2nd printing (2008)

LECTURE NOTES IN MATHEMATICS   ◤ Springer

Edited by J.-M. Morel, F. Takens, B. Teissier, P.K. Maini

Editorial Policy (for the publication of monographs)

1. Lecture Notes aim to report new developments in all areas of mathematics and their applications - quickly, informally and at a high level. Mathematical texts analysing new developments in modelling and numerical simulation are welcome.

   Monograph manuscripts should be reasonably self-contained and rounded off. Thus they may, and often will, present not only results of the author but also related work by other people. They may be based on specialised lecture courses. Furthermore, the manuscripts should provide sufficient motivation, examples and applications. This clearly distinguishes Lecture Notes from journal articles or technical reports which normally are very concise. Articles intended for a journal but too long to be accepted by most journals, usually do not have this "lecture notes" character. For similar reasons it is unusual for doctoral theses to be accepted for the Lecture Notes series, though habilitation theses may be appropriate.

2. Manuscripts should be submitted either to Springer's mathematics editorial in Heidelberg, or to one of the series editors. In general, manuscripts will be sent out to 2 external referees for evaluation. If a decision cannot yet be reached on the basis of the first 2 reports, further referees may be contacted: The author will be informed of this. A final decision to publish can be made only on the basis of the complete manuscript, however a refereeing process leading to a preliminary decision can be based on a pre-final or incomplete manuscript. The strict minimum amount of material that will be considered should include a detailed outline describing the planned contents of each chapter, a bibliography and several sample chapters.

   Authors should be aware that incomplete or insufficiently close to final manuscripts almost always result in longer refereeing times and nevertheless unclear referees' recommendations, making further refereeing of a final draft necessary.

   Authors should also be aware that parallel submission of their manuscript to another publisher while under consideration for LNM will in general lead to immediate rejection.

3. Manuscripts should in general be submitted in English. Final manuscripts should contain at least 100 pages of mathematical text and should always include

   - a table of contents;
   - an informative introduction, with adequate motivation and perhaps some historical remarks: it should be accessible to a reader not intimately familiar with the topic treated;
   - a subject index: as a rule this is genuinely helpful for the reader.

For evaluation purposes, manuscripts may be submitted in print or electronic form, in the latter case preferably as pdf- or zipped ps-files. Lecture Notes volumes are, as a rule, printed digitally from the authors' files. To ensure best results, authors are asked to use the LaTeX2e style files available from Springer's web-server at:

ftp://ftp.springer.de/pub/tex/latex/svmonot1/ (for monographs).

Additional technical instructions, if necessary, are available on request from: lnm@springer.com.

4. Careful preparation of the manuscripts will help keep production time short besides ensuring satisfactory appearance of the finished book in print and online. After acceptance of the manuscript authors will be asked to prepare the final LaTeX source files (and also the corresponding dvi-, pdf- or zipped ps-file) together with the final printout made from these files. The LaTeX source files are essential for producing the full-text online version of the book (see www.springerlink.com/content/110312 for the existing online volumes of LNM).

The actual production of a Lecture Notes volume takes approximately 12 weeks.

5. Authors receive a total of 50 free copies of their volume, but no royalties. They are entitled to a discount of 33.3% on the price of Springer books purchased for their personal use, if ordering directly from Springer.

6. Commitment to publish is made by letter of intent rather than by signing a formal contract. Springer-Verlag secures the copyright for each volume. Authors are free to reuse material contained in their LNM volumes in later publications: a brief written (or e-mail) request for formal permission is sufficient.

Addresses:
Professor J.-M. Morel, CMLA,
École Normale Supérieure de Cachan,
61 Avenue du Président Wilson, 94235 Cachan Cedex, France
E-mail: Jean-Michel.Morel@cmla.ens-cachan.fr

Professor F. Takens, Mathematisch Instituut,
Rijksuniversiteit Groningen, Postbus 800,
9700 AV Groningen, The Netherlands
E-mail: F.Takens@math.rug.nl

Professor B. Teissier, Institut Mathématique de Jussieu,
UMR 7586 du CNRS, Équipe "Géométrie et Dynamique",
175 rue du Chevaleret
75013 Paris, France
E-mail: teissier@math.jussieu.fr

For the "Mathematical Biosciences Subseries" of LNM:

Professor P.K. Maini, Center for Mathematical Biology,
Mathematical Institute, 24-29 St Giles,
Oxford OX1 3LP, UK
E-mail: maini@maths.ox.ac.uk

Springer, Mathematics Editorial I, Tiergartenstr. 17
69121 Heidelberg, Germany,
Tel.: +49 (6221) 487-8259
Fax: +49 (6221) 4876-8259
E-mail: lnm@springer.com